THE THEORY OF GROUPS

Proceedings
of the International Conference on

THE THEORY OF GROUPS

held at the
Australian National University
Canberra, 10 – 20 August, 1965

Edited by

L. G. KOVÁCS and B. H. NEUMANN

GORDON AND BREACH SCIENCE PUBLISHERS

NEW YORK · LONDON · PARIS

Editorial Office for Great Britain:

Gordon and Breach Science Publishers, Ltd.
61 Carey Street
London W.C.2

Editorial Office for France

7–9 rue Emile Dubois
Paris 14ᵉ

Distributed in France by:

Dunod Editeur
92 rue Bonaparte
Paris 6ᵉ

Distributed in Canada by:

The Ryerson Press
299 Queen Street West
Toronto 2ʙ, Ontario

Library of Congress Catalog Card Number: 66–268068

Made and printed in Great Britain by
William Clowes and Sons, Limited, London and Beccles

Preface

The Editors have acted on the principle that authors are responsible for the contents of their papers: style, accuracy, interest, all the things that referees are commonly supposed to judge, are really the author's business. The Editors have, accordingly, paid attention to *minutiae* of printing, to uniformity of bibliographical details, and so on. A few mathematical errors were amended with authors' cooperation.

Canberra, A.C.T.
L.G.K.
B.H.N.

CONTENTS

	Page
PREFACE	v
INTRODUCTION	xi
List of sponsors and donors	xiv
List of chairmen of lectures	xiv
List of study groups, with chairmen and assistant organizers .	xiv
List of participants	xv

PAPERS

L. W. Anderson and *R. P. Hunter*, Groups, homomorphisms, and the Green relations	1
Christine W. Ayoub, On the number of conjugate classes in a group	7
Reinhold Baer, Nilpotency	11
Reinhold Baer, Noetherian soluble groups	17
Reinhold Baer, Noetherian groups	33
Gilbert Baumslag, Finitely presented groups	37
Warren Brisley, On a problem of D. W. Barnes . . .	51
J. L. Britton, On the unrestricted Burnside problem . .	*
Paul F. Conrad, Lateral completions of lattice-ordered groups (T)	57
A. L. S. Corner, Endomorphism rings of torsion-free abelian groups	59
P. J. Cossey, On varieties of A-groups (T) . . .	71
H. S. M. Coxeter, The Lorentz group and the group of homographies	73
W. E. Deskins, On 𝕮-permutable subgroups of finite groups .	79
John D. Dixon, Complements of normal subgroups in infinite groups	81
Walter Feit, On groups with a cyclic Sylow subgroup . .	85
Walter Feit, An analogue of Jordan's Theorem in characteristic p	*
L. Fuchs, On orderable groups	89
T. M. Gagen, On groups with abelian 2-Sylow subgroups .	99

(T) denotes papers presented by title.
  * indicates that no manuscript has been received.

Wolfgang Gaschütz, Nichtabelsche p-Gruppen besitzen äussere p-Automorphismen (read in English) 101

Chander Kanta Gupta, On stability groups of certain nilpotent groups (T) 103

N. D. Gupta, Metabelian groups in the variety of certain two-variable laws 105

N. D. Gupta and *M. F. Newman*, On metabelian groups (T) . 111

Marshall Hall, Jr., Group theory and block designs . . . 115

Trevor Hawkes, Analogues of Prefrattini subgroups . . . 145

Hermann Heineken, Groups with an existence property with respect to commutators 151

D. H. W. Held, Some criteria for the hypercentrality of groups . *

Graham Higman, The orders of relatively free groups . . 153

Graham Higman, Representations of general linear groups and varieties of p-groups 167

K. A. Hirsch, Periodic linear groups 175

R. P. Hunter and *L. W. Anderson*, Certain groups and homomorphisms associated with a semigroup 185

Noboru Ito, On transitive permutation groups of Fermat prime degree 191

Noboru Ito, Two questions on projective planes and block designs *

D. G. James, On the orthogonal groups of ramified lattices over local fields 203

Zvonimir Janko, A characterization of a new simple group . . 205

J. A. Kalman, An intrinsic multiplication in additive l-groups (T) *

Otto H. Kegel, On Huppert's characterization of finite supersoluble groups 209

L. G. Kovács, Varieties and the Hall-Higman paper (T) . . 217

L. G. Kovács and *M. F. Newman*, Just-non-Cross varieties (T). 221

F. Loonstra, Ordering of extensions 225

P. J. Lorimer, Subplanes of projective planes . . . 233

I. D. Macdonald, A theorem about critical p-groups . . . 241

Hanna Neumann, Varieties of groups 251

Sheila Oates, Identical relations in a small number of variables . 261

(T) denotes papers presented by title.
  * indicates that no manuscript has been received.

Sophie Piccard, Quelques problèmes généraux de la théorie des groupes et les groupes libres modulo n (read in English). . 265

K. M. Rangaswamy, Abelian groups with endomorphic images of special type 279

Rimhak Ree, Classification of involutions and centralizers of involutions in certain simple groups 281

John S. Rose, Remarks on system normalizers and Carter subgroups 303

H. Schwerdtfeger, Group-theoretical interpretation of projective incidence theorems 307

Robert Steinberg, On the Galois cohomology of linear algebraic groups 315

A. G. R. Stewart, On the class of certain nilpotent groups . . 321

G. Szekeres, Metabelian groups with two generators . . . 323

Olaf Tamaschke, A generalized character theory on finite groups . 347

G. E. Wall, On Hughes' H_p problem 357

M. A. Ward, Basic commutators for polynilpotent groups . . 363

M. A. Ward, Nilpotent free groups with torsion-free central factor groups . . . . *

Paul M. Weichsel, Critical and basic p-groups 367

Kenneth W. Weston, On a useful theorem for commutator calculation and the theory of associative rings 373

Helmut Wielandt, On the structure of composite groups . . 379

Helmut Wielandt, On automorphisms of doubly transitive permutation groups 389

G. Zappa, Sur les S-partitions de Hall dans les groupes finis . 395

* indicates that no manuscript has been received.

Introduction

The Group Theory Conference took place from 10 to 20 August, 1965. It was sponsored by the International Mathematical Union, the Australian Academy of Science, and, most substantially, by its host, the Australian National University. Of the 89 participants, 45 came from outside Australia, and represented 12 countries: Canada, Eire, Germany, Hungary, India, Italy, Japan, Netherlands, New Zealand, Switzerland, United Kingdom, United States of America. A representative of the U.S.S.R. was unfortunately unable to attend the conference.

The scientific programme of the conference concentrated mainly on non-abelian abstract groups; it consisted of 16 formal lectures, mostly surveys, and 10 informal study groups on special topics, organized informally on an *ad hoc* basis. Perhaps the most important activity of the conference, and certainly the most stimulating, was the informal getting together of small groups of group theorists with cups of coffee and blackboards or paper and pencils to exchange problems and results and to work jointly on some of the problems. An excellent setting for this was provided by Bruce Hall, where the participants from outside Canberra were accommodated.

The nonscientific activities of the conference included a sherry party, a dinner given by the Australian National University, a tour of Canberra, a film show, and visits to Mount Stromlo Observatory (optical astronomy), the Snowy Mountains (water conservation, electricity generation, skiing), the Mills Cross at Hoskingtown (radio astronomy), the Deep Space Instrument Facility at Tidbinbilla (space vehicle tracking and control), and the CSIRO Wild Life Research Station at Gungahlin (kangaroos). Private hospitality and entertainment was provided, among others, by some senior members of the diplomatic missions in Canberra.

This was the first international conference on a special mathematical topic to be held in Australia. It is impossible to assess its value, and it is very difficult to estimate its cost. A rough guess is $50,000* for the cost of bringing the participants together from all over the world and accommodating them during the conference. By far the largest contribution came from the Australian National University; but as it included grants to Visiting Fellows and Visiting Lecturers, most of whom stayed in the University for varying periods before or after the conference, a figure of $6,000 for this contribution represents no more than a wild guess. The International Mathematical Union contributed $3,500 to the fares of overseas delegates, the Australian Academy of Science $200 to the expenses of Australian participants. Industry in Australia donated about $500. Substantial contributions were made by overseas governments, academies, and scientific foundations by contributing to the round-trip fares of some participants; some universities, both in Australia and elsewhere, similarly contributed to the fares of their members.

Governments, etc., generally prefer to spend their money on senior academics rather than junior researchers; in order to ensure the participation also of some of the younger workers in the field, one-term visiting lectureships in Australian universities were arranged for about 12 participants. The visiting lecturers concerned have thus also contributed substantially, by their work, to the notional cost of the conference; but even more valuable was their contribution to the life and liveliness of the conference, and to Australian mathematics. This was made possible by the ready cooperation of their Australian host universities, and in some cases also of their home universities, who gave them the requisite leave of absence.

It is clear from all this that many individuals and organizations have earned our gratitude—keenly felt and gladly expressed—for making the conference possible, and for making it the success which, by common consent, it was. My own very special thanks are due to the key members of the Organizing Committee who did all the hard preparatory chores, as well as the running of the actual conference, and the remaining work, such as paying bills and preparing copy for the PROCEEDINGS: in particular my wife, who acted as Deputy Convener during my

* The uncertainty in the figures given makes it immaterial whether they are interpreted in Australian, Canadian, or U.S. dollars.

absence on study leave, and who conducted an inordinate amount of the initial correspondence; L. G. Kovács, who was the tireless Secretary before the conference, the *spiritus rector* during the conference, and the effective PROCEEDINGS Editor after the conference; and M. F. Newman, who, as Treasurer, had all the money worries. Paul F. Conrad and Zvonimir Janko were members of the Organizing Committee throughout; I. D. Macdonald and I. M. S. Dey took part in the earlier and later deliberations, respectively, and others helped as they became available. The International Mathematical Union nominated as its committee representatives Claude Chevalley and F. E. P. Hirzebruch, whose advice and help proved invaluable.

B. H. Neumann

List of sponsors and donors

Australian Academy of Science
Australian National University
General Motors–Holden Pty Ltd
International Mathematical Union
Remington Rand–Chartres Pty Ltd
The Broken Hill Pty Co Ltd
The Colonial Sugar Refining Co Ltd

List of chairmen of lectures

P. F. Conrad, W. Gaschütz, N. Ito, R. Kochendörffer,
F. Loonstra, I. D. Macdonald, W. Magnus, B. H. Neumann,
Hanna Neumann, Sophie Piccard, R. Ree, H. Schwerdtfeger,
R. Steinberg, G. Szekeres, G. E. Wall, G. Zappa.

List of study groups, with chairmen and assistant organizers

Abelian groups	L. Fuchs; K. M. Rangaswamy
Burnside problem	M. Hall, Jr; P. M. Neumann
Free groups	G. Baumslag; Joan Landman
Geometry and groups	H. S. M. Coxeter; M. A. Ward
Insoluble groups	R. Ree; T. M. Gagen
Nilpotent groups	R. Baer; D. H. W. Held
Ordered groups	P. F. Conrad; R. D. Byrd
Permutation groups	H. Wielandt; J. Wiegold
Soluble groups	K. A. Hirsch; N. D. Gupta
Varieties	Hanna Neumann; P. J. Cossey

List of participants

Dr I. T. A. C. Adamson (Queen's College, Dundee, and University of Western Australia)

Professor L. W. Anderson (Pennsylvania State University)

Mr A. J. Andrews (University of Adelaide)

Mr D. S. Asche (Monash University)

Professor Christine W. Ayoub (Pennsylvania State University)

Professor Reinhold Baer (University of Frankfurt)

Dr D. W. Barnes (University of Sydney)

Professor Gilbert Baumslag (City University of New York)

Mr R. F. Berghout (University of Sydney)

Mr Warren Brisley (University of Newcastle, New South Wales)

Dr J. L. Britton (University of Glasgow and University of Western Australia)

Mr M. S. Brooks (Institute of Advanced Studies, Australian National University)

Mr R. A. Bryce (Institute of Advanced Studies, Australian National University)

Mr R. G. Burns (Institute of Advanced Studies, Australian National University)

Mr R. D. Byrd (Louisiana State University and Institute of Advanced Studies, Australian National University)

Mr J. M. Campbell (School of General Studies, Australian National University)

Mr R. J. Clarke (University of Adelaide)

Dr S. B. Conlon (University of Sydney)

Professor P. F. Conrad (Tulane University and Institute of Advanced Studies, Australian National University)

Mr W. A. Coppel (Institute of Advanced Studies, Australian National University)

Dr A. L. S. Corner (Worcester College, Oxford)

Mr P. J. Cossey (Institute of Advanced Studies, Australian National University)

Professor H. S. M. Coxeter (University of Toronto)

Professor W. E. Deskins (Michigan State University)

Dr I. M. S. Dey (University of Sussex and School of General Studies, Australian National University)

Dr J. D. Dixon (University of New South Wales)

Professor Walter Feit (Yale University)

Professor László Fuchs (University of Budapest and University of New South Wales)

Mr T. M. Gagen (Institute of Advanced Studies, Australian National University)

Professor Wolfgang Gaschütz (University of Kiel)

Mrs C. K. Gupta (Institute of Advanced Studies, Australian National University)

Mr N. D. Gupta (Institute of Advanced Studies, Australian National University)

Professor Marshall Hall, Jr (California Institute of Technology)

Mr T. O. Hawkes (Trinity College, Cambridge, and University of Queensland)

Dr Hermann Heineken (University of Frankfurt and Monash University)

Dr D. H. W. Held (School of General Studies, Australian National University)

Professor Graham Higman (Magdalen College, Oxford)

Professor K. A. Hirsch (Queen Mary College, London)
Professor A. F. Horadam (University of New England)
Professor R. P. Hunter (Pennsylvania State University)
Professor Noboru Ito (Nagoya University)
Dr D. G. James (University of Auckland)
Mr R. K. James (University of Sydney)
Dr Zvonimir Janko (Institute of Advanced Studies, Australian National University)
Professor J. A. Kalman (University of Auckland)
Dr O. H. Kegel (University of Frankfurt and University of Sydney)
Mr B. W. King (Church of England Residential Halls, Sydney)
Professor Rudolf Kochendörffer (University of Rostock and University of Adelaide)
Dr L. G. Kovács (Institute of Advanced Studies, Australian National University)
Dr Joan Landman (Columbia University and School of General Studies, Australian National University)
Professor F. Loonstra ("Technische Hogeschool", Delft)
Dr P. J. Lorimer (University of Canterbury, Christchurch)
Mr Lewis Low (University of Sydney)
Professor I. D. Macdonald (University of Newcastle, New South Wales)
Professor Wilhelm Magnus (Courant Institute, New York University)
Mr Abdul Majeed (Institute of Advanced Studies, Australian National University)
Dr Joachim Neubüser (University of Kiel and School of General Studies, Australian National University)
Professor B. H. Neumann (Institute of Advanced Studies, Australian National University)
Professor Hanna Neumann (School of General Studies, Australian National University)
Mr P. M. Neumann (The Queen's College, Oxford, and Monash University)
Mr W. D. Neumann (University of Adelaide)
Mr Bill Newman (Townsville University College)
Dr M. F. Newman (School of General Studies, Australian National University)
Dr Sheila Oates (St. Hilda's College, Oxford, and University of Queensland)
Mr M. P. O'Donnell (University of Queensland)
Mr K. R. Pearson (University of Adelaide)
Professor Sophie Piccard (University of Neuchâtel)
Mr K. E. Pledger (Victoria University of Wellington)
Mr E. Rahman (University of Sydney)
Dr K. M. Rangaswamy (University of Panjab and Institute of Advanced Studies, Australian National University)
Professor Rimhak Ree (University of British Columbia)
Dr J. S. Rose (University of Mainz and University of Newcastle, England)
Professor Hans Schwerdtfeger (McGill University and University of Adelaide)
Dr R. J. Smith (Townsville University College)
Dr N. F. Smythe (University of New South Wales)

Professor Robert Steinberg (University of California at Los Angeles)

Mr A. G. R. Stewart (School of General Studies, Australian National University)

Professor George Szekeres (University of New South Wales)

Dr Olaf Tamaschke (University of Tübingen and University of Melbourne)

Professor S. J. Tobin (University College Galway and Institute of Advanced Studies, Australian National University)

Mr P. G. Trotter (University of New South Wales)

Professor G. E. Wall (University of Sydney)

Mr J. N. Ward (University of Sydney)

Mr M. A. Ward (Institute of Advanced Studies, Australian National University)

Professor P. M. Weichsel (University of Illinois and Institute of Advanced Studies, Australian National University)

Professor K. W. Weston (University of Notre Dame and University of Western Australia)

Dr James Wiegold (University College of South Wales and Monmouthshire, Cardiff, and Institute of Advanced Studies, Australian National University)

Professor Helmut Wielandt (University of Tübingen)

Professor Guido Zappa (Institute "Ulisse Dini", University of Florence)

Proc. Internat. Conf. Theory of Groups, Austral. Nat. Univ. Canberra, August 1965, pp. 1–5. © Gordon and Breach Science Publishers, Inc. 1967

Groups, homomorphisms, and the Green relations

L. W. ANDERSON* and R. P. HUNTER*

Permutation groups play a vital role in the investigation of the structure of a semigroup. It is the purpose of this note to indicate some of the groups and homomorphisms associated in a natural way with a semigroup.

We let S be a semigroup with identity and recall the Green relations [3]:

$$x\mathscr{L}y \rightleftharpoons Sx = Sy, \qquad \mathscr{H} = \mathscr{L} \cap \mathscr{R},$$
$$x\mathscr{R}y \rightleftharpoons xS = yS, \qquad \mathscr{D} = \mathscr{L} \circ \mathscr{R} = \mathscr{R} \circ \mathscr{L}.$$

If $A, B \subset S$, let

$$A \cdot {}^{\cdot} B = \{x \in S \mid Bx \subset A\},$$
$$A^{\cdot} \cdot B = \{x \in S \mid xB \subset A\},$$
$$A \cdot\cdot B = \{(x, y) \in S \times S \mid xBy \subset A\}.$$

If H is an $\mathscr{H}$-class of S, define the relation $\mathscr{S}$ on $H \cdot {}^{\cdot} H$ as follows:

$$x\mathscr{S}y \rightleftharpoons hx = hy \quad \text{for all} \quad h \in H$$

(equivalently, $x\mathscr{S}y$ if $hx = hy$ for some $h \in H$)

so that $\Gamma = (H \cdot {}^{\cdot} H)/\mathscr{S}$ is a group, called the (right) Schützenberger group of H [4]. We let Γ act upon H as follows: if $t \in \gamma \in \Gamma$, $h \in H$ let $h\gamma = ht$. Then Γ is simply transitive over H and thus it follows that in a compact semigroup H is the underlying space of a compact group. Dually, if we set $\Gamma' = (H^{\cdot} \cdot H)/\mathscr{S}'$, where $\mathscr{S}'$ is the left-hand version of $\mathscr{S}$, then Γ' is the left Schützenberger group of H and Γ' acts simply transitively over H on the left. The groups Γ and Γ' are isomorphic under the following correspondence: fix $h \in H$ and for $\alpha \in \Gamma$, $\beta \in \Gamma'$ let $\beta \leftrightarrow \alpha$ if, and only if, $\beta h = h\alpha$.

* With support of the National Science Foundation Grant GP 4066.

The Schützenberger group has proven to be an extremely useful tool, especially in the theory of compact topological semigroups. To illustrate this point, consider the

Theorem. *If S is a compact connected semigroup with identity and K, the minimal two-sided ideal of S, is finite dimensional and contained in the orbit of some $\mathscr{H}$-class, then K is a group.*

If $K \subset xH$ where H is an $\mathscr{H}$-class, then $K = xH$ and K is a minimal right ideal. Let Γ be the Schützenberger group of H and let $h \in H$. Set $G = \{\alpha \in \Gamma \mid xh\alpha = xh\}$ so that G is a closed normal subgroup of Γ and Γ/G is homeomorphic with K. If $e = e^2 \in K$ then $K = eSe \times (Se \cap E)$ where E is the set of idempotents in S. If $Se \cap E$ is trivial then $K = eSe$ and is a group. If, on the other hand, $Se \cap E$ is not trivial, set $m = \dim(K)$, $n = \dim(eSe)$ and so $m > n$. Since K is the space of a compact group, K has nontrivial cohomology in dimension m. But eSe carries all the cohomology of K, contradicting the assumption that $n < m$. Thus $Se \cap E$ is indeed trivial.

Again, we let S be any semigroup with identity and for $x, y \in S$, set $B_{x,y} = xH_y \cap H_{xy}$ and $A_{x,y} = \{h \in H_y \mid xh \in B_{x,y}\}$. One now shows that $A_{x,y} \cdot {}^{\cdot} A_{x,y} \subset B_{x,y} \cdot {}^{\cdot} B_{x,y}$. Setting $\Gamma(x, y) = A \cdot {}^{\cdot} A/\mathscr{S}_y$ and $\Delta(x, y) = B \cdot {}^{\cdot} B/\mathscr{S}_{xy}$ we have that $\Gamma(x, y)$ and $\Delta(x, y)$ are groups [1] simply transitive over A and B respectively. Further, if $i: A \cdot {}^{\cdot} A \to B \cdot {}^{\cdot} B$ is the identity mapping then the homomorphism $i_{x,y}$

$$\Gamma(x, y) \xrightarrow{\ i_{x,y}\ } \Delta(x, y)$$
$$\uparrow \qquad\qquad \uparrow$$
$$A \cdot {}^{\cdot} A \xrightarrow{\ i\ } B \cdot {}^{\cdot} B$$

induced by i is onto. We note that the kernel of $i_{x,y}$ is given by

$$\ker(i_{x,y}) = (\{[(xy) \cdot {}^{\cdot} x] \cap H_y\} \cdot {}^{\cdot} \{[(xy) \cdot {}^{\cdot} x] \cap H_y\})/\mathscr{S}_y.$$

If $t \in H_y$, then there is an $\alpha \in \Gamma$ such that $t\alpha = y$. An easy computation shows that $\Gamma(x, t) = \alpha\Gamma(x, y)\alpha^{-1}$ and the induced homomorphism

$$\begin{array}{ccc} \Gamma(x, y) & \longrightarrow & \Gamma(x, t) \\ {\scriptstyle i_{x,y}}\downarrow & & \downarrow{\scriptstyle i_{x,t}} \\ \Delta(x, y) & \longrightarrow & \Delta(x, t) \end{array}$$

is an isomorphism.

If $y^2 = y$ then H_y is a group and $A_{x,y}$ is a subgroup of H_y isomorphic with $\Gamma(x, y)$. Similarly, if xy is an idempotent then $B(x, y)$ is a subgroup of H_{xy} isomorphic with $\Delta(x, y)$.

Loosely speaking, a semigroup may be viewed as a collection of groups and homomorphisms. To be more precise, consider the following construction:

Let T be a semigroup and for each $\alpha \in T$ let G_α be a group such that if $\alpha \neq \beta$ then $G_\alpha \cap G_\beta = \emptyset$. Suppose for each pair $\alpha, \beta \in T$ we have a pair of homomorphisms

$$i_{\alpha,\beta\alpha} \colon G_\alpha \to G_{\beta\alpha}$$
$$j_{\beta,\beta\alpha} \colon G_\beta \to G_{\beta\alpha}$$

such that

(i) if $\beta\alpha = \alpha$ then $i_{\alpha,\beta\alpha}$ is the identity and if $\beta\alpha = \beta$ then $j_{\beta,\beta\alpha}$ is the identity;

(ii) $i_{\beta\alpha,\gamma\beta\alpha} \circ i_{\alpha,\beta\alpha} = i_{\alpha,\gamma\beta\alpha}$
$j_{\alpha\beta,\alpha\beta\gamma} \circ j_{\alpha,\alpha\beta} = j_{\alpha,\alpha\beta\gamma}$;

(iii) $i_{\beta\alpha,\gamma\beta\alpha} \circ j_{\beta,\beta\alpha} = j_{\gamma\beta,\gamma\beta\alpha} \circ i_{\beta,\gamma\beta}$.

Now let $S = \bigcup G_\alpha$ and for $a, b \in S$ ($a \in G_\alpha$, $b \in G_\beta$) define

$$ab = j_{\alpha,\alpha\beta}(a) \cdot i_{\beta,\alpha\beta}(b).$$

A tedious calculation shows that this multiplication is associative and hence S is a semigroup.

Now, if S is semigroup and T is a subsemigroup of S such that $T \cap H$ is a single point for each $\mathscr{H}$-class H then S may be constructed in the manner described in the previous paragraph. For each $t \in T$ let $G_t = \Gamma_t$, the right Schützenberger group of H_t. For $s, t \in T$, let $i_{s,ts} \colon G_s \to G_{ts}$ be the homomorphism induced by left multiplication by t. Let Γ_t' be the left Schützenberger group of H_t and $\sigma_t \colon \Gamma_t \to \Gamma_t'$ be the canonical isomorphism, i.e. $\sigma_t(\alpha) = \beta$ if and only if $\beta t = t\alpha$, and let $k_{t,ts} \colon \Gamma_t' \to \Gamma_{ts}'$ be the homomorphism induced by right multiplication by s. Now define

$$j_{t,ts} \colon G_t \to G_{ts} \quad \text{by} \quad j_{t,ts}(\alpha) = \sigma_{ts}^{-1}(k_{t,ts}(\sigma_t(\alpha))).$$

That the collection of homomorphisms $i_{s,ts}$ and $j_{t,ts}$ satisfy conditions (i) and (ii) is easily verified. Condition (iii) is merely a restatement of the fact that multiplication in S is associative. The correspondence between S and $S' = \bigcup G_\alpha$ is given as follows:

If $a \in S$, then there is a unique $t \in T$ such that $a \in H_t$ and there is a unique $\alpha \in \Gamma_t$ such that $a = t\alpha$; thus we let $a \leftrightarrow \alpha$. One readily verifies that this is in fact an isomorphism. This construction generalizes the construction of an inverse semigroup which is the union of groups as given in [2].

If S fails to contain a cross-section for $S/\mathcal{H}$ or if $S/\mathcal{H}$ fails to be a semigroup, of course, the construction given above also fails. However, in this case, we may still obtain a weak sort of representation of S as endomorphisms on a partial group.

A *partial group* is a union of disjoint groups, G_α, in which multiplication of two members is defined only if they are both members of the same group. If P is a partial group, then a function f is called a partial endomorphism on P if the domain of f, $\mathrm{dom}(f)$, is a subpartial group of P, the range of f is contained in P, and if $x, y \in \mathrm{dom}(f)$ such that xy is defined then $f(x)f(y)$ is defined and $f(xy) = f(x)f(y)$. We denote by $\mathcal{P}$ the set of partial endomorphisms on P. Let $\mathcal{P}_0 = \mathcal{P} \cup \{0\}$ and define multiplication on $\mathcal{P}_0$ by:

$$0f = f0 = 0 \quad \text{all} \quad f \in \mathcal{P}_0$$

$$\text{if} \quad f, g \neq 0, \quad \text{let} \quad fg = 0 \quad \text{if} \quad g^{-1}(\mathrm{dom}(f)) = \emptyset,$$

otherwise let $\mathrm{dom}(fg) = g^{-1}(\mathrm{dom}(f))$ and for $x \in \mathrm{dom}(fg)$ set $fg(x) = f(g(x))$. Then $\mathcal{P}_0$ is a semigroup. By a *weak representation* of a semigroup S we mean a function $F: S \to \mathcal{P}_0$, where $\mathcal{P}_0$ is the semigroup of partial endomorphisms on a partial group, such that if $x, y \in S$ then

$$\mathrm{dom}(F(x)F(y)) \subset \mathrm{dom}(F(xy))$$

and

$$F(x)F(y)[t] = F(xy)[t] \quad \text{for all} \quad t \in \mathrm{dom}(F(x)F(y)).$$

A *representation* of S is a weak representation such that

$$\mathrm{dom}(F(x)F(y)) = \mathrm{dom}(F(xy))$$

for all $x, y \in S$.

Let S be a semigroup with identity and let P_s be the partial group consisting of all Schützenberger groups associated with an element of S, i.e.

$$P_s = \{(x, \alpha) \mid \alpha \in \Gamma_x\}$$

with multiplication defined by:

$$(x, \alpha)(y, \beta) \quad \text{not defined if} \quad x \neq y$$

and

$$(x, \alpha)(x, \beta) = (x, \alpha\beta).$$

Setting $\mathscr{P}_0 =$ the set of partial endomorphisms on P_s, we define $F\colon S \to \mathscr{P}_0$ by:

$$\text{if } \quad x \in S, \quad \mathrm{dom}(F(x)) = \{(y, \alpha) \in P_s \mid \alpha \in \Gamma(x, y)\}$$

and

$$F(x)[(y, \alpha)] = (xy, i_{x,y}(\alpha)) \quad \text{if} \quad (y, \alpha) \in \mathrm{dom}(F(x)).$$

An easy calculation shows that if x, y, $t \in S$ then

$$(*) \qquad\qquad x(yH_t \cap H_{yt}) \cap H_{xyt} \subset xyH_t \cap H_{xyt}$$

from which it follows that F is a partial representation and, further, F is a representation if, and only if, equality obtains in condition $(*)$.

References

[1] J. R. Bastida, Grupos y homomorfismos asociados con un semigrupo I, *Bol. Soc. Mat. Mexicana* (2) **8** (1963), 26–45.

[2] A. H. Clifford and G. B. Preston, *The algebraic theory of semigroups*, Vol. I, Math. Surveys No. 7, Amer. Math. Soc., Providence, R.I., 1961.

[3] J. A. Green, On the structure of semigroups, *Ann. of Math.* **54** (1951), 163–172.

[4] M. P. Schützenberger, $\bar{\mathscr{D}}$ représentation des demi-groupes, *C.R. Acad. Sci. Paris* **244** (1957), 1994–1996.

Proc. Internat. Conf. Theory of Groups, Austral. Nat. Univ. Canberra, August 1965, pp. 7–10. © Gordon and Breach Science Publishers, Inc. 1967

On the number of conjugate classes in a group

CHRISTINE W. AYOUB

I became interested in the number of conjugate classes in a finite group quite recently when Professor Chowla asked me the question: Does the class number become infinite as the order of the group tends to infinity? Or, stated more exactly: Let $r(n)$ be the minimum number of classes which a group of order n can possess. Is $\lim_{n \to \infty} r(n) = \infty$?

In looking up the literature I found that this question was posed by Frobenius and answered in the affirmative some sixty years ago by Landau. The proof is very short and I will outline it.

Let G be a group of order n and suppose that G has r conjugate classes; we write

$$(1) \qquad n = 1 + h_2 + \cdots + h_r$$

where the h_i denote the numbers of elements in the nontrivial classes of G, so that $h_i | n$ for $2 \le i \le r$. Putting $n = h_i m_i$ (for $2 \le i \le r$) and dividing by n we obtain:

$$(2) \qquad 1 = \frac{1}{n} + \frac{1}{m_2} + \frac{1}{m_3} + \cdots + \frac{1}{m_r}.$$

Now it is not difficult to show that for any positive integer r the equation (2) has only a finite number of solutions. Thus $\lim_{n \to \infty} r(n) = \infty$.

In what follows I confine my attention to (finite) p-groups.

Let G be a group of order p^n $(n > 2)$ and let

$$E = Z_0 < Z_1 < \cdots < Z_j < \cdots < Z_c = G$$

be the ascending central series of G. It is easy to verify that if $u \in Z_j$, then $\mathrm{Cl}(u) \subseteq Z_{j-1}u$ $(1 \le j \le c)$ where $\mathrm{Cl}(u)$ denotes the class of elements conjugate to u.

For the group G we rewrite the class equation in the form

$$(3) \qquad\qquad p^n = p^m + k_1 p + \cdots + k_s p^s$$

where p^m is the order of the centre of G and k_j denotes the number of classes with p^j elements. Thus if (3) is the class equation for the group, the group contains $p^m + k_1 + \cdots + k_s$ classes. The following simplifies (3):

Theorem 1.   *If* (3) *is the class equation for the group G, then*

(a) *we can take $s \le n - 2$;*

(b) *$p - 1$ divides each k_j.*

Here (a) is well known and is easy to establish. To prove (b) we notice that if x is an element of order p^t with p^i conjugates, then for $(j, p) = 1$ the elements x and x^j have the same centralizer and hence the same number of conjugates. It can be shown that if $j_1, \ldots, j_{p-1}$ are the $p - 1$ reduced residues mod p^t, satisfying $j_\alpha^{p-1} \equiv 1 \pmod{p^t}$ $(1 \le \alpha \le p - 1)$, then $\mathrm{Cl}(x^{j_1}), \ldots, \mathrm{Cl}(x^{j_{p-1}})$ are $p - 1$ distinct classes with p^i elements apiece. Dividing up the classes with p^i elements in this fashion it is not hard to establish that $p - 1$ divides k_i.

In view of Theorem 1 we have that the class equation of a group of order p^n has the form:

$$(4) \qquad\qquad p^n = p^m + \sum_{i=1}^{n-2} l_i (p - 1) p^i$$

and that the number of classes in G is

$$(5) \qquad\qquad r = p^m + \sum_{i=1}^{n-2} l_i (p - 1).$$

Definition.   If (4) is a partition of p^n which makes the expression (5) for r minimal, (4) is called a minimal partition of p^n.

Theorem 2.   *The minimal partition of p^n is*

$$(6) \qquad p^n = p + (p - 1)p + \cdots + (p - 1)p^{n-3} + (p^2 - 1)p^{n-2}.$$

This can be proved quite easily by taking any partition (4) different from (6) and showing that it can be altered so as to yield a second partition of the form (4) with a smaller r.

If there exists a group G with (6) as its class equation, then some element has p^{n-2} conjugates and this implies[1] that G has the (upper and lower) central series

$$E < Z_1 < \cdots < Z_j < \cdots < Z_{n-2} < Z_{n-1} = G$$

where Z_j has order p^j for $1 \leq j \leq n-2$. Thus since for $u \in Z_j \setminus Z_{j-1}$, $\mathrm{Cl}(u) \subseteq Z_{j-1}u$, for (6) to be the class equation we must have $\mathrm{Cl}(u) = Z_{j-1}u$. It is not hard to show that this can happen only if $n = 3$.

Theorem 3. *Let G be a group of order p^n $(n > 3)$ which contains an element with p^{n-2} conjugates, and define*

$$i = \begin{cases} n/2 & \text{if } n \text{ is even,} \\ (n-1)/2 & \text{if } n \text{ is odd.} \end{cases}$$

Let Z_j be the jth term of the ascending central series for G, and let z_j be a generator of Z_j mod Z_{j-1}. If $\mathrm{Cl}(z_j) = Z_{j-1}z_j$ for $i \leq j \leq n-2$, then $\mathrm{Cl}(z_j) = Z_{j-1}z_j$ for $1 \leq j \leq i$. Furthermore, there exist maximal subgroups M_j $(2 \leq j \leq i)$ such that an element v has p^{n-2} conjugates if, and only if, $v \notin M_j$ for $2 \leq j \leq i$.

Thus if the hypotheses of this theorem are fulfilled, then

$$E < Z_1 < \cdots < Z_{n-2} < \begin{matrix} M_2 \\ \vdots \\ M_i \end{matrix} < G,$$

the classes of the elements of $Z_j \setminus Z_{j-1}$ are just the cosets mod Z_{j-1} for $1 \leq j \leq n-2$, the classes of the elements in $G \setminus (\bigcup_{j=2}^{i} M_j)$ are the cosets mod Z_{n-2}, but the cosets of M_j mod Z_{n-2} split into several classes.

Theorem 4. *The minimal partition of p^3 which is the class equation for a group is $p^3 = p + (p^2 - 1)p$. Therefore $r(p^3) = p + p^2 - 1$.*

The minimal partition of p^4 (for $p \geq 3$) which is the class equation for a group is $p^4 = p + (p^2 - 1)p + (p^2 - 1)p^2$. Therefore $r(p^4) = 2p^2 - 1$.

The minimal partition of p^5 (for $p \geq 5$) which is the class equation for a group is $p^5 = p + (p - 1)p + (p^2 - 1)p + (p^2 - p)p^3$. Therefore $r(p^5) = 2p^2 + p - 2$.

---

[1] See W. BURNSIDE, *Theory of groups of finite order*, 2nd edition, Dover, 1955; pp. 124–125.

The minimal partition of p^6 (for $p \geq 5$) which is the class equation for a group is $p^6 = p + (p-1)p + (p-1)p^2 + 2(p^2-p)p^3 + (p^2-2p+1)p^4$. Therefore $r(p^6) = 3p^2 - p - 1$.

This result is obtained by applying the previous theorem and by exhibiting groups with the appropriate class equations.

Remark. We note that in each of the cases considered we have

$$\lim_{p \to \infty} p^{-2} r(p^n) = i = \begin{cases} n/2 & \text{for } n \text{ even,} \\ (n-1)/2 & \text{for } n \text{ odd.} \end{cases}$$

Proc. Internat. Conf. Theory of Groups, Austral. Nat. Univ. Canberra,
August 1965, pp. 11–15. © Gordon and Breach Science Publishers, Inc. 1967

Nilpotency

REINHOLD BAER

It is well known that the class of finite nilpotent groups may be characterized by a great number of equivalent properties. If one drops the requirement that the groups under consideration be finite, then these properties cease to be equivalent and some of them lose part of their significance, if not their meaning. In the following discussion we are going to analyze the interrelations between these various properties, omitting those whose interest appears to be limited to the class of finite groups and compensating for this by the addition of new properties. In the end we shall return part of the way by a discussion of those additional relations between our properties which appear by restriction of our attention to some comparatively small classes like the class of noetherian [or artinian] groups.

The hierarchy of nilpotency properties begins with

Nilpotency of finite class. A group G is nilpotent of finite class if there exists a finite chain of normal subgroups $N(i)$ of G such that

$$1 = N(0), \ N(i) \subseteq N(i+1), \text{ and}$$
$$N(i+1)/N(i) \subseteq \text{center of } G/N(i), \ N(k) = G.$$

Hypercentral groups are groups such that every epimorphic image, not 1, possesses a nontrivial center. It is clear that groups of finite class are hypercentral and simplest examples show the falsity of the converse.

The normalizer condition is satisfied by the group G, if every proper subgroup of G is different from its normalizer:

$$U \subset G \quad \text{implies} \quad U \subset \mathfrak{n}U$$

where $\mathfrak{n}U$ designates the normalizer of U in G.

A normal chain of G consists of subgroups U_σ with $0 \le \sigma \le \beta$ such that U_σ is a normal subgroup of $U_{\sigma+1}$ whenever $\sigma < \beta$ and such that $U_\lambda = \bigcup_{\sigma < \lambda} U_\sigma$ whenever $\lambda \le \beta$ is a limit ordinal and such that finally $U_\beta = G$.

11

The terms of normal chains are called accessible subgroups of G. It is easily seen that the normalizer condition is satisfied by the group G if, and only if, every subgroup of G is accessible.

We note finally the important fact that the normalizer condition is satisfied by every hypercentral group; Kurosh [7; vol II, p. 219].

The weak accessibility condition is satisfied by the group G if every finitely generated subgroup of G is accessible. It is clear from a preceding remark that the normalizer condition implies the weak accessibility condition; the converse is false as may be seen from Baer [3; p. 69, 6.1].

If we term the element a in G an accessible element of G if $\{a\}$ is an accessible subgroup of G, then the following three properties of G are equivalent:

The weak accessibility condition is satisfied by G.

Every element in G is accessible.

G is generated by its accessible elements.

See Baer [3; p. 57, Satz 3.3 and p. 59, Zusatz 3.6] and Gruenberg [6; p. 158, Theorem 2].

Local Hypercentrality is satisfied by the group G if every finitely generated subgroup of G is hypercentral. The weak accessibility condition implies local hypercentrality; see Baer [3; p. 58, Zusatz 3.4] and Gruenberg [p. 158, Theorem 2].

The Engel condition is satisfied by the group G, if for every pair x, y of elements in G almost all the iterated commutators

$$x^{(i)} \circ y = 1.$$

These commutators are defined inductively by the rules

$$x^{(0)} \circ y = y, \qquad x^{(1)} \circ y = x \circ y = x^{-1}y^{-1}xy,$$
$$x^{(i+1)} \circ y = x \circ [x^{(i)} \circ y].$$

It is well known that local hypercentrality implies the Engel condition. The converse is false, as has been shown recently by Golod and Šafarevič [5].

The star condition is satisfied by the group G whenever there exists to every pair of subgroups X, Y of G with

$$X_Y \subset X \subset Y$$

an element y with

$$X \subset \{X, y\} \subseteq Y \quad \text{and} \quad X_Y \subset X \cap X^y.$$

Here the core X_Y of X with respect to Y is defined by

$$X_Y = \bigcap_{y \in Y} X^y \,;$$

it is the product of all the normal subgroups of Y which are part of X.

The star condition is a consequence of the Engel condition; see Baer [4; p. 58, Folgerung 2.2, and p. 59, Zusatz 2.3]. Since a finite group may be shown to be nilpotent—see Baer [4; p. 59, Zusatz 2.3, and p. 63, Hauptsatz 4.1]—if, and only if, it satisfies the star condition, and since the star condition is inherited by all subgroups and epimorphic images, the star condition implies

The most comprehensive nilpotency condition: Every finite epimorphic image of every subgroup is nilpotent.

Nilpotency of finitely generated groups. It is known that finitely generated hypercentral groups are noetherian; see Baer [1; p. 208, Theorem]. Here we term a group noetherian if all its subgroups are finitely generated [and this is equivalent with the validity of the maximum condition for subgroups]. It follows that the following properties of finitely generated groups are equivalent:

nilpotency of finite class; hypercentrality; normalizer condition; weak accessibility condition; local hypercentrality.

On the other hand it is shown by the example of Golod and Šafarevič [5] that finitely generated Engel groups need not be hypercentral. Since noetherian Engel groups are known to be nilpotent of finite class— see Baer [2; p. 257, Satz N]—it follows that the group G is locally hypercentral if, and only if, it is a locally noetherian Engel group.

Certain ramifications of this hierarchy arise from the concept of Engel element. Here we term an element e of the group G an Engel element of G, more precisely a left Engel element of G, if for every x in G almost all the commutators

$$e^{(i)} \circ x = 1.$$

It is clear then that the Engel condition is satisfied by the group G if, and only if, every element of G is an Engel element of G. Consequently each of the following two conditions is a consequence of the Engel condition:

(E.1) G is generated by its Engel elements.

(E.2)   Every subgroup of G which is generated by two elements is generated by its Engel elements.

These two conditions are prima facie weaker than the Engel condition.

It is known that a noetherian group is nilpotent of finite class if, and only if, it is generated by its Engel elements; see Baer [2; p. 257, Satz N]. Thus the group G is locally hypercentral if, and only if, G is locally noetherian and generated by its Engel elements. Furthermore G is an Engel group if every subgroup which is generated by two elements is noetherian and generated by its Engel elements.

To find a common consequence of the star condition and the conditions (E.i) we introduce the concept of a

Star system.   This is a subset $\mathfrak{S}$ of the group G meeting the following three requirements:

$\mathfrak{S}$ is not empty;

$\mathfrak{S} = \mathfrak{S}^G$ is normal;

if X, Y are subgroups of G with $X_Y \subset X \subset Y$, if $\mathfrak{S} \cap Y$ is not part of X and $\mathfrak{S} \cap X$ is not part of X_Y, then there exists an element y in $\mathfrak{S} \cap Y$ which does not belong to X and an element x in $\mathfrak{S} \cap X$ which does not belong to X_Y such that x^y belongs to X.

One may prove that the set of all the Engel elements of a group G is a star system; see Baer [4; p. 58, Lemma 2.1]. Furthermore the star condition is satisfied by the group G if, and only if, the set of elements in G is a star system; see Baer [4; p. 59, Zusatz 2.3].

It follows now that the condition (E.i) implies the condition (S.i) phrased as follows:

(S.1) G is generated by a star system.

(S.2)   Every subgroup of G which is generated by two elements is generated by a star system.

It is furthermore seen that (S.1) is a consequence of the star condition.

A group may be termed a star group if it is generated by one of its star systems. Then the following comprehensive result may be proven— see Baer [4; p. 63, Hauptsatz 4.1]:

The following properties of the group G are equivalent.

(1) G is hypercentral and finitely generated.

(2) G is a noetherian star group.

(3) G is a star group whose abelian subgroups are finitely generated.

(4) $\begin{cases} \text{(a) Subgroups of } G\text{, generated by two elements, are star groups.} \\ \text{(b) Abelian subgroups of } G \text{ are finitely generated.} \end{cases}$

(5)
- (a) G is finitely generated.
- (b) Epimorphic images of G are star groups.
- (c) If $H \neq 1$ is an epimorphic image of G, then there exists a normal subgroup $N \neq 1$ of H such that the abelian subgroups of N' are finitely generated.

(6)
- (a) G is finitely generated.
- (b) Epimorphic images of G are star groups.
- (c) If $H \neq 1$ is an epimorphic image of G, then there exists a normal subgroup $N \neq 1$ of H such that H induces a noetherian group of automorphisms in N'.

(7)
- (a) G is finitely generated.
- (b) Every finite epimorphic image of a subgroup of G is a star group.
- (c) Every epimorphic image, not 1, of G possesses an abelian normal subgroup, not 1.

(8)
- (a) Every abelian subgroup of G is finitely generated.
- (b) Every finite epimorphic image of a subgroup of G is a star group.
- (c) Every epimorphic image, not 1, of G possesses an abelian normal subgroup, not 1.

Similar results on artinian star groups have recently been obtained by Dr Dieter Held.—A detailed account of our results may be found in Baer [4].

References

[1] REINHOLD BAER, The hypercenter of a group, *Acta Math.* **89** (1953), 165–208.
[2] REINHOLD BAER, Engelsche Elemente noetherscher Gruppen, *Math. Ann.* **133** (1957), 256–270.
[3] REINHOLD BAER, Erreichbare und engelsche Gruppenelemente, *Abh. Math. Sem. Univ. Hamburg* **27** (1964), 44–74.
[4] REINHOLD BAER, Die Sternbedingung: Eine Erweiterung der Engelbedingung, *Math. Ann.* **162** (1965), 54–73.
[5] E. S. GOLOD and I. R. ŠAFAREVIČ, On class field towers, *Izv. Akad. Nauk SSSR Ser. Mat.* **28** (1964), 261–272; E. S. GOLOD, On nil-algebras and finitely approximable p-groups, *ibid.* 273–276 [Russian].
[6] K. W. GRUENBERG, The Engel elements of a soluble group, *Illinois J. Math.* **3** (1959), 151–168.
[7] A. G. KUROSH, *The theory of groups*, Chelsea, New York, 1960.
[8] B. I. PLOTKIN, Generalized soluble and generalized nilpotent groups, *Uspehi Mat. Nauk (N.S.)* **13** (1958), 89–172 [Russian]; *Amer. Math. Soc. Transl.* (2) **17** (1961), 29–115.

Proc. Internat. Conf. Theory of Groups, Austral. Nat. Univ. Canberra,
August 1965, pp. 17–32. © Gordon and Breach Science Publishers, Inc. 1967

Noetherian soluble groups

REINHOLD BAER

A group is termed *polycyclic* if all its subgroups are finitely generated
and almost all its derivatives equal 1. Mal'cev has shown that the group
G is polycyclic if, and only if, every abelian subgroup of G is finitely
generated and almost all derivatives of G equal 1; and we have shown
elsewhere that a group is polycyclic if, and only if, all its abelian subgroups are finitely generated and all its epimorphic images, not 1,
possess abelian normal subgroups, not 1; see Baer [2; p. 173, Hauptsatz
4]. In the present note we give a further improvement of these results
by showing that a group is polycyclic if, and only if, all its abelian
subnormal subgroups are finitely generated and all its epimorphic
images, not 1, possess abelian subnormal subgroups, not 1 [Theorem 4.2].
Furthermore we obtain the following extension of this result: The group
G is noetherian and possesses a polycyclic subgroup of finite index if,
and only if, every abelian subnormal subgroup of G is noetherian,
independent sets of subnormal subgroups of epimorphic images of G
are finite, and infinite epimorphic images of G possess abelian subnormal
subgroups, not 1 [Theorem 5.8].

Notations

   $\{\ldots\}$ = subgroup, generated by the elements enclosed in the brackets
$A \subset B :=:A$ is a proper subgroup of B
$A \cap B$ = intersection of A and B
   $a \circ b = a^{-1}b^{-1}ab$
$A \circ B$ = subgroup, generated by all the $a \circ b$ for a in A and b in B
   $G' = G \circ G, \quad G^{(0)} = G, \quad G^{(i+1)} = [G^{(i)}]'$
   $\mathfrak{z}G$ = center of G
   $\mathfrak{c}X$ = centralizer of X in G

Terminology

The group G is *hypercentral*, if $_\delta H \neq 1$ for every epimorphic image $H \neq 1$ of G; and G is *nilpotent of finite class*, if there exists a finite central chain of G connecting 1 and G.

The group G is *hyperabelian*, if every epimorphic image, not 1, of G possesses an abelian normal subgroup, not 1; and G is *soluble of finite degree*, if $G^{(i)} = 1$ for at least one i.

The group G is *noetherian*, if all its subgroups are finitely generated; and this is equivalent with the validity of the maximum condition for subgroups.

$$\text{Polycyclic} = \text{noetherian} + \text{hyperabelian}$$
$$= \text{noetherian} + \text{soluble of finite degree.}$$

Subnormal subgroup = member of a finite normal chain.

$\mathfrak{H}G$ = product of all hypercentral normal subgroups of G.

$\mathfrak{K}G$ = product of all hyperabelian normal subgroups of G.

1. We begin by deriving a simple criterion for hypercentrality that will be needed in the sequel.

Proposition 1.1. *The group G is hypercentral if, and only if, it is locally hypercentral and every epimorphic image, not 1, of G possesses a noetherian and abelian normal subgroup, not 1.*

The necessity of the conditions is quite obvious. The proof of their sufficiency will be preceded by the proofs of some auxiliary propositions, involving weaker hypotheses.

(1.1.a)   *If $F \neq 1$ is a finite normal subgroup of the locally hypercentral group L, then $F \cap {}_\delta L \neq 1$.*

This is an immediate consequence of McLain [10; p. 7, Theorem 2.2].

(1.1.b)   *If the normal subgroup $F \neq 1$ of the locally hypercentral group L is free abelian of finite rank, then $F \cap {}_\delta L \neq 1$.*

Proof.   Every normal subgroup X of L with $1 \subset X \subseteq F$ is free abelian of finite positive rank. Hence there exists among these X one M of minimal rank. Then M is a normal subgroup of L with $1 \subset M \subseteq F$. It is a free abelian group of finite positive rank. If Y is a normal subgroup of L with $1 \subset Y \subseteq M$, then we deduce from the minimality of the rank of M that the ranks of M and Y are equal. Since the rank of M is finite, this implies the finiteness of M/Y. Thus we have shown:

(∗)   If Y is a normal subgroup of L with $1 \subset Y \subseteq M$, then M/Y is finite.

Assume now by way of contradiction that $M \not\subseteq_{\mathfrak{z}} L$. This is equivalent to $1 \neq M \circ L$. Hence $M \circ L$ is a normal subgroup of L with $1 \subset M \circ L \subseteq M$. We deduce from $(*)$ the finiteness of $M/(M \circ L)$ so that the index $[M : (M \circ L)]$ is a positive integer. Consequently there exists a prime p which does not divide $[M : (M \circ L)]$.

It follows from our choice of M that M is a normal subgroup of L and free abelian of finite positive rank. The characteristic subgroup M^p of the normal subgroup M of L is a normal subgroup of L; and M/M^p is a finite elementary abelian p-group, not 1. Among all the normal subgroups T of L with $M^p \subseteq T \subset M$—note that M^p is one of these T—there exists because of the finiteness of M/M^p a maximal one, say S. Then M/S is a minimal normal subgroup of L/S. Since L/S is with L locally hypercentral, and since M/S is a finite normal subgroup, not 1, of L/S, we may apply (1.1.a). Hence $1 \neq (M/S) \cap_{\mathfrak{z}} (L/S)$. Since M/S is a minimal normal subgroup of L/S, this implies $M/S \subseteq_{\mathfrak{z}} (L/S)$. This is equivalent with

$$M \circ L \subseteq S \subset M.$$

From $M^p \subseteq S \subset M$ we deduce that the prime p divides the finite index $[M : S]$. But $[M : S]$ is a divisor of $[M : (M \circ L)]$ so that p too divides $[M : (M \circ L)]$, contradicting our choice of the prime p. This contradiction proves that $M \subseteq_{\mathfrak{z}} L$. Hence

$$1 \subset M \subseteq F \cap_{\mathfrak{z}} L.$$

(1.1.c)   *If the normal subgroup $F \neq 1$ of the locally hypercentral group L is noetherian and abelian, then $F \cap_{\mathfrak{z}} L \neq 1$.*

Proof.   If F is torsionfree, then F is a free abelian normal subgroup of finite rank and application of (1.1.b) shows that $F \cap_{\mathfrak{z}} L \neq 1$. If F is not torsionfree, then the torsion subgroup T of the noetherian abelian group F is a finite characteristic subgroup, not 1, of F and hence a finite normal subgroup, not 1, of L. Application of (1.1.a) shows that

$$1 \subset T \cap_{\mathfrak{z}} L \subseteq F \cap_{\mathfrak{z}} L;$$

and thus we have verified (1.1.c) in either case.

Proof of Proposition 1.1.   As we pointed out before, we have to prove only the sufficiency of our conditions. Thus we assume that the group G is locally hypercentral and that every epimorphic image, not 1, of G possesses an abelian and noetherian normal subgroup different from 1. Consider any epimorphic image $H \neq 1$ of G. Then H likewise is locally hypercentral. By hypothesis, there exists a normal subgroup $F \neq 1$ of H which is abelian and noetherian. Application of (1.1.c) shows that

$$1 \subset F \cap {}_3H \subseteq {}_3H.$$

Thus G is hypercentral.

Remark 1.2. It is well known that there exist groups G with the following properties:

G is infinite; $G'' = {}_3G = 1$;

every finitely generated subgroup of G is a finite p-group.

See, for instance, Baer [5; p. 69, 6.1].

This shows that locally hypercentral groups which are in a fairly strong sense soluble need not be hypercentral.

Corollary 1.3. *The group G is hypercentral, if it is the product of the normal subgroups X of G with the property:*

(∗) *X is hypercentral and ${}_3Y$ is noetherian for every epimorphic image Y of X.*

Proof. Products of hypercentral normal subgroups are locally hypercentral; see, for instance, Baer [6; Proposition D.2]. Hence G is locally hypercentral. If $H \neq 1$ is an epimorphic image of G, then H too is the product of its normal subgroups with the property (∗). Hence there exists a normal subgroup $A \neq 1$ of H with the property (∗). It follows in particular that ${}_3A$ is noetherian and different from 1. But ${}_3A$ is a characteristic subgroup of the normal subgroup A of H and hence a normal subgroup of H. Thus we may apply Proposition 1.1 to show the hypercentrality of G.

Remark 1.4. It follows in particular from Corollary 1.3 that products of noetherian and hypercentral normal subgroups are hypercentral. The example mentioned in Remark 1.2 shows that products of normal subgroups which are nilpotent of finite class need not be hypercentral.

2. Our objective in the present section is the derivation of the following criterion.

Proposition 2.1. *A hypercentral group is noetherian if [and only if] its abelian normal subgroups are noetherian.*

Proof. Suppose that G is a hypercentral group and that every abelian normal subgroup of G is noetherian. There exists—as in every

NOETHERIAN SOLUBLE GROUPS

group—a maximal abelian normal subgroup A of G. Then A is noetherian and $A \subseteq cA$. Assume by way of contradiction that $A \subset cA$. Since cA is, with A, a normal subgroup of G, it follows that cA/A is a normal subgroup, not 1, of G/A. Since G is hypercentral, this implies

$$1 \neq [cA/A] \cap \mathfrak{z}[G/A].$$

Consequently there exists a subgroup B of G with $A \subset B \subseteq cA$ and cyclic $B/A \subseteq \mathfrak{z}[G/A]$. Because of the last condition B is a normal subgroup of G; because of the cyclicity of B/A we have $B = \{A, b\}$. Since b belongs to the centralizer of A, we have found an abelian normal subgroup B of G with $A \subset B$, contradicting the maximality of A. Thus we have shown $A = cA$ so that G/A is essentially the same as the group of automorphisms induced in A by G. Since G is hypercentral, this group of automorphisms is likewise hypercentral and *a fortiori* hyperabelian. But A is an abelian and noetherian group and every hyperabelian group of automorphisms of such a group is known to be noetherian; see Baer [2; p. 171, Satz 2]. Thus A and G/A are noetherian, proving that G itself is noetherian.

3. It will be convenient to denote by $\mathfrak{H}G$ *the product of all the hypercentral normal subgroups of the group* G. This is a well determined characteristic subgroup of G.

Proposition 3.1. *$\mathfrak{H}G$ is noetherian and hypercentral if, and only if, every hypercentral normal subgroup of G is noetherian.*

Proof. Hypercentral normal subgroups of G are subgroups of $\mathfrak{H}G$ so that they are noetherian in case $\mathfrak{H}G$ is noetherian. If conversely every hypercentral normal subgroup of G is noetherian, then we deduce from Corollary 1.3 and Remark 1.4 that $\mathfrak{H}G$ as a product of noetherian hypercentral normal subgroups is a hypercentral subgroup; and this implies, by hypothesis, that $\mathfrak{H}G$ is noetherian, too.

Theorem 3.2. *The following properties of the hyperabelian group G are equivalent:*

(i) *G is noetherian.*

(ii) *Every abelian subnormal subgroup of G is noetherian.*

(iii) *Every abelian normal subgroup of a hypercentral normal subgroup of G is noetherian.*

(iv) *Every hypercentral normal subgroup of G is noetherian.*

Proof. It is clear that (i) implies (ii) and that (ii) implies (iii); and it is a consequence of Proposition 2.1 that (iv) is a consequence of (iii).

Assume finally by way of contradiction that the group G meets requirement (iv) without being noetherian. We let $H = \mathfrak{H}G$. It is a consequence of (iv) and Proposition 3.1 that the characteristic subgroup H of G is noetherian and hypercentral. The group of automorphisms, induced in H by G, is essentially the same as G/cH. Thus it is hyperabelian. It is known that hyperabelian groups of automorphisms of noetherian and hyperabelian groups are noetherian; see Baer [2; p. 172, Hauptsatz 3]. Application of this result shows that G/cH is noetherian.

$cH \subseteq H$ would imply that H and the epimorphic image G/H of G/cH are noetherian. Hence G itself would be noetherian, contradicting our assumption. Hence $cH \nsubseteq H$ and this is equivalent to $HcH/H \neq 1$. Since G is hyperabelian, this implies the existence of an abelian normal subgroup, not 1, of G/H which is part of HcH/H; see Baer [4; p. 17, Lemma 3.2]. Consequently there exists a normal subgroup A of G with

$$A' \subseteq H \subset A \subseteq HcH.$$

We consider the descending central chains A_i and H_i of A and H respectively which are defined inductively by the following rules:

$$A_0 = A, \qquad A_{i+1} = A \circ A_i,$$
$$H_0 = H, \qquad H_{i+1} = H \circ H_i.$$

Clearly $A_1 = A' \subseteq H = H_0$. Thus we may make the inductive hypothesis: $A_i \subseteq H_{i-1}$. It follows that

$$A_{i+1} = A \circ A_i \subseteq [HcH] \circ H_{i-1} \subseteq [H \circ H_{i-1}][cH \circ H] = H_i,$$

completing our inductive proof of $A_j \subseteq H_{j-1}$.

Since H is noetherian and hypercentral, H is nilpotent of finite class. Hence $H_k = 1$ for at least one k and this implies $A_{k+1} = 1$. Thus A is nilpotent of finite class and as such A is a hypercentral normal subgroup. By definition of $\mathfrak{H}$ we have

$$A \subseteq \mathfrak{H}G = H \subset A,$$

a contradiction. Thus (iv) implies (i), proving the equivalence of conditions (i)–(iv).

Example 3.3. McLain [9] has constructed a group M with the following properties:

(a) M is not abelian.

(b) M is the product of its abelian normal subgroups.

(c) 1 and M are the only characteristic subgroups of M.

Since M is in particular locally hypercentral, though not hypercentral, application of Corollary 1.3 shows that

(d) not all abelian normal subgroups of M are noetherian.

Denote by G the holomorph of M. It is a consequence of (c) that M is a minimal normal subgroup of G and that M is part of every normal subgroup, not 1, of G. Since M is not hypercentral, 1 is the only hypercentral normal subgroup of G. Thus G is a group with the following two properties:

every hypercentral normal subgroup of G is noetherian;

there exist abelian subnormal subgroups of G which are not noetherian.

Hence condition (iv) of Theorem 3.2 is really weaker than condition (ii).

Example 3.4. Denote by N a group, generated by the elements c, a_i, b_i for $i = 0, \pm 1, \pm 2, \ldots$, subject to the following relations:

$$\{c, a_0, a_{\pm 1}, a_{\pm 2}, \ldots\} \quad \text{and} \quad \{c, b_0, b_{\pm 1}, b_{\pm 2}, \ldots\}$$

are elementary abelian p-groups for p an odd prime;

$$a_i \circ b_j = \begin{cases} c & \text{if } i = j, \\ 1 & \text{if } i \neq j. \end{cases}$$

We note that $N' = {}_3 N = \{c\}$.

This group N possesses automorphisms α, β, σ defined by the following rules:

$$c^\alpha = c^\beta = c^{-1}, \quad a_i^\alpha = a_i^{-1}, \quad b_i^\alpha = b_i, \quad a_i^\beta = a_i, \quad b_i^\beta = b_i^{-1},$$
$$c^\sigma = c, \quad a_i^\sigma = a_{i+1}, \quad b_i^\sigma = b_{i+1}.$$

The group $\Sigma = \{\alpha, \beta, \sigma\}$ of automorphisms is abelian; it is the direct product of the elementary abelian group $\{\alpha, \beta\}$ of order 4 and the infinite cyclic group $\{\sigma\}$.

Denote finally by ω the automorphism of N meeting the following requirements:

$$a_i^\omega = b_i, \quad b_i^\omega = a_i, \quad c^\omega = c^{-1}.$$

This automorphism ω has order 2 and satisfies

$$\alpha^\omega = \beta, \quad \beta^\omega = \alpha, \quad \omega\sigma = \sigma\omega.$$

Thus Σ is a normal subgroup of index 2 of the group $\Gamma = \{\Sigma, \omega\}$ of automorphisms of N, so that in particular $\Gamma'' = 1$.

We form the product $G = N\Gamma$ in the holomorph of N. Then $\{c\}$ is the only abelian normal subgroup, not 1, of G. Furthermore $G'''' = 1$ and $G = \{a_0, b_0, \alpha, \beta, \sigma, \omega\}$. Thus G is finitely generated; its abelian normal subgroups are finite; it is soluble in a very strict sense; but G is not noetherian. This shows the impossibility of substituting in Theorem 3.2 for condition (ii) or (iv) the weaker condition:

every abelian normal subgroup of G is noetherian.

4. In this section we are going to investigate the groups whose abelian subnormal subgroups are noetherian. It will be convenient to denote by $\mathfrak{K}G$ *the product of all the hyperabelian normal subgroups of G.* This is a well determined characteristic subgroup of G and clearly $\mathfrak{H}G \subseteq \mathfrak{K}G$.

Proposition 4.1. *If every abelian subnormal subgroup of the group G is noetherian, then*

(a) *$\mathfrak{H}G$ contains every hypercentral subnormal subgroup of G,*

(b) *$\mathfrak{K}G$ is polycyclic,*

(c) *every hyperabelian subnormal subgroup of G is part of $\mathfrak{K}G$,*

(d) *1 is the only abelian subnormal subgroup of $G/\mathfrak{K}G$ and $\mathfrak{K}[G/\mathfrak{K}G] = 1$.*

Proof. If S is a hyperabelian subnormal subgroup of G, then every abelian subnormal subgroup of S is an abelian subnormal subgroup of G and hence noetherian. Application of Theorem 3.2 shows that S is noetherian and hence we have shown

(e) *every hyperabelian subnormal subgroup is polycyclic.*

$\mathfrak{K}G$ is, by (e) and its definition, the product of its polycyclic normal subgroups and the same is true of every epimorphic image $H \neq 1$ of $\mathfrak{K}G$. Hence there exists a polycyclic normal subgroup $J \neq 1$ of H. There exists a positive integer j such that $1 = J^{(j)} \subset J^{(j-1)}$ and $J^{(j-1)}$ is, as a characteristic subgroup of the normal subgroup J of H, an abelian normal subgroup of H. This shows that $\mathfrak{K}G$ is hyperabelian and by (e) polycyclic, proving (b).

Consider now a finite number of subgroups $S(i)$ of G such that $S(i)$ is a normal subgroup of $S(i+1)$ and $S(n) = G$. Then every $S(i)$ is a subnormal subgroup of G and every abelian subnormal subgroup of $S(i)$ is a subnormal subgroup of G and as such it is by hypothesis noetherian. Proposition 3.1 is applicable by (e). Hence $\mathfrak{H}S(i)$ is noetherian and hypercentral and application of (b) shows that $\mathfrak{K}S(i)$ is polycyclic. As characteristic subgroups of the normal subgroup $S(i)$ of $S(i+1)$ both $\mathfrak{H}S(i)$ and $\mathfrak{K}S(i)$ are normal subgroups of $S(i+1)$. Application of the definitions of $\underline{\mathfrak{H}}$ and $\mathfrak{K}$ shows

$$\mathfrak{H}S(i) \subseteq \mathfrak{H}S(i+1) \quad \text{and} \quad \mathfrak{K}S(i) \subseteq \mathfrak{K}S(i+1).$$

It follows by complete induction that

$$\mathfrak{H}S(0) \subseteq \mathfrak{H}S(n) = \mathfrak{H}G \quad \text{and} \quad \mathfrak{K}S(0) \subseteq \mathfrak{K}S(n) = \mathfrak{K}G;$$

and (a) and (c) are immediate consequences of these inequalities.

If $S/\mathfrak{K}G$ is a hyperabelian subnormal subgroup of $G/\mathfrak{K}G$, then S is a subnormal subgroup of G. Since $\mathfrak{K}G$ is, by (b), polycyclic, S is itself hyperabelian. We deduce $S \subseteq \mathfrak{K}G$ from (c) so that $S/\mathfrak{K}G = 1$. Hence

(d*) 1 *is the only hyperabelian subnormal subgroup of* $G/\mathfrak{K}G$.

It is clear that (d) is a consequence of (d*).

Theorem 4.2. *The group G is polycyclic if, and only if,*

(a) *every abelian subnormal subgroup of G is noetherian and*

(b) *every epimorphic image, not 1, of G possesses an abelian subnormal subgroup, not 1.*

Proof. We note first that G is polycyclic if, and only if, G is noetherian and hyperabelian. This puts into evidence the necessity of (a) and (b).

If conversely conditions (a) and (b) are satisfied by G, then we deduce from Proposition 4.1, (b) and (d) that firstly $\mathfrak{K}G$ is polycyclic and that secondly $G = \mathfrak{K}G$, proving the sufficiency of our conditions.

Remark 4.3. It is interesting to contrast Theorem 4.2 with the following result—see Baer [7; p. 164, Satz 1.1]:

The group G is polycyclic if, and only if,

(A) *every abelian subgroup of G is noetherian and*

(B) *every epimorphic image, not 1, of G possesses a locally hypercentral normal subgroup, not 1.*

Noting that abelian subnormal subgroups are always contained in locally hypercentral normal subgroups one sees that (B) is a consequence of condition (b) of Theorem 4.2 whereas condition (a) of Theorem 4.2 is clearly weaker than (A).

5. Following common usage we term a group *almost polycyclic* if it possesses a polycyclic subgroup of finite index; for some general properties of this class of groups see Baer [2; p. 149, Satz 1]. The characterization of polycyclic groups, obtained in Section 4, shall now be extended to almost polycyclic groups.

Lemma 5.1. *If 1 is the only abelian normal subgroup of G, then the following properties of G are equivalent:*

(i) *The maximum condition is satisfied by the finite normal subgroups of G.*

(ii) *The product of all finite normal subgroups of G is finite.*

(iii) *Independent sets of finite normal subgroups of G are finite.*

Terminological reminder. The set Σ of subgroups is *independent* if the subgroup generated by the subgroups in Σ is the direct product of the subgroups in Σ. One verifies that Σ is independent if, and only if, different subgroups in Σ centralize each other and

$$X \cap \prod_{\substack{Y \neq X \\ Y \in \Sigma}} Y = 1 \quad \text{for every } X \text{ in } \Sigma.$$

Proof. In the presence of (i) there exists a maximal finite normal subgroup M of G. If X is any finite normal subgroup of G, then MX is a finite normal subgroup of G. Because of the maximality of M we conclude $X \subseteq MX = M$. Hence M is the product of all the finite normal subgroups of G; and we have deduced (ii) from (i).

It is immediately obvious that both (i) and (iii) are consequences of (ii).

Assume finally the existence of an infinite properly ascending chain of finite normal subgroups N_i of G. Then we have

$$N_1 \subset \cdots \subset N_i \subset N_{i+1} \subset \cdots \subset \bigcup_{i=1}^{\infty} N_i = N;$$

and it is clear that the join N of all the N_i is an infinite normal subgroup of G. Next we note that cN_i is a normal subgroup of G and that G/cN_i is essentially the same as the group of automorphisms induced in N_i

by G. Since N_i is finite, so is G/cN_i. Since 1 is the only abelian normal subgroup of G, we have

$$1 = {}_3N_i = N_i \cap cN_i.$$

Let $N_i^* = N \cap cN_i$. This is a normal subgroup of G with the following two properties:

(a) $1 = N_i \cap N_i^*$.

(b) $N/N_i^* \simeq NcN_i/cN_i \subseteq G/cN_i$ is finite.

Now we are going to construct by complete induction finite normal subgroups F_i of G with the following properties:

(1) $1 \subset F_i \subset N$.

(2) $F_1, \ldots, F_n$ are independent for every n.

To begin our construction we let $F_1 = N_1$. Suppose that we have already constructed $F_1, \ldots, F_n$ for some positive n. Then $T_n = \prod_{i=1}^n F_i$ is a finite normal subgroup of G with $T_n \subseteq N$. Since T_n is finite and $N = \bigcup_{i=1}^{\infty} N_i$, there exists a positive integer m with

$$T_n \subseteq N_m.$$

Since N/N_m^* is, by (b), finite, and since N is by construction infinite, there exists a positive integer k with

$$1 \neq N_k \cap N_m^*.$$

Let $F_{n+1} = N_k \cap N_m^*$. Then F_{n+1} is a finite normal subgroup of G, since N_k is finite, which is part of N. Thus (1) is satisfied. Because of (a) we have

$$\prod_{i=1}^n F_i \cap F_{n+1} = T_n \cap F_{n+1} \subseteq N_m \cap N_m^* = 1.$$

Hence $F_1, \ldots, F_n, F_{n+1}$ are, by (2), independent.

This completes the inductive construction of the F_i, showing the existence of an infinite independent set of finite normal subgroups of G. Hence (iii) is false if (i) is false. This completes the proof of the equivalence of conditions (i)–(iii).

Remark 5.2. The hypothesis that 1 is the only abelian normal subgroup of G has been used only in our derivation of condition (i) from (iii). But here it is indispensable as may be seen from the example of Prüfer's group of type p^{∞} which meets requirement (iii), but neither (i) nor (ii).

Lemma 5.3. *If 1 is the only finite normal subgroup of G, if 1 is the only finite abelian subnormal subgroup of G, and if independent sets of [finite] subnormal subgroups of G are finite, then 1 is the only finite subnormal subgroup of G.*

Proof. If this were false, then there would exist a minimal finite subnormal subgroup M of G. Clearly M is a finite simple group and not abelian. Application of a theorem of Wielandt [11; p. 463, (1.a)] shows that M normalizes every subnormal subgroup of G. This implies:

If A and B are two different subgroups of G, conjugate to M in G, then A and B centralize each other.

Now one verifies without difficulty that the normal subgroup $\{M^G\}$ of G, spanned by M, is the direct product of all the subgroups in M^G. Hence M^G is an independent set of subnormal subgroups of G and as such M^G is, by hypothesis, finite. Consequently $\{M^G\}$ is the direct product of finitely many finite groups and as such $\{M^G\}$ is a finite normal subgroup of G. This implies by our first hypothesis

$$1 \subset M \subseteq \{M^G\} = 1,$$

a contradiction that proves the validity of our lemma.

Theorem 5.4. *The group G is almost polycyclic if, and only if,*

(a) *every abelian subnormal subgroup of G is noetherian,*

(b) *every epimorphic image, not 1, of G, possesses a subnormal subgroup, not 1, with finite commutator subgroup and*

(c) *independent sets of subnormal subgroups of epimorphic images of G are finite.*

Proof. If G is almost polycyclic, then G is noetherian, proving the validity of (a) and (c). If furthermore $H \neq 1$ is an epimorphic image of G, then there exists a polycyclic subgroup S of H with finite index $[H:S]$. In particular S possesses but a finite number of conjugate subgroups in H. Their intersection J is, by Poincaré's Theorem, a polycyclic normal subgroup of H with finite H/J. If firstly $J = 1$, then H is finite. If secondly $J \neq 1$, then there exists a positive integer n with

$$1 = J^{(n)} \subset J^{(n-1)}.$$

Clearly $J^{(n-1)}$ is as a characteristic subgroup of J an abelian normal subgroup, not 1, of H. This shows the validity of (b)—and much more.

Assume conversely the validity of (a), (b), (c). Because of (a) we may apply Proposition 4.1. Consequently $\mathfrak{K}G$ is a polycyclic characteristic subgroup of G and 1 is the only abelian subnormal subgroup of $H = G/\mathfrak{K}G$.

Because of (c) we may apply Lemma 5.1 upon H. It follows that the product P of all finite normal subgroups of H is finite. If we let $J = H/P$, then J is an epimorphic image of G.

Every finite normal subgroup of J has the form X/P. Here X is a normal subgroup of H which is finite, since P and X/P are finite. It follows that $X \subseteq P$ because of our construction of P. Hence $X/P = 1$ so that

(1)   1 is the only finite normal subgroup of J.

Suppose that $j \neq 1$ is a subnormal element of J. This is equivalent to saying that $\{j\}$ is a subnormal subgroup of J. It follows from Baer [3; p. 418, Satz 2] that the normal subgroup $T = \{j^J\}$ of J, spanned by j, consists of subnormal elements only. $T \neq 1$, since $j \neq 1$. Hence T is infinite by (1).

Naturally T has the form $T = Q/P$ where Q is a normal subgroup of H. We note that

$$1 = {}_3P = P \cap \mathfrak{c}P,$$

since 1 is the only abelian subnormal subgroup of H. Since P is finite, so is $H/\mathfrak{c}P$ as $H/\mathfrak{c}P$ is essentially the same as the group of automorphisms, induced in P by H.

If $Q \cap \mathfrak{c}P = 1$, then

$$Q = Q/[Q \cap \mathfrak{c}P] \simeq Q\,\mathfrak{c}P/\mathfrak{c}P \subseteq H/\mathfrak{c}P$$

is finite. But Q is infinite, since $Q/P = T$ is infinite. Hence $Q \cap \mathfrak{c}P \neq 1$. Consider an element $a \neq 1$ in $Q \cap \mathfrak{c}P$. Then Pa is an element in T and as such Pa is a subnormal element in J. It follows that $\{P, a\}$ is a subnormal subgroup of H. Hence $\{P, a\} \cap \mathfrak{c}P$ is likewise a subnormal subgroup of H. Because of $\{a\} \subseteq \mathfrak{c}P$ we may apply Dedekind's modular law. It follows that

$$\{P, a\} \cap \mathfrak{c}P = P\{a\} \cap \mathfrak{c}P = \{a\}[P \cap \mathfrak{c}P] = \{a\} \neq 1$$

is a cyclic subnormal subgroup of H. This is impossible. Hence 1 is the only subnormal element of J and this implies:

(2)   1 is the only abelian subnormal subgroup of J.

Because of (1), (2), and (c) we may apply Lemma 5.3 on J. It follows that

(3) 1 is the only finite subnormal subgroup of J.

Assume now by way of contradiction that $J \neq 1$. Then there exists by (b) a subnormal subgroup $S \neq 1$ of J with finite S'. With S its commutator subgroup S' is subnormal and (3) implies $S' = 1$. Hence S is abelian and $S = 1$ is a consequence of (2). This is a contradiction proving $J = 1$. Hence $H = P$ is finite so that $G/\Re G$ is finite. Since $\Re G$ is polycyclic, G is almost polycyclic, as we wanted to show.

Since condition (b) has been used in the last paragraph only, our proof of Theorem 5.4 is easily seen to contain a proof of the following

Proposition 5.5. *If every abelian subnormal subgroup of G is noetherian and if independent sets of subnormal subgroups of epimorphic images of G are finite, then*

(1) *$\Re G$ is polycyclic,*

(2) *the product P of all finite normal subgroups of $H = G/\Re G$ is finite,*

(3) *1 is the only finite and the only abelian subnormal subgroup of H/P.*

To extend this result we denote by $\mathfrak{A}\mathfrak{P}G$ the *product of all almost polycyclic normal subgroups of G.* This is in any case a well determined characteristic subgroup of G.

Corollary 5.6. *If every abelian subnormal subgroup of G is noetherian, and if independent sets of subnormal subgroups of epimorphic images of subnormal subgroups of G are finite, then*

(1) *$\mathfrak{A}\mathfrak{P}G$ is almost polycyclic,*

(2) *$\mathfrak{A}\mathfrak{P}G$ contains every almost polycyclic subnormal subgroup of G,*

(3) *1 is the only almost polycyclic subnormal subgroup of $G/\mathfrak{A}\mathfrak{P}G$ and*

(4) *$\mathfrak{A}\mathfrak{P}[G/\mathfrak{A}\mathfrak{P}G] = 1$.*

Proof. Application of Proposition 5.5 shows that $\Re G$ is polycyclic and that the product P of all finite normal subgroups of $G/\Re G$ is finite. There exists one and only one subgroup Q of G with $\Re G \subseteq Q$ and $P = Q/\Re G$. Then Q is a characteristic and almost polycyclic subgroup. In particular $Q \subseteq \mathfrak{A}\mathfrak{P}G$.

Suppose now that X is an almost polycyclic normal subgroup of G. Then there exists a polycyclic subgroup Y of X with finite index $[X : Y] = j$. Since X is noetherian, there exists only a finite number of subgroups of index j in X; see, for instance, Baer [1; p. 331, Folgerung 3]. The intersection J of all S with $[X : S] = j$ is consequently a characteristic subgroup of X whose index $[X : J]$ is finite by Poincaré's Theorem. As a characteristic subgroup of a normal subgroup J is a normal subgroup of G. By construction $J \subseteq Y$. Hence J is polycyclic. It follows that $J \subseteq \Re G$. Furthermore

$$X \; \Re G / \Re G \; \simeq \; X / [X \cap G \Re]$$

is an epimorphic image of X/J, since $J \subseteq \Re G \cap X$. Hence $X \; \Re G / \Re G$ is a finite normal subgroup of $G / \Re G$ so that $X \; \Re G / \Re G \subseteq P$ and $X \subseteq X \; \Re G \subseteq Q$. Thus we have shown that $\mathfrak{A} \mathfrak{P} G \subseteq Q$. Hence $\mathfrak{A} \mathfrak{P} G = Q$ is almost polycyclic proving (1).

If T is an almost polycyclic subnormal subgroup of G, then there exist finitely many subgroups T_i with $T = T_0$, T_i is a normal subgroup of T_{i+1} and $T_n = G$. Naturally $T = \mathfrak{A} \mathfrak{P} T_0$. Since T_i is a subnormal subgroup of G, the abelian subnormal subgroups of T_i are subnormal subgroups of G and hence noetherian. Thus (1) is applicable so that $\mathfrak{A} \mathfrak{P} T_i$ is an almost polycyclic characteristic subgroup of the normal subgroup T_i of T_{i+1}. Hence $\mathfrak{A} \mathfrak{P} T_i \subseteq \mathfrak{A} \mathfrak{P} T_{i+1}$. Consequently

$$T = \mathfrak{A} \mathfrak{P} T_0 \subseteq \cdots \subseteq \mathfrak{A} \mathfrak{P} T_i \subseteq \mathfrak{A} \mathfrak{P} T_{i+1} \subseteq \cdots \subseteq \mathfrak{A} \mathfrak{P} T_n = \mathfrak{A} \mathfrak{P} G,$$

proving (2).

If U is an almost polycyclic subnormal subgroup of $G / \mathfrak{A} \mathfrak{P} G$, then there exists one and only one subgroup V of G with $\mathfrak{A} \mathfrak{P} G \subseteq V$ and $U = V / \mathfrak{A} \mathfrak{P} G$. Clearly V is a subnormal subgroup of G and an extension of the almost polycyclic group $\mathfrak{A} \mathfrak{P} G$ by the almost polycyclic group U. It is easily verified that V is an almost polycyclic subnormal subgroup of G. Application of (2) shows $V \subseteq \mathfrak{A} \mathfrak{P} G$ so that $U = 1$, proving (3); and (4) is an immediate consequence of (3).

Remark 5.7. Denote by S a non-abelian, finite, simple group. If G is the direct product of a countable infinity of groups isomorphic to S, then 1 is the only abelian subnormal subgroup of G. Thus G meets requirements (a), (b) of Theorem 5.4, though G is not noetherian. Hence requirement (c) of Theorem 5.4 is indispensable. Similarly one sees that the second hypothesis of Proposition 5.5 is indispensable for the validity of Proposition 5.5 (2).

Theorem 5.8. *The group G is almost polycyclic if, and only if,*

(a) *every abelian subnormal subgroup of G is noetherian,*

4—K.

(b) *independent sets of subnormal subgroups of epimorphic images of G are finite and*

(c) *infinite epimorphic images of G possess abelian subnormal subgroups, not 1.*

The necessity of conditions (a)–(c) is almost obvious; their sufficiency is an immediate consequence of Corollary 5.6.

References

[1] REINHOLD BAER, Das Hyperzentrum einer Gruppe, III, *Math. Z.* **59** (1953), 299–338.

[2] REINHOLD BAER, Auflösbare Gruppen mit Maximalbedingung, *Math. Ann.* **129** (1955), 139–173.

[3] REINHOLD BAER, Nilgruppen, *Math. Z.* **62** (1955), 402–437.

[4] REINHOLD BAER, Gruppen mit Minimalbedingung, *Math. Ann.* **150** (1963), 1–44.

[5] REINHOLD BAER, Erreichbare und engelsche Gruppenelemente, *Abh. Math. Sem. Univ. Hamburg* **27** (1964), 44–74.

[6] REINHOLD BAER, Local and global hypercentrality and supersolubility, *Indaz, Math.* **28** (1966), 93–126.

[7] REINHOLD BAER, Noethersche Gruppen, II, *Math. Ann.* **165** (1966), 163–180.

[8] MARSHALL HALL, JR., *The theory of groups*, Macmillan, New York, 1959.

[9] D. H. McLAIN, A characteristically-simple group, *Proc. Cambridge Philos. Soc.* **50** (1954), 641–642.

[10] D. H. McLAIN, On locally nilpotent groups, *Proc. Cambridge Philos. Soc.* **52** (1956), 5–11.

[11] HELMUT WIELANDT, Über den Normalisator der subnormalen Untergruppen, *Math. Z.* **69** (1958), 463–465.

Proc. Internat. Conf. Theory of Groups, Austral. Nat. Univ. Canberra,
August 1965, pp. 33–36. © Gordon and Breach Science Publishers, Inc. 1967

Noetherian groups

REINHOLD BAER

More than a quarter of a century ago Charles Hopkins discovered that
in sufficiently many rings the maximum condition is a consequence of
the minimum condition. No such phenomenon has been discussed
within the theory of groups. In the following discussion we aim at
showing that certain aspects of the theory of groups with minimum
condition are consequences of the corresponding facts of the theory of
groups with maximum condition.

We shall term a group $\left\{ \begin{array}{c} artinian \\ noetherian \end{array} \right\}$ if the $\left\{ \begin{array}{c} minimum \\ maximum \end{array} \right\}$ condition is

satisfied by its subgroups. The following two theorems are known since
more than a decade. They show that a phenomenon similar to the
one discovered by Hopkins in ring theory exists in group theory; and
they will constitute the backbone of our investigation.

Theorem N.A. *Hyperabelian
groups of automorphisms of noetherian abelian groups are noetherian.*

Theorem A.A. *Torsion groups of
automorphisms of artinian abelian
groups are finite.*

Here we term a group *hyperabelian* if its epimorphic images, not 1,
possess abelian normal subgroups, not 1.

We are going to state next criteria for a group to be soluble and
noetherian or soluble and artinian; and these criteria will either be
strictly parallel or else the artinian criterion will arise from the
noetherian one principally by substituting, in a fashion similar to the
above theorems, finite for noetherian.

Theorem N.B. *The following properties of the group G are equivalent:*

Theorem A.B. *The following properties of the group G are equivalent:*

33

(1) *G is noetherian and* $G^{(i)} = 1$ *for almost all* i.

(2) *Abelian subnormal subgroups of* G *are noetherian; epimorphic images, not* 1, *possess abelian subnormal subgroups, not* 1.

(3) *Hyperabelian subnormal subgroups induce noetherian groups of automorphisms in their abelian normal subgroups; epimorphic images, not* 1, *possess abelian subnormal subgroups, not* 1; *and* $_3\mathfrak{A}G$ *is noetherian.*

(4) *Hypercentral normal subgroups are noetherian;* G *is hyperabelian.*

(1) *G is artinian and* $G^{(i)} = 1$ *for almost all* i.

(2) *Abelian subnormal subgroups of* G *are artinian; epimorphic images, not* 1, *possess abelian subnormal subgroups, not* 1; *G is a torsion group.*

(3) *Hyperabelian subnormal subgroups induce finite groups of automorphisms in their abelian normal subgroups; epimorphic images, not* 1, *possess abelian subnormal subgroups, not* 1; *and* $_3\mathfrak{A}G$ *is artinian.*

(4) *Hypercentral normal subgroups are artinian;* G *is a hyperabelian torsion group.*

Here $\mathfrak{A}G$ is the product of all abelian normal subgroups of G and $_3\mathfrak{A}G$ is the center of $\mathfrak{A}G$; and we term a group *hypercentral* if its epimorphic images, not 1, have center, not 1.

To obtain a more striking form of these two theorems we consider the classes of groups instead of their individual members. For this consideration we need a group theoretical property $\mathfrak{k}$, meeting the following requirements:

A. Subnormal subgroups and epimorphic images of $\mathfrak{k}$-groups are $\mathfrak{k}$-groups.

B. Extensions of $\mathfrak{k}$-groups by $\mathfrak{k}$-groups are $\mathfrak{k}$-groups.

C. If G is a $\mathfrak{k}$-group, then every epimorphic image, not 1, of G possesses an abelian subnormal subgroup, not 1.

Theorem N.C. $\mathfrak{k} = polycyclicity$ *if, and only if,* $\mathfrak{k}$-*groups of automorphisms of abelian* $\mathfrak{k}$-*groups are noetherian and not all* $\mathfrak{k}$-*groups are torsion groups.*

Theorem A.C. *Every* $\mathfrak{k}$-*group is artinian if, and only if,* $\mathfrak{k}$-*groups of automorphisms of abelian* $\mathfrak{k}$-*groups are finite.*

The similarity between these two theorems suggests the search for a common source of these two results. This common root will be a criterion essentially for $G/_3\mathfrak{A}G$ to be noetherian. As in the artinian case $G/_3\mathfrak{A}G$ will be finite, this produces in a way a subsumption of artinian theory under the noetherian one.

Theorem C. *Let* e *be a group theoretical property, meeting the following requirements:*

I. *Subgroups and epimorphic images of* e-*groups are* e-*groups.*

II. *Extensions of* e-*groups by* e-*groups are* e-*groups.*

III. e-*groups are noetherian.*

IV. *Hyperabelian groups of automorphisms of hypercentral* e-*groups are* e-*groups.*

(C.1)   *Then* $G/_3\mathfrak{A}G$ *is an* e-*group and* $G^{(i)} = 1$ *for almost all* i *if, and only if,*

G is hyperabelian,

e-*groups of automorphisms are induced by* G *in its abelian normal subgroups and*

every hypercentral normal subgroup induces an e-*group of automorphisms in at least one of its maximal abelian normal subgroups.*

(C.2) $S/_3\mathfrak{A}S$ *is an* e-*group for every subnormal subgroup* S *of* G *and* $G^{(i)} = 1$ *for almost all* i *if, and only if,*

every hyperabelian subnormal subgroup induces in all its abelian normal subgroups e-*groups of automorphisms and*

every epimorphic image, not 1, *of* G *possesses an abelian subnormal subgroup, not* 1.

The applicability of Theorem C in the proof of Theorems N.B and A.B stems from the fact that both the property of being noetherian and the property of being finite meet the requirements I–IV. Beyond that we need naturally the Theorems N.A and A.A and the following general principle which permits to substitute often for hyperabelian the following condition:

epimorphic images, not 1, possess abelian subnormal subgroups, not 1.

Theorem D. $G^{(i)} = 1$ *for almost all* i *if, and only if,*

$X^{(i)} = 1$ *for almost all* i *and every hyperabelian subnormal subgroup* X *of* G *and epimorphic images, not* 1, *of G possess abelian subnormal subgroups not* 1.

A detailed account of these investigations will be published in the *Mathematische Annalen*.

Bibliography

[1] REINHOLD BAER, Finite extensions of Abelian groups with minimum condition, *Trans. Amer. Math. Soc.* **79** (1955), 521–540.
[2] REINHOLD BAER, Auflösbare Gruppen mit Maximalbedingung, *Math. Ann.* **129** (1955), 139–173.
[3] REINHOLD BAER, Auflösbare, artinsche, noethersche Gruppen, *Math. Ann.* [forthcoming].

Proc. Internat. Conf. Theory of Groups, Austral. Nat. Univ. Canberra, August 1965, pp. 37–50. © Gordon and Breach Science Publishers, Inc. 1967

Finitely presented groups

GILBERT BAUMSLAG*

1. Introduction

1.1. Many problems in analysis, geometry, and topology give rise to problems in associated finitely presented groups. It is therefore not surprising that the study of finitely presented groups is so diverse and difficult. In particular there seems to be no single main theme in the entire theory. In this brief survey we shall proceed along six main lines: subgroups, algorithms, invariants, constructions, properties, and unsolved problems.

1.2. I would like to point out that this account is in no way complete— there is a great body of material which is not even mentioned (see, e.g., H. S. M. Coxeter and W. O. J. Moser [16]).

2. Subgroups

2.1. The first comprehensive subgroup theorems seem to have arisen in the theory of automorphic functions. One such theorem concerns the so-called Fuchsian groups, which are special quotient groups of discrete groups of 2×2 real matrices with determinant 1. Here a group G of 2×2 real matrices is discrete if it does not contain any infinitesimal substitutions; in other words there is no sequence

$$g_1, g_2, g_3, \ldots \quad (g_1 \neq 1, g_2 \neq 1, g_3 \neq 1, \ldots)$$

of elements of G which converges to 1. If the discrete group G contains J, the subgroup generated by -1, then the quotient group G/J is termed Fuchsian.

The monumental work of R. Fricke and F. Klein [21] contains the following

* The author gratefully acknowledges help from the National Science Foundation.

Theorem 1 (Fricke and Klein, 1897). *A finitely generated group F is Fuchsian if and only if it can be presented in the form*[1]

$$F = (a_1, b_1, \ldots, a_k, b_k, c_1, \ldots, c_l, d_1, \ldots, d_m; c_1^{e_1}, \ldots, c_l^{e_l},$$
$$a_1^{-1}b_1^{-1}a_1b_1\cdots a_k^{-1}b_k^{-1}a_kb_k \cdot c_1\cdots c_l \cdot d_1\cdots d_m).$$

Obviously subgroups of Fuchsian groups are Fuchsian; so we have the following immediate but somewhat surprising consequence of Theorem 1.

Corollary A. *Let*

$$F = (a_1, b_1, \ldots, a_k, b_k, c_1, \ldots, c_l, d_1, \ldots, d_m; c_1^{e_1}, \ldots, c_l^{e_l},$$
$$a_1^{-1}b_1^{-1}a_1b_1\cdots a_k^{-1}b_k^{-1}a_kb_k \cdot c_1\cdots c_l \cdot d_1\cdots d_m).$$

If G is a finitely generated subgroup of F then G can be presented in the form

$$G = (w_1, x_1, \ldots, w_h, x_h, y_1, \ldots, y_i, z_1, \ldots, z_j; y_1^{f_1}, \ldots, y_i^{f_i},$$
$$w_1^{-1}x_1^{-1}w_1x_1\cdots w_h^{-1}x_h^{-1}w_hx_h \cdot y_1\cdots y_i \cdot z_1\cdots z_j).$$

This corollary has many interesting specializations. For example, if

$$G_k = (a_1, b_1, \ldots, a_k, b_k; a_1^{-1}b_1^{-1}a_1b_1\cdots a_k^{-1}b_k^{-1}a_kb_k)$$

then every subgroup of finite index in G_k is a G_l for some $l \geq k$. A direct algebraic proof of this fact seems out of the question.

2.2. By examining the geometry involved in the proof of Theorem 1 L. Greenberg [25] has proved a number of interesting theorems about Fuchsian groups. For example Greenberg has shown that the intersection of two finitely generated subgroups of a Fuchsian group is again finitely generated, thereby generalizing the corresponding result due to A. G. Howson [32] for free groups.

This theorem about free groups prompts us to point out another specialization of Corollary A, namely that a finitely generated subgroup of a free group is free. The first proof that *every* subgroup of a free group is free is due to J. Nielsen [51], 1921. The next proof, by O. Schreier [59] in 1927, gave rise to the so-called Reidemeister–Schreier method for finding generators and defining relations for a subgroup H of a group G given by generators and defining relations

---

[1] We do not exclude the possibility that k or l or m is zero; it is to be interpreted to mean that the relevant generators are absent.

(K. Reidemeister [58], 1932). This method involves choosing a system of representatives of the (right or left) cosets of H in G; in many applications this turns out to be a major stumbling block. In the theory of groups with a single defining relation in particular, however, it can be used to great advantage. Perhaps the most important theorem in this theory is the Freiheitssatz of W. Magnus [40]:

Theorem 2 (W. Magnus, 1930). *Let G be a group with a single defining relator r:*

$$G = (a_1, a_2, \ldots, a_n; r).$$

If r involves all the generators $a_1, a_2, \ldots, a_n$ and is cyclically reduced, then $a_1, a_2, \ldots, a_{n-1}$ freely generate a free subgroup of G.

One consequence of Theorem 2 is the torsion-freeness of groups with a single defining relation when the relator is not a proper power (A. Karrass, W. Magnus and D. Solitar [33], 1960).

2.3. The last result was prompted by the following question of C. D. Papakyriakopoulos about elements of finite order in certain groups arising in connection with the Poincaré Conjecture: if w is an element in the derived group of the free group on $a_1, b_1, \ldots, a_k, b_k$, is

$$G = (a_1, b_1, \ldots, a_k, b_k; a_1^{-1}b_1^{-1}a_1b_1 \cdots a_k^{-1}b_k^{-1}a_kb_k, a_1^{-1}w^{-1}b_1^{-1}a_1b_1w)$$

torsion-free? This question was answered affirmatively by E. S. Rapaport [57] in 1964. It brings to mind a remarkable application of topology to group theory involving knot groups. A knot is a circle embedded as a 1-dimensional subcomplex K of the 3-sphere S^3 (see, e.g., L. P. Neuwirth [50]). The group of the knot K is the fundamental group of the residual space $S^3 - K$. E. Artin [2] proved in 1926 that a group G is a knot group if and only if it can be presented in the form

$$G = (t_1, \ldots, t_n; T_1t_1T_1^{-1}t_2^{-1}, \ldots, T_{n-1}t_{n-1}T_{n-1}^{-1}t_n^{-1}, T_nt_nT_n^{-1}t_1^{-1})$$

where $T_1, \ldots, T_n$ are words in $t_1, \ldots, t_n$ satisfying the relation

$$T_1t_1T_1^{-1} \cdots T_nt_nT_n^{-1} = t_2t_3 \cdots t_nt_1.$$

One of the consequences of Papakyriakopoulos' work in [55] is the

Theorem 3 (Papakyriakopoulos, 1957). *Every knot group is torsion-free.*

No algebraic proof of Theorem 3 exists. Conceivably the Reidemeister–Schreier method might yield a reasonable set of generators and defining relations for the derived group G_2 of a knot group G since G/G_2 is infinite cyclic. The combinatorial difficulties are, however, enormous.

2.4. The subgroup theorems I have mentioned so far are concerned with rather special classes of finitely presented groups. Thinking along somewhat more general lines one might ask whether a finitely generated subgroup of a finitely presented group can be infinitely related. Actually it was only in 1937 that B. H. Neumann [46] proved that there *are* finitely generated infinitely related groups. The first example of a finitely presented group with a finitely generated but infinitely related subgroup appeared over twenty years later (G. Baumslag, W. W. Boone, and B. H. Neumann [7], 1959). A variation of their example is

$$G = (a, b, c, d;\, a^{-1}c^{-1}ac,\, a^{-1}d^{-1}ad,\, b^{-1}c^{-1}bc,\, b^{-1}d^{-1}bd),$$

the direct product of two free groups of rank two. The two-generator subgroups of G are either free or free abelian, but the subgroup

$$H = \mathrm{gp}(ac, b, d)$$

generated by ac, b, and d is infinitely related.

A general survey of all the finitely generated subgroups of finitely presented groups was not even dreamed of in 1959. But in 1961 G. Higman [29] settled this problem in the most amazing way by proving

Theorem 4 (G. Higman, 1961). *A finitely generated group can be embedded in a finitely presented group if and only if it can be defined by a recursively enumerable set of defining relations.*

It follows from this theorem that the (restricted) direct product of all finite groups, one for each isomorphism class, is a subgroup of a finitely presented group. It is therefore little wonder that finitely presented groups are so complicated.

3. Algorithms

3.1. Let f be a recursive function whose range is not a recursive set and let G be the group generated by a, b, c and d subject to the defining relations

$$b^{-f(n)}ab^{f(n)} = d^{-f(n)}cd^{f(n)} \qquad (n = 0, 1, 2, \ldots).$$

This group has an unsolvable word problem. By Theorem 4, G can be effectively embedded in a finitely presented group H; so H has an unsolvable word problem! This is G. Higman's approach to the word problem and is the culmination of over fifty years work on this and related problems.

3.2. The search for an algorithmic solution of the word problem, the conjugacy problem and the isomorphism problem was started by M. Dehn [17] in 1911 when he settled all three problems positively for the groups G_k (see 2.1).

The next main advance was in 1926 when E. Artin [2] provided a positive solution of the word problem for the braid groups B_n, where

$$B_n = (s_1, \ldots, s_{n-1}; s_i s_{i+1} s_i = s_{i+1} s_i s_{i+1}, s_i s_k = s_k s_i$$

$$(1 \leq i < k-1 \leq n-1, \quad k < n))$$

for $n = 2, 3, \ldots$.

The first real breakthrough in the word problem occurred in 1932 when W. Magnus [41] proved by an ingenious use of the Freiheitssatz the

Theorem 5 (Magnus, 1932). *Every group with a single defining relator has a solvable word problem.*

Seventeen years later V. A. Tartakovski [62], [63] made another big breakthrough by proving

Theorem 6 (V. A. Tartakovski, 1949). *If the group G is given by a system of generators and defining relations, where the defining relators are in some sense independent, then G has a solvable word problem.*

Theorem 6 has been rather hazily put since the actual conditions are somewhat complicated. We refer the reader to the survey by M. Hall [27] for the details. Tartakovski's work was extended shortly afterwards by J. L. Britton [12] in 1956. Later, in 1960, M. Greendlinger [26] solved the conjugacy problem for a related class of groups.

3.3. The first examples of finitely presented groups with unsolvable word problems were discovered by P. S. Novikov [54] and W. W. Boone [10] (see also [11]) at about the same time.

Theorem 7 (P. S. Novikov, 1955; W. W. Boone, 1954–57). *There is a finitely presented group with an unsolvable word problem.*

Shortly afterwards J. L. Britton [13], 1958, proceeding along slightly different lines constructed further finitely presented groups with unsolvable word problems (see also J. L. Britton [14], 1963).

In 1960 A. A. Fridman [22] solved the word problem for a further class of finitely presented groups. Among the groups in this class is one which P. S. Novikov [53] has shown has an unsolvable conjugacy problem. So *there exist finitely presented groups with unsolvable conjugacy problems but having solvable word problems.*

3.4. The existence of a finitely presented group with an unsolvable word problem leads to a host of negative results about finitely presented groups. Perhaps the most impressive of these is summarized by the following

Theorem 8 (S. I. Adyan [1], 1955; M. O. Rabin [56], 1958). *Let P be a property of groups inherited by isomorphic images and finitely presented subgroups. If there is a finitely presented group with P and a finitely presented group without P, then there is no general and effective procedure whereby one can determine whether any finitely presented group has P.*

In particular it follows that there is no general and effective procedure whereby one can determine whether any finitely presented group is of order one! This settles the isomorphism problem negatively in a rather spectacular way.

A similar theorem about elements exists (G. Baumslag, W. W. Boone, and B. H. Neumann [7], 1959).

In 1958 A. A. Markov [43] utilized the negative solution of the isomorphism problem, showing that the homeomorphism problem for three-manifolds is effectively undecidable. More recently J. Stallings (1963, unpublished) has proved that there is no general and effective procedure whereby one can determine whether any finitely presented group is a knot group.

3.5. In closing this section I should mention the solution of the word problem for finitely presented residually finite groups by J. C. C. McKinsey [44], 1943 (see also V. H. Dyson [18], 1964), the solution of the conjugacy problem for finitely generated nilpotent groups by N.

Blackburn [9], 1965, and the solution of the conjugacy problem for the free product of two free groups with a cyclic subgroup amalgamated by S. Lipschutz (unpublished).

4. Invariants

4.1. In view of the negative solution of the isomorphism problem it is perhaps not surprising that invariants are hard to find. Of course for any finitely presented group the various factor groups by the terms of the lower central series are computable invariants (see, e.g., K. T. Chen, R. H. Fox, and R. C. Lyndon [15], 1958).

4.2. A more interesting type of invariant is the deficiency of a finitely presented group G, which is the maximum of the deficiencies of the finite presentations of G. The deficiency of a finite presentation

$$G = (a_1, a_2, \ldots, a_n; r_1, r_2, \ldots, r_m)$$

is $n - m$. The deficiency of a finite group is non-positive. In 1904 I. Schur [60] investigated finite groups with zero deficiency; these turn out to have trivial Schur-multiplicator. In modern terminology the Schur-multiplicator of a group G is the second cohomology group $H^2(G, C^*)$ of G with coefficients in the multiplicative group C^* of non-zero complex numbers, where the action of G on C^* is trivial.

4.3. We have been led inevitably to a further collection of invariants, the various homology and cohomology groups of a group. Despite the fact that the precise role of homology and cohomology in group theory is little understood their usefulness is undisputed. In this regard I would like to mention the rather surprising work of W. Ledermann and B. H. Neumann [34], [35], 1956, on the number of automorphisms of a finite group, in which cohomology plays an important part.

More recently J. L. Mennicke [45], 1965, on the one hand, and H. Bass, M. Lazard and J-P. Serre on the other ([3], 1964) proved independently, and at approximately the same time, the following

Theorem 9 (J. L. Mennicke, 1965; H. Bass, M. Lazard, J-P. Serre, 1964). *Let $n > 2$ and let $SL(n, Z)$ be the group of $n \times n$ matrices of determinant 1 over the ring Z of rational integers. Then every normal subgroup of $SL(n, Z)$ is either finite or of finite index.*

Mennicke's proof of Theorem 9 is achieved by some simple, but ingenious algebraic manipulations; the proof by Bass *et al.* is largely based on some complicated cohomological calculations.

Another application of cohomology to group theory is the very recent proof by W. Gaschütz [23], 1965, that a finite nonabelian p-group has an outer automorphism.

4.4. Cohomology groups are, in general, very difficult to compute. The cohomology of free groups and cyclic groups have long been known (see S. Eilenberg and S. MacLane [20], 1947).

In 1950 R. C. Lyndon computed the cohomology of groups with a single defining relation. In particular he proved

Theorem 10 (R. C. Lyndon [38], 1950). *The cohomological dimension of a group with a single defining relator r which is not a proper power is at most two.*

The content of Theorem 10 is simply that the cohomology groups of such groups with a single defining relation are zero in every dimension greater than two, irrespective of the coefficient groups involved. Groups of cohomological dimension two crop up in topology as the fundamental groups of aspherical spaces (see S. MacLane [39], p. 137). Now Papakyriakopoulos proved in [55] that the residual space $S^3 - K$ of a knot K is aspherical. So we have then the remarkable

Theorem 11 (C. D. Papakyriakopoulos [55], 1957). *The cohomological dimension of a knot group is at most two.*

5. Constructions

5.1. The existence of a free product of groups with an amalgamated subgroup was first proved by O. Schreier [59] in 1927. W. Magnus [41] was first to recognize its value, making applications to the theory of groups with a single defining relation in 1932. The usefulness of the notion of a generalized free product for calculating the fundamental groups of certain spaces was underlined by E. R. van Kampen [64] in 1933.

The real power of this construction was made apparent by the embedding theorems for countable groups by G. Higman, B. H. Neumann, and H. Neumann [30], 1949. Since then their approach has led to a remarkably straightforward example of a finitely presented group with an unsolvable word problem by J. L. Britton [14], 1963; it has also been of great use in G. Higman's investigations into the subgroups of finitely presented groups [29], 1961.

Of course the direct and the free product of two finitely presented groups is again finitely presented; however it is easy to see that the generalized free product of two finitely presented groups is finitely presented only if the amalgamated subgroup is finitely generated (G. Baumslag [4], 1962).

5.2. Another important way of making finitely presented groups is by forming extensions. One need only observe that an extension of a finitely presented group G by a finitely presented group H is again finitely presented.

6. Properties

6.1. In making an extension of one finitely presented group G by a second finitely presented group H one needs, among other things, information about the automorphism group of G and the factor groups of H.

Now the automorphism group of a finitely presented group is an awkward group to work out. There is no general information available and, indeed, very little special information. It is however known that the automorphism group of a finitely generated free group is finitely presented (J. Nielsen [52], 1924) and also that the automorphism group of a finitely generated abelian group is finitely presented (W. Magnus [42], 1934). More recently P. Gold [24], 1961, has proved that the automorphism groups of the groups G_k are finitely generated; whether they are also finitely related seems still unknown.

6.2. Despite slight evidence to the contrary there are finitely presented groups whose automorphism groups are not finitely presented—this has recently been established by J. Lewin [37], 1964. In fact Lewin has shown that the group

$$G = (a, b, c, d; c^{-1}aca^{-2}, c^{-1}bcb^{-2}, a^{-1}b^{-1}ab, d^{-1}adb^{-1}a^{-1},$$
$$d^{-1}bdb^{-1}, d^{-1}c^{-1}dc)$$

has an *infinitely generated* automorphism group.

This group G is residually finite. Now the automorphism group of a finitely generated residually finite group is again residually finite

(G. Baumslag [5], 1963); however we see from the group G above that finiteness of presentation need not persist in the automorphism group of even a residually finite, finitely presented group.

It has been known for a long time now that free groups are residually finite (F. W. Levi [36], 1933). The list of finitely presented residually finite groups now includes polycyclic groups (K. A. Hirsch [31], 1946), the free product of two free groups with a cyclic subgroup amalgamated (G. Baumslag [6], 1963), and the free product of two finitely generated torsion-free nilpotent groups amalgamating an isolated (closed) subgroup of each factor (G. Baumslag [6], 1963).

6.3.   Finitely generated residually finite groups are hopfian, i.e., every epi-endomorphism is an automorphism. The first example of a finitely generated non-hopfian group was constructed by B. H. Neumann [47], 1950; this was followed by an example of a finitely presented non-hopfian group (G. Higman [28], 1951). A strikingly simple example of a non-hopfian group is provided by the group

$$G = (a, b; a^{-1}b^2ab^{-3});$$

indeed the epi-endomorphism defined by

$$a \to a, \qquad b \to b^2$$

has a nontrivial kernel (G. Baumslag and D. Solitar [8], 1962).

7. Unsolved problems

7.1.   The unsolved problems presented here are mainly new. In a way they are natural outgrowths of the line of development adopted in this survey. Consequently they are grouped into five classes.

7.2.   *Subgroup problems*

(i) Does every finitely presented group have a finitely generated center?

(ii) Is every finitely generated subgroup of a group with a single defining relation finitely related?

(iii) Is every finitely generated subgroup of a knot group finitely related?

7.3.   *Algorithmic problems*

(i) Does a finitely presented residually finite group have a solvable conjugacy problem?

(ii) Does there exist a computable complete set of invariants for a group with a single defining relation?

(iii) Does every knot group have a solvable word problem?

7.4. *Invariants problems*

(i) If G coincides with its derived group and $G \neq 1$, is there always some G-module A for which the second cohomology group $H^2(G, A)$ of G with coefficients in A is nontrivial?

(ii) If

$$G = (t_1, t_2, \ldots, t_n; T_1 t_1 T_1^{-1} t_2^{-1}, \ldots, T_n t_n T_n^{-1} t_1^{-1}),$$

where the T_i are words in the t_i, is the cohomological dimension of G less than three?

(iii) Let F be a free group, let G be a residually nilpotent group and suppose that

$$G/G_i \cong F/F_i \qquad (i = 1, 2, \ldots),$$

where X_i denotes the ith term of the lower central series of the group X. Is then the cohomological dimension of G at most two?

7.5. *Constructions problems*

(i) Is every knot group a generalized free product of two free groups? (This question is due to L. P. Neuwirth.)

7.6. *Properties problems*

(i) Is every knot group residually finite? (This question is due to S. Abhyankar.)

(ii) Is the automorphism group of a polycyclic group finitely presented?

(iii) Is a group with a single defining relation which is not torsion free, residually finite?

References

[1] S. I. ADYAN, Algorithmic unsolvability of problems of recognition of certain properties of groups, *Dokl. Akad. Nauk SSSR* **103** (1955), 533–535 [Russian].

[2] E. ARTIN, Theorie der Zöpfe, *Abh. Math. Sem. Univ. Hamburg* **4** (1926), 47–72.

[3] H. BASS, M. LAZARD, and J-P. SERRE, Sous-groupes d'indice fini dans $SL(n, Z)$, *Bull. Amer. Math. Soc.* **70** (1964), 385–392.

[4] GILBERT BAUMSLAG, A remark on generalized free products, *Proc. Amer. Math. Soc.* **13** (1962), 53–54.

[5] GILBERT BAUMSLAG, Automorphism groups of residually finite groups, *J. London Math. Soc.* **38** (1963), 117–118.

[6] GILBERT BAUMSLAG, On the residual finiteness of generalized free products of nilpotent groups, *Trans. Amer. Math. Soc.* **106** (1963), 193–209.

[7] GILBERT BAUMSLAG, W. W. BOONE, and B. H. NEUMANN, Some unsolvable problems about elements and subgroups of groups, *Math. Scand.* **7** (1959), 191–201.

[8] GILBERT BAUMSLAG and DONALD SOLITAR, Some two-generator one-relator non-hopfian groups, *Bull. Amer. Math. Soc.* **68** (1962), 199–201.

[9] NORMAN BLACKBURN, Conjugacy in nilpotent groups, *Proc. Amer. Math. Soc.* **16** (1965), 143–148.

[10] WILLIAM W. BOONE, Certain simple, unsolvable problems of group theory, I, II, *Nederl. Akad. Wetensch. Proc. Ser. A.* **57** (1954), 231–237, 492–497; III, IV, ibid. **58** (1955), 252–256, 571–577; V, VI, ibid. **60** (1957), 22–27, 227–232.

[11] WILLIAM W. BOONE, The word problem, *Ann. of Math.* (2) **70** (1959), 207–265.

[12] J. L. BRITTON, Solution of the word problem for certain types of groups, I, II, *Proc. Glasgow Math. Assoc.* **3** (1956–57), 45–54, 68–90.

[13] J. L. BRITTON, The word problem for groups, *Proc. London Math. Soc.* (3) **8** (1958), 493–506.

[14] JOHN L. BRITTON, The word problem, *Ann. of Math.* (2) **77** (1963), 16–32.

[15] K. T. CHEN, R. H. FOX, and R. C. LYNDON, Free differential calculus, IV. The quotient groups of the lower central series, *Ann. of Math.* (2) **68** (1958), 81–95.

[16] H. S. M. COXETER and W. O. J. MOSER, *Generators and relations for discrete groups*, Springer, Berlin-Göttingen-Heidelberg, 1957, 1966.

[17] M. DEHN, Über unendliche diskontinuierliche Gruppen, *Math. Ann.* **71** (1911), 116–144.

[18] V. H. DYSON, The word problem and residually finite groups, *Notices Amer. Math. Soc.* **11** (1964), 743.

[19] SAMUEL EILENBERG and SAUNDERS MACLANE, Relations between homology and homotopy groups of spaces, *Ann. of Math.* (2) **46** (1945), 480–509.

[20] SAMUEL EILENBERG and SAUNDERS MACLANE, Cohomology theory in abstract groups, I, *Ann. of Math.* (2) **48** (1947), 51–78.

[21] R. FRICKE and F. KLEIN, *Vorlesungen über der Theorie der automorphen Funktionen*, Teubner, Leipzig–Berlin, 1926.

[22] A. A. FRIDMAN, On inter-relations between the identity problem and the conjugacy problem in finitely presented groups, *Trudy Moskov. Mat. Obšč.* **9** (1960), 329–356 [Russian].

[23] WOLFGANG GASCHÜTZ, Kohomologische Trivialitäten und äussere Automorphismen von p-Gruppen, *Math. Z.* **88** (1965), 432–433.

[24] PHILIP JOHN GOLD, *The mapping class and symplectic modular groups*, Ph.D. thesis, New York University, 1961.

[25] LEON GREENBERG, Discrete groups of motions, *Canad. J. Math.* **12** (1960), 415–426.

[26] MARTIN GREENDLINGER, On Dehn's algorithms for the conjugacy and word problems, with applications, *Comm. Pure Appl. Math.* **13** (1960), 641-677.

[27] MARSHALL HALL, JR., Generators and relations in groups—the Burnside problem, pp. 42–49 in T. L. Saaty's *Lectures on modern mathematics*, vol. II, Wiley, New York, 1964.

[28] GRAHAM HIGMAN, A finitely related group with an isomorphic proper factor group, *J. London Math. Soc.* **26** (1951), 59–61.

[29] G. HIGMAN, Subgroups of finitely presented groups, *Proc. Roy. Soc. London Ser. A.* **262** (1961), 455–575.

[30] GRAHAM HIGMAN, B. H. NEUMANN, and HANNA NEUMANN, Embedding theorems for groups, *J. London Math. Soc.* **24** (1949), 247–254.

[31] K. A. HIRSCH, On infinite soluble groups, III, *Proc. London Math. Soc.* (2) **49** (1946), 184–194.

[32] A. G. HOWSON, On the intersection of finitely generated free groups, *J. London Math. Soc.* **29** (1954), 428–434.

[33] A. KARRASS, W. MAGNUS, and D. SOLITAR, Elements of finite order in groups with a single defining relation, *Comm. Pure Appl. Math.* **13** (1960), 57–66.

[34] W. LEDERMANN and B. H. NEUMANN, On the order of the automorphism group of a finite group, I, *Proc. Roy. Soc. London Ser. A.* **233** (1956), 494–506.

[35] W. LEDERMANN and B. H. NEUMANN, On the order of the automorphism group of a finite group, II, *Proc. Roy. Soc. London Ser. A.* **235** (1956), 235–246.

[36] FRIEDRICH LEVI, Über die Untergruppen der freien Gruppen, II, *Math. Z.* **37** (1933), 90–97.

[37] JAQUES LEWIN, *Residual properties of loops and rings*, Ph.D. thesis, New York University, 1964.

[38] R. C. LYNDON, Cohomology theory of groups with a single defining relation, *Ann. of Math.* (2) **52** (1950), 650–665.

[39] SAUNDERS MACLANE, *Homology*, Academic Press, New York, 1963.

[40] WILHELM MAGNUS, Über diskontinuierliche Gruppen mit einer definierenden Relation. (Der Freiheitssatz), *J. Reine Angew. Math.* **163** (1930), 141–165.

[41] W. MAGNUS, Das Identitätsproblem für Gruppen mit einer definierenden Relation, *Math. Ann.* **106** (1932), 295–307.

[42] WILHELM MAGNUS, Über n-dimensionale Gittertransformationen, *Acta Math.* **64** (1934), 353–367.

[43] A. A. MARKOV, Insolubility of the problem of homeomorphy, *Proc. Internat. Congress Math. 1958*, pp. 300–306 [Russian]; Cambridge Univ. Press, Cambridge etc. 1960.

[44] J. C. C. MCKINSEY, The decision problem for some classes of sentences without quantifiers, *J. Symb. Logic* **8** (1943), 61–76.

[45] J. L. MENNICKE, Finite factor groups of the unimodular group, *Ann. of Math.* (2) **81** (1965), 31–37.

[46] B. H. NEUMANN, Some remarks on infinite groups, *J. London Math. Soc.* **12** (1937), 120–127.

[47] B. H. NEUMANN, A two-generator group isomorphic to a proper factor group, *J. London Math. Soc.* **25** (1950), 247–248.

[48] HANNA NEUMANN, Generalized free products with amalgamated subgroups, *Amer. J. Math.* **70** (1948), 590–625.

[49] HANNA NEUMANN, Generalized free products with amalgamated subgroups, II, *Amer. J. Math.* **71** (1949), 491–540.

[50] L. P. NEUWIRTH, *Knot groups*, Ann. of Math. Studies No. 56, Princeton Univ. Press, Princeton, N.J., 1965.

[51] J. NIELSEN, Om Regning med ikke-kommutative Faktorer og dens Anvendelse i Gruppeteorien, *Mat. Tidsskrift B* **1921**, 77–94.

[52] JAKOB NIELSEN, Die Isomorphismengruppe der freien Gruppen, *Math. Ann.* **91** (1924), 169–209.

[53] P. S. NOVIKOV, Unsolvability of the conjugacy problem in the theory of groups, *Izv. Akad. Nauk SSSR Ser. Mat.* **18** (1954), 485–524 [Russian].

[54] P. S. NOVIKOV, *On the algorithmic unsolvability of the word problem in group theory*, Trudy Mat. Inst. im. Steklov. No. 44, Izdat. Akad. Nauk SSSR, Moscow, 1955 [Russian].

[55] C. D. PAPAKYRIAKOPOULOS, On Dehn's Lemma and the asphericity of knots, *Ann. of Math.* (2) **66** (1957), 1–26.

[56] MICHAEL O. RABIN, Recursive unsolvability of group theoretic problems, *Ann. of Math.* (2) **67** (1958), 172–194.

[57] ELVIRA STRASSER RAPAPORT, Proof of a conjecture of Papakyriakopoulos, *Ann. of Math.* (2) **79** (1964), 506–513.

[58] K. REIDEMEISTER, *Einführung in die kombinatorische Topologie*, F. Vieweg, Braunschweig, 1932.

[59] O. SCHREIER, Die Untergruppen der freien Gruppen, *Abh. Math. Sem. Univ. Hamburg* **5** (1927), 161–183.

[60] I. SCHUR, Über die Darstellungen der endlichen Gruppen durch gebrochene lineare Substitutionen, *J. Reine Angew. Math.* **127** (1904), 20–50.

[61] V. A. TARTAKOVSKIĬ, The sieve method in group theory, *Mat. Sb.* N.S. **25** (**67**) (1949), 3–50 [Russian].

[62] V. A. TARTAKOVSKIĬ, Application of the sieve method to the solution of the word problem for certain types of groups, *Mat. Sb.* N.S. **25** (**67**) (1949), 251–274 [Russian].

[63] V. A. TARTAKOVSKIĬ, Solution of the word problem for groups with a k-reduced basis for $k > 6$, *Izv. Akad. Nauk SSSR Ser. Mat.* **13** (1949), 483–494 [Russian].

[64] EGBERT R. VAN KAMPEN, On the connection between the fundamental groups of some related spaces, *Amer. J. Math.* **55** (1933), 261–267.

Proc. Internat. Conf. Theory of Groups, Austral. Nat. Univ. Canberra,
August 1965, pp. 51–56. © Gordon and Breach Science Publishers, Inc. 1967

On a problem of D. W. Barnes

WARREN BRISLEY

The problem is to determine whether the variety generated by all the proper factors of a critical group G is maximal in the variety generated by G, and, if not, to say how far removed from maximality it may be. This question seems to have originated at the Fourth Summer Research Institute of the Australian Mathematical Society (Sydney, 1964); it has been traced through B. H. Neumann and Hanna Neumann back to D. W. Barnes.

We shall prove:

Theorem. *Given any positive integer n, there exists a critical (finite) metabelian p-group G, with $p > n+2$, such that the variety of proper factors of G is at least n steps removed from maximality in the variety of G.*

Proof. We construct a group $G = G_0$, and a set of groups G_k, $k = 1, 2, \ldots, n$, such that

$$V_0 \supset V_1 \supset \cdots \supset V_k \supset V_{k+1} \supset \cdots \supset V_n \supset \mathrm{Var}((\mathsf{QS}-\mathrm{I})G_0)$$

where V_k denotes the variety generated by G_k for each k, and $\mathrm{Var}((\mathsf{QS}-\mathrm{I})G_0)$ is the variety of proper factors of G_0.

The following notation is adopted: $\Phi(G)$ denotes the Frattini subgroup of G, $\gamma_m(G)$ the mth term of the lower central series of G, and G^p the group $\mathrm{Gp}\{g^p \mid g \in G\}$. Left normed commutators with repeated entries are defined inductively by

$$(x, y, 0z) = (x, y) = x^{-1}y^{-1}xy, \qquad (x, y, iz) = ((x, y, (i-1)z), z),$$
$$i = 1, 2, \ldots.$$

Similar meanings are given to the symbols (x, y, ix, jy) and (x, y, ix, jy, kz) where i, j, k are nonnegative integers. We use x^y to denote $y^{-1}xy$.

The metabelian p-group G_0 is constructed as a split extension of the metacyclic group

$$\text{Gp}\{a, c \mid a^{p^{n+2}} = 1, \; c^a = c^{1+p}, \; a^{p^{n+1}} = c^{p^n}\}$$

by the cyclic group $\text{Gp}\{b \mid b^{p^{n+1}} = 1\}$:

$$G_0 = \text{Gp}\{a, b \mid a^{p^{n+2}} = 1, \; c^a = c^{1+p}, \; a^{p^{n+1}} = c^{p^n},$$
$$b^{p^{n+1}} = 1, \; a^b = ac, \; cb = bc\}.$$

The relations between a and b imply

$$(a, b, ia) = c^{p^i}, \quad (a, b, ia)^{p^{n+1-i}} = 1, \quad \text{for } i = 0, 1, \ldots, n+1.$$

G_0 can be proved to have order p^{3n+3} and class $n+2$; a normal form $x = a^\lambda b^\mu c^\nu$ (with $0 \le \lambda, \mu, \nu < p^{n+1}$) may be adopted for its elements. Since $\gamma_m(G_0) = \text{Gp}\{c^{p^{m-2}}\}$ for $m \ge 2$, some of the laws of G_0, and hence of V_0, are:

$$x^{p^{n+2}} = 1, \quad \text{and} \quad (x_1, x_2, \ldots, x_m)^{p^{n+3-m}} = 1 \quad \text{for } m = 2, 3, \ldots, n+2.$$

The law $x^{p^{n+2}} = 1$ holds because G_0 is a regular p-group generated by elements of order at most p^{n+2}; note that p exceeds the class of G_0.

Let $x = a^\lambda b^\mu c^\nu$ be central. Then

$$1 = (x, b) = (a^\lambda, b) = c^q \quad \text{with} \quad q = \sum_{s=0}^{n} p^s \binom{\lambda}{s+1}.$$

Since $p^{n+1} \mid q$, we have $\lambda = 0$. Also,

$$1 = (x, a) = (b^\mu, a)(c^\nu, a) = c^{-\mu}(c, a)^\nu = c^{p\nu - \mu}.$$

Thus $p\nu - \mu \equiv 0 \bmod p^{n+1}$, and the center of G is $\text{Gp}\{b^p c\}$. By Theorem 4.1 of [1], G_0 is therefore critical.

The group G_1 is constructed as a factor group of a subdirect product of two copies, H_1 and H_2, of G_0. Take $H_1 = \text{Gp}\{a_1, b_1\}$, $H_2 = \text{Gp}\{a_2, b_2\}$, $K = \text{Gp}\{a', b'\} \subset H_1 \times H_2$ where $a' = a_1 a_2^2$, $b' = b_1 b_2$. Then:

$$(a', b') = c_1(a_2^2, b_2) = c_1 c_2^{2+p} \quad \text{where } c_j = (a_j, b_j), \quad j = 1, 2;$$

$$(a', b', ia') = c_1^{p^i} c_2^{p^i(2+p)^{i+1}}, \quad i = 1, 2, \ldots, n.$$

Clearly $(a', b', na') = c_1^{p^n} c_2^{2^{n+1} p^n}$ is central in K. Setting $G_1 \cong K/\text{Gp}\{(a', b', na')\}$, we obtain

$$G_1 = \text{Gp}\{a, b \mid a^{p^{n+2}} = b^{p^{n+1}} = (a, b, b) = 1, \; a^{p^{n+1}} = (a, b)^{p^n},$$
$$(a, b, ia)^{p^{n+1-i}} = 1 \text{ for } i = 1, 2, \ldots, n-1;$$
$$(a, b, na) = 1\}.$$

G_1 is of order $p^{(n+1)(n+6)/2 - 1}$, and since its class is $n+1$ we have $V_0 \supset V_1$.

The groups $G_2, G_3, \ldots, G_n$ are constructed by defining $G_{k+1} = G_k/\mathrm{Gp}\{(a, b, (n-k)a)^p\}$ for $k = 1, 2, \ldots, n-1$, so that

$$G_k = \mathrm{Gp}\{a, b \mid a^{p^{n+2}} = b^{p^{n+1}} = (a, b, b) = 1, a^{p^{n+1}} = (a, b)^{p^n},$$
$$(a, b, ia)^{p^{n+1-i}} = 1, \qquad i = 0, 1, 2, \ldots, n-k,$$
$$(a, b, ja)^{p^{n-j}} = 1, \qquad j = n-k+1, \ldots, n\},$$

and G_k is of order $p^{(n+1)(n+6)/2 - k}$ for $k = 1, 2, \ldots, n$. Since $\gamma_{n-k+2}(G_k) = \mathrm{Gp}\{(a, b, sa) \mid s = n-k, \ldots, n-1\}$, each G_{k+1} has the law

$$(x_1, x_2, \ldots, x_{n-k+2})^{p^k} = 1,$$

while G_k does not satisfy this law; and since by construction $V_k \supseteq V_{k+1}$ we have

$$V_k \supset V_{k+1}, \qquad k = 0, 1, 2, \ldots, n-1.$$

It remains to show that $V_n \supset \mathrm{Var}((\mathsf{QS}-\mathsf{I})G_0)$, and this is established by examining the laws of G_n, and groups which generate $\mathrm{Var}((\mathsf{QS}-\mathsf{I})G_0)$.

The maximal factor group of G_0 is $F = G_0/\mathrm{Gp}\{c^{p^n}\}$, since $\mathrm{Gp}\{c^{p^n}\}$ is the unique minimal normal subgroup of G_0. F clearly satisfies the laws:

$$x^{p^{n+1}} = 1 \quad \text{and} \quad (x_1, x_2, \ldots, x_j)^{p^{n+2-j}} = 1 \quad \text{for} \quad j = 2, 3, \ldots, n+2.$$

Each maximal subgroup of G_0 is generated by one element of G_0 together with $\Phi(G_0)$. Consider such a maximal subgroup $M = \mathrm{Gp}\{g, \Phi(G_0)\}$ of G_0. Then since $\Phi(G_0) = \mathrm{Gp}\{\gamma_2(G_0), G_0^p\}$, any two elements x_1, x_2 of M may be written $x_1 = g^s f_1$, $x_2 = g^t f_2$, with $f_1, f_2 \in \Phi(G_0)$. Since

$$(x_1, x_2) = (g^s, f_2)(g^s, f_2, f_1)(f_1, f_2)(f_1, g^t)(f_1, g^t, f_2),$$

it follows that $\gamma_2(M) = \mathrm{Gp}\{(\gamma_2(G_0))^p, \gamma_3(G_0)\}$ and that the order of (x_1, x_2) divides p^n. Similarly M satisfies the laws

$$(x_1, x_2, \ldots, x_j)^{p^{n+2-j}} = 1 \qquad \text{for } j = 2, 3, \ldots, n+2.$$

We then have that the groups generating $\mathrm{Var}((\mathsf{QS}-\mathsf{I})G_0)$ satisfy the laws

$$x^{p^{n+2}} = 1, \qquad (x_1, x_2, \ldots, x_j)^{p^{n+2-j}} = 1 \quad \text{for } j = 2, 3, \ldots, n+2.$$

We note that F needs only two generators, and that the maximal subgroups of G_0 are $N = \mathrm{Gp}\{a^p, b, c\}$ and $N_\beta = \mathrm{Gp}\{ab^\beta, b^p, c\}$, $\beta = 0, 1, 2, \ldots, p-1$, each requiring at most three generators.

We now consider the laws of G_n; we already know that G_n satisfies the set S of laws

$$S: \; x^{p^{n+2}} = 1, \quad (x_1, x_2)^{p^{n+1}} = 1, \quad (x_1, x_2, \ldots, x_j)^{p^{n+2-j}} = 1,$$
$$j = 3, 4, \ldots, n+2.$$

We establish, by the following argument, that *all 2-variable laws and all 3-variable laws of G_n are consequences of the set S.*

Consider first the 2-variable laws; since G_n is metabelian, the relations (1) hold:

$$(1) \begin{cases} \text{(i) } (x, y, z)(y, z, x)(z, x, y) = 1; \\[2mm] \text{(ii) } (x_1, x_2, x_3, \ldots, x_t) = (x_1, x_2, x_{i_3}, \ldots, x_{i_t}) \text{ where } i_3, i_4, \ldots, i_t \text{ is} \\ \quad \text{any rearrangement of } 3, 4, \ldots, t; \\[2mm] \text{(iii) } g_1, g_2 \in \gamma_2(G_n) \Rightarrow (g_1 g_2, g) = (g_1, g)(g_2, g) \text{ for all } g \in G_n. \end{cases}$$

By using these relations, any 2-variable law which is not a consequence of S may be written in the form

$$(2) \qquad \prod_{m=0}^{n-1} \prod_{s+t=m} (x, y, sx, ty)^{\beta_{st}} = 1$$

where $0 \le \beta_{00} < p^{n+1}$, $0 \le \beta_{st} < p^{n-m}$ for $m > 0$. Then (2) must be true in $G_n/\gamma_3(G_n)$, and hence $\beta_{00} = 0$. We continue by induction on h: suppose $\beta_{st} = 0$ for all s, t such that $s + t \le h < n - 1$. Then since the law (2) holds in $G_n/\gamma_{h+4}(G_n)$, we have:

$$\prod_{s+t=h+1} (x, y, sx, ty)^{\beta_{st}} = 1$$

in $G_n/\gamma_{h+4}(G_n)$. By substituting $(ab)^\alpha$ for x and a for y it follows that

$$(a, b, (h+1)a)^{\xi(\alpha)} = 1, \quad \xi(\alpha) = \sum_{s=0}^{h+1} \alpha^s \beta_{st}, \quad s+t = h+1.$$

By taking $\alpha = 1, 2, \ldots, h+2$, and using S, it follows that

$$\begin{pmatrix} 1 & 1 & \cdots & 1 \\ 2 & 2^2 & & 2^{h+2} \\ \vdots & \vdots & & \vdots \\ h+2 & (h+2)^2 & & (h+2)^{h+2} \end{pmatrix} \begin{pmatrix} \beta_{0, h+1} \\ \beta_{1, h} \\ \vdots \\ \beta_{h+1, 0} \end{pmatrix} \equiv 0 \quad (\bmod \; p^{n-(h+1)}).$$

Since the determinant of this matrix is $(h+2)!B$, where B is a product of differences of distinct pairs from the set $\{1, 2, \ldots, h+2\}$, and $p > n+1 \geq h+2$, then $\beta_{st} = 0$ for all s, t such that $s+t = h+1$. Thus, all β_{st} in the proposed law are zero, and all 2-variable laws of G_n are consequences of S.

Any 3-variable law of G_n, not a consequence of S, can be reduced to the following form by repeated application of the relations (1):

$$(3) \qquad (x, y)^{\alpha_1} (y, z)^{\alpha_2} (z, x)^{\alpha_3} \prod_{m=1}^{n-1} A_m B_m C_m W_m(x, y, z) = 1$$

where:

(i) $W_m(x, y, z) = \prod_{i,j,k} (x, y, ix, jy, kz)^{\xi_{ijk}} \prod_{r,s,t} (x, z, rx, sy, tz)^{\eta_{rst}}$

with $k > 0$, $s > 0$, $i+j+k = r+s+t = m$, and $0 \leq \xi_{ijk}, \eta_{rst} < p^{n-m}$;

(ii) A_m, B_m, C_m are products of powers of commutators of weight $m+2$ whose entries consist only of x and y, and y and z, and z and x, respectively.

Substituting $z = 1$ in this law, we have that $(x, y)^{\alpha_1} \prod_{m=1}^{n-1} A_m = 1$ is a 2-variable law, and hence is a consequence of S, and so these terms can be removed from the product. Setting $y = 1$ and $x = 1$, we can remove also $(z, x)^{\alpha_3} \prod C_m$ and $(y, z)^{\alpha_2} \prod B_m$, leaving $\prod_{m=1}^{n-1} W_m(x, y, z) = 1$ in place of (3).

This law must hold in $G_n/\gamma_4(G_n)$, so that

$$(x, y, z)^{\xi_{001}} (x, z, y)^{\eta_{010}} = 1.$$

Putting $z = x$ yields $(x, y, x)^{\xi_{001}} = 1$, giving $\xi_{001} = 0$, and putting $y = x$ gives $\eta_{010} = 0$. We continue by induction on h: suppose $\xi_{ijk} = \eta_{rst} = 0$ for all i, j, k, r, s, t such that $i+j+k = r+s+t \leq h < n-1$. Then since the remaining law holds in $G_n/\gamma_{h+4}(G_n)$, we have

$$W_{h+1}(x, y, z) = 1$$

in $G_n/\gamma_{h+4}(G_n)$. Put $z = x^\alpha$: then

$$\prod_{i,j,k} (x, y, ix, jy, kx)^{\alpha^k \xi_{ijk}} = 1, \qquad i+j+k = h+1,$$

i.e., using relations (1),

$$\prod_{t,j} (x, y, tx, jy)^{\xi_{tj}(\alpha)} = 1$$

where

$$\zeta_{tj}(\alpha) = \sum_{k=1}^{t} \alpha^k \xi_{ijk}, \qquad i+k = t, \quad t+j = h+1.$$

Since this is a 2-variable law, it is a consequence of S, and $\zeta_{tj}(\alpha) \equiv 0$ (mod $p^{n-(h+1)}$) for each t, j pair. By taking $\alpha = 1, 2, \ldots, t$, this implies

$$\begin{pmatrix} 1 & 1 & \cdots & 1 \\ 2 & 2^2 & \cdots & 2^t \\ \vdots & \vdots & & \vdots \\ t & t^2 & \cdots & t^t \end{pmatrix} \begin{pmatrix} \xi_{t-1,j,1} \\ \xi_{t-2,j,2} \\ \vdots \\ \xi_{0,j,t} \end{pmatrix} \equiv 0 \pmod{p^{n-(h+1)}}$$

for each t, j pair, $t+j = h+1$. Since $p > n+1 > h+1 \geq t$, it follows that $\xi_{ijk} = 0$ for each admissible i, j, k triplet, $i+j+k = h+1$, $k > 0$. Similarly, putting $y = x^\alpha$, and $\alpha = 1, 2, \ldots, h+1$ gives $\eta_{rst} = 0$ for each admissible r, s, t triplet, $r+s+t = h+1$, $s > 0$. Thus, by induction, there is no 3-variable law which is not a consequence of S.

We now use a result of Hanna Neumann (see [2]) that an n-generator group belongs to a variety if it satisfies the n-variable laws of that variety. As it has been established that all maximal factor groups and maximal subgroups of G_0 are either 2- or 3-generator groups satisfying the set S, whose consequences include all 2- or 3-variable laws of G_n, then all these groups lie in V_n; however, they all satisfy the law $(x_1, x_2)^{p^n} = 1$, which G_n does not satisfy. Thus $V_n \supset \mathrm{Var}((\mathsf{QS}-\mathrm{I})G_0)$, and the chain of varieties between the variety generated by G_0 and the variety of its proper factors has length at least n.

This completes the proof of the theorem.

Acknowledgements are due to I. D. Macdonald for continuing help and advice throughout this work.

References

[1] PAUL M. WEICHSEL, On critical p-groups, *Proc. London Math. Soc.* (3) **14** (1964), 83–100.
[2] HANNA NEUMANN, On varieties of groups and their associated near-rings, *Math. Z.* **65** (1956), 36–69.

Proc. Internat. Conf. Theory of Groups, Austral. Nat. Univ. Canberra, August 1965, pp. 57–58. © Gordon and Breach Science Publishers, Inc. 1967

Lateral completions of lattice-ordered groups

PAUL F. CONRAD

A lattice ordered group ("l-group") is said to be *laterally complete* if each set of positive pairwise disjoint elements has a least upper bound. Holland [4] has shown that any l-group can be embedded in the laterally complete l-group of all order preserving permutations of a suitably chosen totally ordered set. Lorenzen [5] has shown that an l-group G such that

(∗) $a \wedge (-x+a+x) = 0$ implies $a = 0$ for all a, x in G

can be embedded in a large cardinal sum S of totally ordered groups, and clearly S is laterally complete. An l-group that satisfies (∗) will be called *representable*. In [3] it is shown that an abelian l-group can be embedded in a laterally complete abelian l-group of real valued functions. Thus each of the main embedding theorems for l-groups is actually an embedding into a laterally complete l-group. We prove that two important classes of l-groups have the property that each group in the class admits a unique minimal lateral completion. In each case we obtain a refinement of the known embedding theorem.

For the remainder of this summary we assume that G and H are l-groups, and that G is a subgroup and a sublattice ("l-subgroup") of H. H is said to be a *lateral completion* of G if

(i) H is laterally complete,

(ii) no proper l-subgroup of H which contains G is laterally complete, and

(iii) $C \cap G \neq 0$ for all nonzero convex l-subgroups C of H.

There exist l-groups that do not admit a lateral completion, but if we omit property (iii) from our definition, things get pretty wild.

57

Theorem. *A lateral completion of an abelian l-group is abelian.*

Theorem. *If G is a representable l-group with zero radical, then G has a unique lateral completion H. Moreover, H is representable with zero radical.*

Theorem. *If G is an l-group with a basis S, then G has a unique lateral completion H. S is also a basis for H, and H is representable if and only if G is representable.*

For the definition and properties of the radical of G see [2], and for the definition and properties of a basis for G see [1]. The following questions are open:

(1) Find an example of an l-group with two nonequivalent lateral completions.

(2) If G is an abelian laterally complete l-group, does G have zero radical?

Added in proof (December 1966): A representable l-group has a unique lateral completion (which is necessarily representable), and an l-group with zero radical has a unique lateral completion (which necessarily has zero radical). This is a much stronger version of the second theorem and it follows that the answer to question (2) is no.

References

[1] PAUL CONRAD, Some structure theorems for lattice-ordered groups, *Trans. Amer. Math. Soc.* **99** (1961), 212–240.
[2] PAUL CONRAD, The relationship between the radical of a lattice-ordered group and complete distributivity, *Pacific J. Math.* **14** (1964), 493–499.
[3] PAUL CONRAD, JOHN HARVEY, and CHARLES HOLLAND, The Hahn embedding theorem for abelian lattice-ordered groups, *Trans. Amer. Math. Soc.* **108** (1963), 143–169.
[4] CHARLES HOLLAND, The lattice-ordered group of automorphisms of an ordered set, *Michigan Math. J.* **10** (1963), 399–408.
[5] PAUL LORENZEN, Über halbgeordnete Gruppen, *Math. Z.* **52** (1949), 483–526.

Proc. Internat. Conf. Theory of Groups, Austral. Nat. Univ. Canberra, August 1965, pp. 59–69. © Gordon and Breach Science Publishers, Inc. 1967

Endomorphism rings of torsion-free abelian groups

A. L. S. CORNER

The problem of finding necessary and sufficient conditions for a ring to be isomorphic with the (full) endomorphism ring of an abelian group presents considerable difficulties, and it is probably too much to hope for more than partial results. In the torsion case significant progress has recently been made by Pierce, who has completely characterized the endomorphism rings of abelian p-groups without elements of infinite height [10]. The present paper contains a summary of some results that I have obtained on the endomorphism rings of reduced torsion-free abelian groups; details of the most recent results are being prepared for publication elsewhere [5], [6]. It may be of interest to note that both Pierce's results and mine make use of topological notions.

The paper is in three sections. The first two summarize results and give an account of some applications, mainly teratological; Section 1 deals with the countable case, while Section 2 is concerned with generalizations to the uncountable case. The third section is an appendix to the first, and is devoted to a proof of the main result of Section 1, namely a characterization of the (finitely topologized) endomorphism rings of countable reduced torsion-free abelian groups; although this is contained in the more general results discussed in Section 2, it is much easier to establish, and it seems worthwhile to make available a proof which is free from the complications needed in the wider situation.

In the absence of firm indication to the contrary, every group occurring in this paper is to be assumed abelian, torsion-free, and additively written; without exception, every ring is associative and has an identity element. Maps are written on the right.

1. Endomorphism rings of countable, reduced, torsion-free groups

In a previous paper [2] I have shown that every reduced torsion-free ring $\mathbf{A}$ of finite or countable rank n is isomorphic with the endomorphism ring $\mathbf{E}(G)$ of a reduced torsion-free group G of rank $2n$. (A group is *reduced* if its divisible subgroup is trivial; we attribute to a ring properties of its additive group in cases where no confusion can arise.) It is immediate from the finite-rank case of this result that the class of endomorphism rings of reduced torsion-free groups of finite rank may be characterized up to isomorphism as the class of reduced torsion-free rings of finite rank.[1] However, the countable-rank case is less satisfactory: there are many reduced torsion-free groups of countable rank whose endomorphism rings are uncountable, and these do not fall within the scope of the theorem.

In order to obtain a more satisfactory result it appears to be necessary, or at least highly convenient, to make use of a topology first introduced in the context of abelian groups by Szele (see [7], Section 65). The *finite topology* on the endomorphism ring $\mathbf{E}(G)$ of a group G is defined by taking as a basis of neighborhoods of 0 the family of subsets

$$(1.1) \qquad N(X) = \{\varphi \in \mathbf{E}(G) \mid X\varphi = 0\},$$

where X ranges over the set of all finite subsets of G. It is clear that the $N(X)$ are right ideals of $\mathbf{E}(G)$; and it is well known, and easily established, that in its finite topology $\mathbf{E}(G)$ is a complete Hausdorff topological ring. It is also clear that, *for every finite subset X of G, the quotient group $\mathbf{E}(G)/N(X)$ is isomorphic with a subgroup of the direct sum of a finite number of copies of G;* for $N(X)$ is the kernel of the homomorphism $\mathbf{E}(G) \to G^X$ given by $\varphi \to (x\varphi)_{x \in X}$.

Theorem 1.1. *A topological ring $\mathbf{A}$ is isomorphic with the endomorphism ring $\mathbf{E}(G)$ of a countable reduced torsion-free group G if and only if $\mathbf{A}$ is complete and Hausdorff and admits as a basis of neighborhoods of 0 a descending sequence of right ideals N_k ($k = 1, 2, \ldots$) with the property that each quotient group $\mathbf{A}/N_k$ is countable, reduced, and torsion-free.*

The necessity of the conditions is easily established: we have only to enumerate the elements of an arbitrary countable reduced torsion-free group G as $g_1, g_2, \ldots$, write $N_k = N(g_1, \ldots, g_k)$ for every positive

---

[1] Zassenhaus has given conditions under which a torsion-free ring of finite rank n may be realized as the endomorphism ring of a torsion-free group of the same rank n. See [12].

integer k, and apply the observations of the preceding paragraph. It is less trivial that the conditions are sufficient; the proof is given in Section 3.

Using Theorem 1.1 we can clarify the status of the result mentioned in the opening paragraph of this section. Let us call a basis of neighborhoods of 0 satisfying the conditions of Theorem 1.1 a *good* basis. Then if $\mathbf{A}$ is a discrete ring which is realizable as the endomorphism ring of a countable reduced torsion-free group, almost all the terms of a good basis N_k $(k = 1, 2, \ldots)$ of $\mathbf{A}$ must be trivial; therefore $\mathbf{A}$ is countable, reduced, and torsion-free, because its additive group is isomorphic with $\mathbf{A}/N_k$ for large k. Conversely, if $\mathbf{A}$ is a discrete ring whose additive group is countable, reduced, and torsion-free, then $\mathbf{A}$ is trivially complete and Hausdorff, and the sequence $0, 0, \ldots$ is a good basis. Hence

Theorem 1.2. *A discrete ring is isomorphic with the finitely topologized endomorphism ring of a countable reduced torsion-free group if and only if its additive group is countable, reduced, and torsion-free.*

We remark further in this connection that it is not difficult to show that the endomorphism ring of a countable group is discrete if and only if it is countable.

As a first application we deduce two closure properties for the class of endomorphism rings of countable reduced torsion-free groups:

Theorem 1.3. (i) *Let G be a countable reduced torsion-free group, and let $\mathbf{A}'$ be a closed subring of $\mathbf{E}(G)$; then there exists a countable reduced torsion-free group G' such that $\mathbf{E}(G') \cong \mathbf{A}'$.*

(ii) *Let G_n $(n = 1, 2, \ldots)$ be a sequence of countable reduced torsion-free groups; then there exists a countable reduced torsion-free group G such that $\mathbf{E}(G) \cong \prod_n \mathbf{E}(G_n)$, where the direct product is taken in the product topology.*

Proof. Since closed subspaces and cartesian products of complete Hausdorff spaces are complete and Hausdorff, we have only to establish the existence of good bases in the two cases.

(i) Let N_k $(k = 1, 2, \ldots)$ be a good basis for $\mathbf{A} = \mathbf{E}(G)$, and write $N_k' = \mathbf{A}' \cap N_k$. Then the N_k' form a basis at 0 for the induced topology on $\mathbf{A}'$; and, since $\mathbf{A}'/N_k' \cong (\mathbf{A}' + N_k)/N_k \subseteq \mathbf{A}/N_k$, this basis is good.

(ii) For each positive integer n, let N_{nk} ($k = 1, 2, \ldots$) be a good basis for $\mathbf{A}_n = \mathbf{E}(G_n)$. Write

$$N_k = N_{1k} \times \cdots \times N_{kk} \times \prod_{n > k} \mathbf{A}_n.$$

The N_k form a basis at 0 for the product ring, and

$$\left(\prod_n \mathbf{A}_n \right) / N_k \cong \prod_{n \le k} (\mathbf{A}_n / N_{nk}),$$

which is countable, reduced, and torsion-free; so the basis is good.

A number of direct-decomposition pathologies can be shown to occur by constructing suitable rings and applying Theorem 1.2.

Theorem 1.4. ([2]) *There exists a countable torsion-free group whose direct summands are all (directly) decomposable. In particular, G is not expressible as a direct sum of indecomposable groups.*

Theorem 1.5. ([3]) *There exists a countable torsion-free group G which is isomorphic with its direct cube but not with its direct square,*

$$G \cong G \oplus G \oplus G \not\cong G \oplus G.$$

Theorem 1.6. ([4]) *There exists a countable hopfian group G whose direct square $G \oplus G$ is not hopfian.*

(A group G is *hopfian* if every endomorphism of G onto itself is an automorphism of G.)

J. D. Reid [11] has applied the finite-rank case of the theorem with which this section opens to construct a group of finite rank whose endomorphism ring is not commutative but all of whose endomorphic images are fully invariant; this provides a negative solution to a problem of Szele and Szendrei ([7], Problem 47). (See also A. Orsatti, Su di un problema di T. Szele e J. Szendrei, *Rend. Sem. Mat. Univ. Padova* **35** (1965), 171–175.)

I end this section with a problem which should respond to similar treatment: Does there exist a torsion-free group of finite rank with infinitely many nonisomorphic direct summands?

2. Generalization to the uncountable case

It can be shown that the direct sum of two copies of the ring of p-adic integers cannot be realized as an endomorphism ring for any prime p; in consequence, we cannot extend Theorem 1.1 to the uncountable case by simply suppressing the countability restrictions. In order to exclude what might be called p-adic trouble, we introduce the following definitions.

Let C be a reduced torsion-free group of cardinal $< 2^{\aleph_0}$; such a group we call a *control group*. A group G is C-*controlled* if every subgroup of finite rank in G is isomorphic with a subgroup of C. A topological ring $\mathbf{A}$ is *residually C-controlled* if it admits a basis of neighborhoods of 0 consisting of a family N_i ($i \in I$) of right ideals such that the corresponding quotient groups $\mathbf{A}/N_i$ are all C-controlled; such a basis N_i ($i \in I$) is $\mathfrak{m}$-*good*, where $\mathfrak{m}$ is an infinite cardinal, if in addition

$$(2.1) \qquad \sum_{i \in I} |\mathbf{A}/N_i| \le \mathfrak{m}.$$

Finally, a group (ring) is *controlled* (*residually controlled*) if it is C-controlled (residually C-controlled) for some control group C.

It is trivial that a direct sum of C-controlled groups is C'-controlled, where C' is the direct sum of countably many copies of C. Using this fact we easily obtain

Proposition 2.1. *The endomorphism ring $\mathbf{E}(G)$ of a controlled group G is a complete, Hausdorff, residually controlled ring admitting a $|G|$-good basis.*

The converse can be proved under an accessibility restriction on the cardinals that are involved; it is undecided whether it is true in general.

Theorem 2.2. *Let $\mathbf{A}$ be a complete Hausdorff residually controlled ring admitting an $\mathfrak{m}$-good basis, where $\mathfrak{m}$ is an infinite cardinal $< \aleph_\iota$ (the first strongly inaccessible cardinal). Then there exist $2^{\mathfrak{m}}$ nonisomorphic controlled groups G of cardinal $\mathfrak{m}$ such that $\mathbf{E}(G) \cong \mathbf{A}$.*

The proof of Theorem 2.2 proceeds by a transfinite induction based on Fuchs's construction[2] of "large" indecomposable groups [9]; indeed by taking $\mathbf{A}$ to be the discretely topologized ring of rational integers, we deduce

Theorem 2.3. *For every infinite cardinal $\mathfrak{m} < \aleph_\iota$ there exist $2^{\mathfrak{m}}$ nonisomorphic indecomposable groups of cardinal $\mathfrak{m}$.*

It may be remarked in passing that there is very little group theory in the proof of Theorem 2.2; the greater part of the argument is devoted

---

[2] Note that a set-theoretical flaw in Fuchs's induction makes it fail for exorbitantly large cardinals: see [7].

to establishing a theorem in Linear Algebra, of which perhaps the most interesting consequence is

Theorem 2.4. *Let V be a vector space of arbitrary infinite dimension $(< \aleph_\iota)$ over an arbitrary field. Then there exists a finite set of subspaces $U_1, \ldots, U_n$ of V with the property that every endomorphism φ of V such that $U_i \varphi \subseteq U_i$ $(i = 1, \ldots, n)$ is a scalar multiple of the identity map.*

The countable-dimensional case of this theorem and its finite-dimensional analogues are due to Sheila Brenner [1], who also raised the question whether the theorem was true in the uncountable-dimensional case. She has shown further that in the finite-dimensional case one can always take $n = 4$. In the infinite-dimensional case, $n = 5$ always suffices, but it is not known, even in the countable case, whether this is best-possible.

We quote without proof two simple applications of Theorem 2.2 to the problem of determining which (not necessarily abelian) groups occur as automorphism groups of torsion-free groups. (The problem of determining which finite groups occur has been considered by Hallett and Hirsch [9].)

Theorem 2.5. *Let U_λ $(\lambda \in \Lambda)$ be a family of finite groups, each of which is the automorphism group of a torsion-free group, and suppose that the indexing set Λ is of cardinal $< \aleph_\iota$. Then the complete direct product $\prod_{\lambda \in \Lambda} U_\lambda$ can be realized as the automorphism group of a torsion-free group.*

Theorem 2.6. *Let U be a group of cardinal $< \aleph_\iota$ which admits a total order, for example an absolutely free group or a torsion-free abelian group. Then there exists a torsion-free group whose automorphism group is isomorphic with the direct product $U \times \mathscr{C}(2)$, where $\mathscr{C}(2)$ is the group of order 2.*

It is perhaps worth pointing out that a given abstract ring may be realizable as an endomorphism ring of a controlled group in more than one topology. For example, if $\mathbf{A}$ is the direct product of $\mathfrak{m}$ copies of the ring of rational integers, then Theorem 2.2 shows that $\mathbf{A}$ is realizable both in the discrete topology and also in the product of the discrete topologies on the factors. However, it can be shown by a slenderness argument that if the endomorphism rings of two countable reduced torsion-free groups are isomorphic they are necessarily topologically isomorphic.

We mention in closing the section that Theorem 2.2 can be extended to a yet wider class of rings. Call a topological ring **A** *residually controlled in rank r* if it admits a basis of neighborhoods of 0 consisting of a family N_i $(i \in I)$ of right ideals such that for some control group C every subgroup of rank at most r in each quotient $\mathbf{A}/N_i$ is isomorphic with a subgroup of C. Then it can be proved that *every complete Hausdorff ring which is residually controlled in rank 4 and of cardinal $< \aleph_i$ can be realized as the endomorphism ring of a reduced torsion-free group*; cardinality requirements analogous to those of Theorem 2.2 can be exacted. It is not known whether the 4 of the statement is best-possible, but the example of the direct product of two copies of the ring of p-adic integers shows that it cannot be replaced by a 1.

3. Proof of Theorem 1.1

The necessity of the conditions has already been established; their sufficiency remains to be proved.

Let **A** be a complete Hausdorff ring admitting a good basis N_k $(k = 1, 2, \ldots)$. Note that the Hausdorff condition is

$$(3.1) \qquad \bigcap_k N_k = 0;$$

here, as throughout this section, k runs over the positive integers. To start with, ignore the topology on **A** and form the direct sum C of the quotients $\mathbf{A}/N_k$. Clearly, C is a countable reduced torsion-free group. Moreover, since the N_k are right ideals of **A**, C is canonically a right **A**-module; and if we write

$$(3.2) \qquad e_k = 1 \bmod N_k \quad (\in \mathbf{A}/N_k),$$

then e_k is a generator of the cyclic submodule $\mathbf{A}/N_k$, so that

$$(3.3) \qquad C = \sum_k e_k \mathbf{A}.$$

The proof is based on the observation, implicit in what follows, that the canonical representation $\mathbf{A} \to \mathbf{E}(C)$ embeds **A** as a topological subring of $\mathbf{E}(C)$; we shall embed C in a larger **A**-module G in such a way that the cyclic submodules of C become fully invariant subgroups of G.

The construction makes use of the methods of [2]. We topologize the group C by taking as a basis of neighborhoods of 0 the sequence of subgroups qC $(q=1, 2, \ldots)$, and form the completion $\hat{C}$. By [2], Lemma 1.4, $\hat{C}$ is a reduced torsion-free group containing C as a pure subgroup, and it is canonically a $\hat{\mathbf{Z}}$-module, where $\hat{\mathbf{Z}}$ is the corresponding completion of the ring $\mathbf{Z}$ of rational integers; moreover, since every endomorphism of C extends to a unique $\hat{\mathbf{Z}}$-endomorphism of $\hat{C}$, $\hat{C}$ is canonically also a right $\mathbf{A}$-module. By [2], Lemma 1.5, the ring $\hat{\mathbf{Z}}$ contains a (unital) subring $\mathbf{P}$ of cardinal $2^{\aleph_0}$, every element of which is the product of a rational integer and a unit of $\mathbf{P}$; and the argument of the first part of [2] Section 2 shows that there is a countable subring $\mathbf{\Pi}$ of $\mathbf{P}$ such that $\mathbf{P}$ is *linearly disjoint from C over* $\mathbf{\Pi}$, in the sense that if $\gamma_1, \ldots, \gamma_n$ are elements of $\mathbf{P}$ that are linearly independent over $\mathbf{\Pi}$, and if $c_1, \ldots, c_n$ are elements of C such that $\gamma_1 c_1 + \cdots + \gamma_n c_n = 0$ in $\hat{C}$, then $c_1 = \cdots = c_n = 0$. (The existence of the subring $\mathbf{\Pi}$ is established in [2] for a countable reduced torsion-free ring rather than a countable reduced torsion-free group, but it is easy to see that the multiplicative structure of the ring is not used.)

Since the subring $\mathbf{\Pi}$ is countable, the integral domain $\mathbf{P}$ is of transcendence degree $2^{\aleph_0}$ over $\mathbf{\Pi}$; so there certainly exists a countable family α_c $(c \in C)$ of elements of $\mathbf{P}$ that are algebraically independent over $\mathbf{\Pi}$. For later use we note that formally distinct monomials in the α_c are elements of $\mathbf{P}$ that are linearly independent over $\mathbf{\Pi}$. We define G to be the pure subgroup of $\hat{C}$ generated by C and the $\alpha_c(c\mathbf{A})$ $(c \in C)$,

$$(3.4) \qquad\qquad G = \{C, \alpha_c(c\mathbf{A}) \, (c \in C)\}_* .$$

As a pure subgroup of a reduced torsion-free group generated by countably many countable subgroups, G is countable, reduced, and torsion-free; and, since C and the $\alpha_c(c\mathbf{A})$ are sub-$\mathbf{A}$-modules of $\hat{C}$, G is a right $\mathbf{A}$-module. It is easy to see that G is faithful as an $\mathbf{A}$-module. For it is immediate from (3.2) that

$$(3.5) \qquad\qquad N_k = \{a \in \mathbf{A} \mid e_k a = 0\};$$

and it follows that if $a \in \mathbf{A}$ is such that $Ga = 0$, then $a \in N_k$ for all k, whence $a = 0$ by (3.1). Thus, ignoring topology, we may identify $\mathbf{A}$ in a natural way with a subring of $\mathbf{E}(G)$.

To prove that $\mathbf{A}$ exhausts $\mathbf{E}(G)$, consider an arbitrary endomorphism φ of G. Fix an element $c \in C$. Then $c\varphi$, $(\alpha_c c)\varphi \in G$; so by (3.4) there exist a positive integer q and elements $x, y \in C$, $x_d, y_d \in \mathbf{A}$ $(d \in C)$, almost all zero, such that

(3.6)
$$q(c\varphi) = y + \sum_d \alpha_d(dy_d),$$
$$q((\alpha_c c)\varphi) = x + \sum_d \alpha_d(dx_d).$$

Now G is a dense topological subgroup of $\hat{C}$. Therefore $\hat{G} = \hat{C}$, so that φ extends to a unique $\hat{\mathbf{Z}}$-endomorphism of $\hat{C}$. Consequently $(\alpha_c c)\varphi = \alpha_c(c\varphi)$, whence

(3.7)
$$x + \sum_d \alpha_d(dx_d) = \alpha_c\left(y + \sum_d \alpha_d(dy_d)\right).$$

But the elements 1, α_d, $\alpha_c\alpha_d$ $(d \in C)$ of $\mathbf{P}$ are linearly independent over $\mathbf{\Pi}$, so by linear disjointness we may equate their coefficients in (3.7) to deduce that

$$x = 0, \qquad dy_d = 0 \quad (d \in C),$$
$$cx_c = y, \qquad dx_d = 0 \quad (d \in C, d \neq c).$$

Substituting these values in (3.6) and simplifying our notation slightly, we conclude that *for each $c \in C$ there exist a positive integer $q = q(c)$ and an element $x = x(c) \in \mathbf{A}$ such that*

(3.8)
$$q(c\varphi) = cx.$$

Now fix an element $a \in \mathbf{A}$. It follows from (3.8) that, for each k, we have $q_k(e_k a)\varphi = e_k a_k$, where $q_k = q(e_k a)$ and $a_k = x(e_k a)$. But then $e_k a_k$ belongs to the intersection $q_k \hat{C} \cap e_k \mathbf{A}$, which is equal to $q_k e_k \mathbf{A}$ by the evident purity of $e_k \mathbf{A}$ in $\hat{C}$. Therefore we may always choose $q_k = 1$,

(3.9)
$$(e_k a)\varphi = e_k a_k.$$

Applying (3.8) a second time with $c = (e_k + e_{k+1})a$, we find that

$$q(e_k a + e_{k+1} a)\varphi = e_k(ax) + e_{k+1}(ax),$$

where $q = q(c)$, $x = x(c)$. But, by (3.9),

$$(e_k a + e_{k+1} a)\varphi = e_k a_k + e_{k+1} a_{k+1}.$$

Comparison of the last two displayed equations shows with the help of (3.5) that

$$qa_k - ax \in N_k, \qquad qa_{k+1} - ax \in N_{k+1}.$$

But $N_{k+1} \subseteq N_k$, by hypothesis; therefore $q(a_k - a_{k+1}) \in N_k$, whence

(3.10) $a_k - a_{k+1} \in N_k$,

because the quotient $\mathbf{A}/N_k$ is torsion-free. Now (3.10) means that (a_k) is a Cauchy sequence in the given topology on $\mathbf{A}$. Therefore completeness guarantees the existence of an element $a^* \in \mathbf{A}$ such that $a_k - a^* \in N_k$. Thus, by (3.5) and (3.9),

(3.11) $(e_k a)\varphi = e_k a^*$,

where a^* depends only on a and not on k. Now let h be an arbitrary positive integer. By the continuity of the ring operations there exists a positive integer k such that

(3.12) $aN_k + N_k \subseteq N_h$.

Apply (3.8) for a third time with $c = e_k + e_{k+1}a$: we have

$$q(e_k + e_{k+1}a)\varphi = e_k x + e_{k+1}(ax),$$

where $q = q(c)$, $x = x(c)$. But (3.11) shows that

$$(e_k + e_{k+1}a)\varphi = e_k 1^* + e_{k+1}a^*,$$

and comparing the last two displayed equations we deduce at once with the help of (3.5) that

$$x - q1^* \in N_k, \qquad ax - qa^* \in N_{k+1} \subseteq N_k.$$

Therefore $q(a1^* - a^*) \in aN_k + N_k \subseteq N_h$, whence $a1^* - a^* \in N_h$. Since h was an arbitrary positive integer, (3.1) implies that $a^* = a1^*$. Thus (3.11) simplifies further to

$$(e_k a)\varphi = (e_k a)1^*.$$

Since every element of C is a finite sum of terms of the form $e_k a$, this proves that φ operates on C as 1^*; and it follows at once by the density of C in G that φ operates as 1^* on the whole of G. This completes the proof that $\mathbf{A}$ exhausts $\mathbf{E}(G)$.

It now only remains for us to prove that the given topology on $\mathbf{A}$ coincides with the finite topology on $\mathbf{E}(G)$. (i) Since $\mathbf{A} = \mathbf{E}(G)$, (3.5) shows that $N_k = N(e_k)$; so the basic neighborhoods of 0 in the given topology are open in the finite topology. (ii) To prove that the basic neighborhoods of 0 in the finite topology are open in the given topology, it is clearly enough to show that $N(g)$ is open in the given topology for each $g \in G$. Now the continuity of multiplication in the given topology implies at once that for each k and each $a \in \mathbf{A}$ the right ideal

$$N(e_k a) = \{b \in \mathbf{A} \mid ab \in N_k\}$$

is open in the given topology. Since each element of C is a finite sum of elements in distinct direct summands $e_k\mathbf{A}$, it follows that $N(c)$ is open in the given topology for each $c \in C$. Consider an element $g \in G$. By (3.4) we may write

$$qg = x + \sum \alpha_c(cx_c),$$

where q is a positive integer, $x \in C$, and the x_c $(c \in C)$ are elements of $\mathbf{A}$, almost all zero. Then an element $a \in \mathbf{A}$ belongs to $N(g)$ if and only if

$$xa = 0, \qquad (cx_c)a = 0 \quad (c \in C).$$

Thus $N(g)$ is the intersection of a finite number of right ideals of the form $N(d)$ $(d \in C)$; as such, it is open in the given topology.

This completes the proof of the theorem.

References

[1] SHEILA BRENNER, Endomorphism algebras of vector spaces with distinguished sets of subspaces [in preparation].

[2] A. L. S. CORNER, Every countable reduced torsion-free ring is an endomorphism ring, *Proc. London Math. Soc.* (3) **13** (1963), 687–710.

[3] A. L. S. CORNER, On a conjecture of Pierce concerning direct decompositions of abelian groups, *Proc. Colloq. Abelian Groups (Tihany,* 1963), pp. 43–48; Akadémiai Kiadó, Budapest, 1964.

[4] A. L. S. CORNER, Three examples on hopficity in torsion-free abelian groups, *Acta Math. Acad. Sci. Hungar.* **16** (1965), 304–310.

[5] A. L. S. CORNER, Endomorphism algebras of large modules with distinguished submodules [in preparation].

[6] A. L. S. CORNER, Torsion-free abelian groups with prescribed finitely topologized endomorphism rings [in preparation].

[7] L. FUCHS, *Abelian groups,* Akadémiai Kiadó, Budapest, 1958.

[8] L. FUCHS, The existence of indecomposable abelian groups of arbitrary power, *Acta Math. Acad. Sci. Hungar.* **10** (1959), 453–457.

[9] J. T. HALLETT and K. A. HIRSCH, Torsion-free groups having finite automorphism groups, I, *J. Algebra* **3** (1965), 287–298.

[10] R. S. PIERCE, Endomorphism rings of primary abelian groups, *Proc. Colloq. Abelian Groups (Tihany,* 1963), pp. 125–137; Akadémiai Kiadó, Budapest, 1964.

[11] J. D. REID, On subcommutative rings, *Acta Math. Acad. Sci. Hungar.* **16** (1965), 23–26.

[12] H. ZASSENHAUS, Orders as endomorphism rings of modules of the same rank, *J. London Math. Soc.* [to appear].

Proc. Internat. Conf. Theory of Groups, Austral. Nat. Univ. Canberra, August 1965, p. 71. © Gordon and Breach Science Publishers, Inc. 1967

On varieties of A-groups

P. J. COSSEY

As usual, we denote by $\mathfrak{A}_p$ the variety of abelian groups of exponent dividing p. By an A-group we mean a soluble locally finite group whose Sylow subgroups are all abelian. A group is called monolithic if the intersection of its nontrivial normal subgroups is nontrivial.

Theorem. *Any variety of A-groups which cannot be generated by a finite group contains the product variety $\mathfrak{A}_p\mathfrak{A}_q\mathfrak{A}_r$ for some set of three distinct primes p, q, r.*

This is proved in two steps:

(A)   Any infinite set of (isomorphism classes of) finite monolithic groups of $\mathfrak{A}_p\mathfrak{A}_q\mathfrak{A}_r$ generates $\mathfrak{A}_p\mathfrak{A}_q\mathfrak{A}_r$ if p, q, r are distinct primes.

(B)   If a variety of A-groups cannot be generated by a finite group, then it contains an infinite set of (isomorphism classes of) finite monolithic groups of $\mathfrak{A}_p\mathfrak{A}_q\mathfrak{A}_r$, for some set of three distinct primes p, q, r.

71

Proc. Internat. Conf. Theory of Groups, Austral. Nat. Univ. Canberra, August 1965, pp. 73–77. © Gordon and Breach Science Publishers, Inc. 1967

The Lorentz group and the group of homographies

H. S. M. COXETER

The object of this paper is to describe some geometric aspects of the isomorphism (first noticed by Liebmann [7], p. 54) between the Lorentz group and the group of homographies. Artin ([1], p. 204) expressed it as follows:

$\Omega_4 \cong PSL_2(R(i))$, the projective unimodular group of two dimensions over the field of complex numbers; Ω_4 is ... a simple group which is called the Lorentz group.

From his use of the word "simple," it is clear that Ω_4 means the group of real homogeneous linear transformations of determinant 1 leaving invariant the indefinite quaternary form

$$x_1^2 + x_2^2 + x_3^2 - x_4^2$$

and the sign of x_4 ([12], Section 8(c)). When interpreted in affine 4-space, these are sense-preserving affine transformations leaving invariant each of the two sheets of the hyperboloid

$$(1) \qquad x_1^2 + x_2^2 + x_3^2 - x_4^2 = -1,$$

which is a unit sphere of timelike radius in the Lorentz–Minkowskian 4-space ([9], p. 405).

When the x's are interpreted as homogeneous coordinates, the null cone (or isotropic cone)

$$(2) \qquad x_1^2 + x_2^2 + x_3^2 - x_4^2 = 0$$

becomes a nonruled quadric in the real projective 3-space. This serves as the absolute for a hyperbolic space whose points, being interior to the quadric, satisfy

$$x_1^2 + x_2^2 + x_3^2 - x_4^2 < 0,$$

and can be normalized so as to satisfy (1).

The substitution

$$\xi = \frac{x_1}{x_4}, \quad \eta = \frac{x_2}{x_4}, \quad \zeta = \frac{x_3}{x_4}$$

enables us to replace the quadric (2) by a sphere

$$\xi^2 + \eta^2 + \zeta^2 = 1$$

in Euclidean 3-space. Then stereographic projection from $(0, 0, 1)$ yields the complex number plane whose points $z = x + iy$ are given by

$$z = \frac{\xi + i\eta}{1 - \zeta}, \quad \xi = \frac{z + \bar{z}}{1 + z\bar{z}}, \quad i\eta = \frac{z - \bar{z}}{1 + z\bar{z}}, \quad \zeta = \frac{-1 + z\bar{z}}{1 + z\bar{z}}$$

([6], p. 34). In other words, the points of (2), which are the points at infinity of the hyperbolic space, can be expressed in terms of a complex parameter z by the equations

$$x_1 = z + \bar{z}, \quad x_2 = (z - \bar{z})/i, \quad x_3 = z\bar{z} - 1, \quad x_4 = z\bar{z} + 1,$$

with the convention that $z = \infty$ when $(x_1, x_2, x_3, x_4) = (0, 0, 1, 1)$.

Since the circles of the inversive plane [8] represent the planes of hyperbolic space ([2], p. 265), the general homography

$$z' = \frac{az + b}{cz + d},$$

which is the product of inversions in two or four circles, represents the product of reflections in two or four planes, which is the general displacement in hyperbolic space. Since

$$z = \frac{x_1 + ix_2}{x_4 - x_3},$$

this displacement is the Lorentz transformation

$$(x_1', x_2', x_3', x_4') = (x_1, x_2, x_3, x_4)L$$

([10], p. 119), where

$$
L = \tfrac{1}{2}
\begin{pmatrix}
a\bar{d}+d\bar{a} & -i(a\bar{d}-d\bar{a} & a\bar{b}+b\bar{a} & a\bar{b}+b\bar{a} \\
+b\bar{c}+c\bar{b} & +b\bar{c}-c\bar{b}) & -c\bar{d}-d\bar{c} & +c\bar{d}+d\bar{c} \\
i(a\bar{d}-d\bar{a} & a\bar{d}+d\bar{a} & i(a\bar{b}-b\bar{a} & i(a\bar{b}-b\bar{a} \\
-b\bar{c}+c\bar{b}) & -b\bar{c}-c\bar{b} & -c\bar{d}+d\bar{c}) & +c\bar{d}-d\bar{c}) \\
a\bar{c}+c\bar{a} & -i(a\bar{c}-c\bar{a} & a\bar{a}-b\bar{b} & a\bar{a}-b\bar{b} \\
-b\bar{d}-d\bar{b} & -b\bar{d}+d\bar{b}) & -c\bar{c}+d\bar{d} & +c\bar{c}-d\bar{d} \\
a\bar{c}+c\bar{a} & -i(a\bar{c}-c\bar{a} & a\bar{a}+b\bar{b} & a\bar{a}+b\bar{b} \\
+b\bar{d}+d\bar{b} & +b\bar{d}-d\bar{b}) & -c\bar{c}-d\bar{d} & +c\bar{c}+d\bar{d}
\end{pmatrix}.
$$

In particular, the loxodromic homography $z' = e^{\alpha+i\beta}z$ represents

$$
\begin{pmatrix}
\cos\beta & \sin\beta & 0 & 0 \\
-\sin\beta & \cos\beta & 0 & 0 \\
0 & 0 & \cosh\alpha & \sinh\alpha \\
0 & 0 & \sinh\alpha & \cosh\alpha
\end{pmatrix},
$$

and the parabolic homography $z' = z+2$ represents

$$
\begin{pmatrix}
1 & 0 & 2 & 2 \\
0 & 1 & 0 & 0 \\
-2 & -1 & 0 & -2 \\
2 & 0 & 2 & 3
\end{pmatrix}.
$$

The group of homographies plays an interesting role in a theorem of Kerékjártó [5]: The (topological) sphere is the only compact surface that admits a triply transitive continuous topological group.

Abandoning continuity, we notice that almost all these results still hold when we replace the real field (to which the x's belong) and the complex field (to which a, b, c, d, and z belong) by the Galois fields $GF(q)$ and $GF(q^2)$ respectively, where q is a power of an odd prime. Of course, the complex conjugate $\bar{z}$ must then be replaced by z^q, and i by the $\tfrac{1}{4}(q^2-1)$th power of a primitive root of $GF(q^2)$.

In this way we obtain two geometrical interpretations for the group $PSL(2, q^2)$: the inversive plane of order q, which has recently been axiomatized very simply by Christoph Hering, and hyperbolic 3-space over $GF(q)$.

In the real inversive plane, any three noncoaxal circles determine a *bundle* consisting of the smallest set of circles which contains these three and, for every two members of the set, the whole of the coaxal pencil to which those two belong. The bundle is said to be *hyperbolic, parabolic,* or *elliptic* according as the three generating circles (and consequently all the circles in the bundle) have a common orthogonal circle, a common point, or neither. According to Liebmann, this inversive plane, with its circles, point-pairs, points, and elliptic bundles, represents real hyperbolic 3-space, with its planes, lines, "ends" (or "points at infinity"), and proper points. (Dembowski and Hughes [4], p. 173, use the term "bundle" in a different sense.)

The finite inversive plane of order q consists of $q^2 + 1$ points and $q(q^2 + 1)$ circles, with $q + 1$ points on each circle and, through each point, $q(q + 1)$ circles (corresponding to the $q(q + 1)$ lines of the affine plane $EG(2, q)$). Applying Liebmann's correspondence, we see that the finite hyperbolic space contains $q^2 + 1$ ends, $q(q^2 + 1)$ planes and $\binom{q^2 + 1}{2}$ lines, with $q + 1$ planes through each line and $\binom{q + 1}{2}$ lines in each plane. However, although this kind of hyperbolic space contains planes and lines, it contains no proper points; for, the absolute polar plane of such a point would be an ultra-infinite plane carrying an elliptic polarity, but over a finite field no such polarity exists. In other words, over a finite field a ternary quadratic form cannot be definite ([1], p. 144; [11], p. 4; [3], p. 78). It follows that, *in the finite inversive plane, there are no elliptic bundles; every three circles have either a common point or a common orthogonal circle.*

In view of the lack of proper points we naturally replace the finite hyperbolic space by a finite *exterior-hyperbolic* space or three-dimensional de Sitter's world ([3], pp. 83, 88), which contains $q(q^2 + 1)$ points or "events" (the poles of the $q(q^2 + 1)$ proper planes), $\frac{1}{2}q^2(q^2 + 1)$ "timelike" lines (already mentioned before as lines in the ordinary hyperbolic space, but now each contains $q - 1$ points as well as two ends), $\frac{1}{2}q^2(q^2 + 1)$ "spacelike" lines (which are the polar lines of the timelike lines, and each contain $q + 1$ points), $q^2 + 1$ isotropic planes (which, being tangent planes to the absolute quadric, are affine planes each containing q^2 points), and $(q + 1)(q^2 + 1)$ "null" lines (each containing q points).

These results can be checked by adding the numbers of lines of the three kinds to make $(q^2 + q + 1)(q^2 + 1)$, which is the number of lines in the projective space $PG(3, q)$.

References

[1] EMIL ARTIN, *Geometric algebra*, Interscience, New York, 1957.
[2] H. S. M. COXETER, *Non-euclidean geometry*, 5th ed., Univ. of Toronto Press, Toronto, 1965.
[3] H. S. M. COXETER, *Geometry*, in T. L. Saaty's *Lectures on modern mathematics*, vol. III, Wiley, New York, 1965.
[4] P. DEMBOWSKI and D. R. HUGHES, On finite inversive planes, *J. London Math. Soc.* **40** (1965), 171–182.
[5] BÉLA KERÉKJÁRTÓ, Sur le caractère topologique du groupe homographique de la sphère, *Acta Math.* **74** (1941), 311–341.
[6] FELIX KLEIN, *Lectures on the icosahedron and the solution of equations of the fifth degree*, 2nd ed., Kegan Paul, London, 1913.
[7] HEINRICH LIEBMANN, *Nichteuklidische Geometrie*, Göschen, Leipzig, 1905.
[8] MARIO PIERI, Nuovi principii di geometria delle inversioni, *Giorn. Mat. Battaglini* (3) **49** (1911), 49–96; **50** (1912), 106–140.
[9] A. A. ROBB, *Geometry of time and space*, Cambridge Univ. Press, Cambridge, 1936.
[10] HANS SCHWERDTFEGER, *Geometry of complex numbers*, Univ. of Toronto Press, Toronto, 1962.
[11] BENIAMINO SEGRE, Le geometrie di Galois, *Ann. Mat. Pure Appl.* (4) **48** (1959), 1–96.
[12] HERMANN WEYL, *Gruppentheorie und Quantenmechanik*, Hirzel, Leipzig, 1928.

Proc. Internat. Conf. Theory of Groups, Austral. Nat. Univ. Canberra,
August 1965, pp. 79–80. © Gordon and Breach Science Publishers, Inc. 1967

On ℭ-permutable subgroups
of finite groups

W. E. DESKINS

Some of the theorems of Ore [4], Kegel [3], Itô and Szép [2], and
Deskins [1] on quasinormal subgroups are generalized here to wider
classes of subgroups. Only finite groups are treated.

A collection ℭ of subgroups of the group G is a cover for G provided

(i) for each primary element y of an arbitrary maximal subgroup
M of G there exists in ℭ an element H such that $y^H \subseteq H \cap M$,
and

(ii) ℭ is closed under the inner automorphisms of G.

A cover ℭ is inductive if the set $\mathℭ_K = \{H \cap K : H \text{ in } \mathℭ\}$ is a cover for
each subnormal subgroup K of G.

Theorem 1. *If ℭ is an inductive cover for G and L is a subgroup of G
which permutes with each element of ℭ, then L is subnormal in G.*

This is proved by showing that a subgroup which is maximal in the
set of all proper ℭ-permutable subgroups of G is normal in G when ℭ
is a cover.

Let $\mathscr{P}$ denote a property of groups which satisfies

(iii) $\mathscr{P}$ is inherited by subgroups and quotient groups;

(iv) each group G has a unique $\mathscr{P}$-radical (a normal subgroup R
minimal with respect to the property that G/R is a $\mathscr{P}$-group);
and

(v) an extension of a $\mathscr{P}$-group by a $\mathscr{P}$-group is a $\mathscr{P}$-group.

Theorem 2. *If ℭ is an inductive cover of G by $\mathscr{P}$-groups and L is a
ℭ-permutable subgroup of G, then the $\mathscr{P}$-radical of L lies in the core of L.*

Covers of an arithmetic type yield similar results.

7—ᴋ.

A set Π of subgroups of G of prime power orders is a prime cover for G provided

   (vi) each primary element of G belongs to at least one element of Π, and

   (vii) Π is closed under the inner automorphisms of G.

Theorem 3. *If Π is a prime cover for G and L is a subgroup of G which permutes with each element of Π then L is subnormal in G and $L/core(L)$ is nilpotent.*

References

[1] W. E. DESKINS, On quasinormal subgroups of finite groups, *Math. Z.* **82** (1963), 125–132.

[2] N. ITÔ and J. SZÉP, Über die Quasinormalteiler von endlichen Gruppen, *Acta Sci. Math. Szeged* **23** (1962), 168–170.

[3] OTTO H. KEGEL, Sylow-Gruppen und Subnormalteiler endlicher Gruppen, *Math. Z.* **78** (1962), 205–221.

[4] OYSTEIN ORE, Contributions to the theory of groups, *Duke Math. J.* **5** (1939), 431–460.

Proc. Internat. Conf. Theory of Groups, Austral. Nat. Univ. Canberra, August 1965, pp. 81–83. © Gordon and Breach Science Publishers, Inc. 1967

Complements of normal subgroups in infinite groups

JOHN D. DIXON

The theorems of Schur and Zassenhaus (see [6]), which concern the existence and conjugacy of complements of a normal Hall subgroup in a finite group, prove very useful in examining the structure of finite groups. The present paper discusses the validity of an analogous theorem for a certain class of infinite groups.

We first note that the theorem cannot be extended without some quite severe restrictions. In fact, there exists a metabelian group G of exponent 6 (and order the power of the continuum), which has a normal abelian Sylow 3-group with no complement in G. (See [3].) Similar examples show that even when complements exist they need not be conjugate. We therefore begin by imposing a finiteness condition.

An *f-chain* for a group G is a well-ordered chain

$$1 = G_0 \subseteq G_1 \subseteq \cdots \subseteq G_\alpha \subseteq \cdots \subseteq G_\lambda = G$$

of subgroups of G such that $|G_\alpha : G_{\alpha-1}|$ is finite if α is not a limit ordinal, and $G_\alpha = \bigcup_{\beta < \alpha} G_\beta$ when α is a limit ordinal. A group with an f-chain will be called an *f-group*. It is easily shown that any f-group is locally finite. On the other hand, periodic solvable groups, periodic FC-groups, and countable locally solvable groups are all f-groups.

We can now state a generalized Schur–Zassenhaus theorem as follows.

Theorem. *Let G be an f-group and let A be a normal abelian Sylow π-group (for some set π of primes). If the centralizer $C_G(A)$ has finite index in G, then*

(i) *A has a complement in G, and*

(ii) *all complements of A in G are conjugate in G.*

The proof of this theorem is independent of the earlier proofs of the original Schur–Zassenhaus theorem. The principal idea is to apply results of the conjugacy of complements to deduce existence of complements, and this is done using elementary properties of the wreath product. The main steps in the proof are as follows.

(1)   Let A be a normal abelian subgroup of a group G, and suppose that H and K are subgroups of G such that $G = AH = AK$ and $A \cap K = 1$. If the index $|K : K \cap H|$ is finite and equals n, and A possesses unique nth roots, then $a^{-1}Ka \subseteq H$ for some $a \in A$. (Compare Theorem 27 of [6].)

(2)   Let W be the (unrestricted) wreath product $N \operatorname{Wr} F$ of two groups N and F, and let M be the normal subgroup N^F of W. If H is a subgroup of W containing M, and U and V are complements of M in H, then $c^{-1}Uc = V$ for some $c \in M$. (This is proved by a short computation. See Theorem 10.1 of [5].)

(3)   Let H be a subgroup of index n in a group G. Let A be a normal abelian subgroup of G contained in H such that A possesses unique nth roots. Then

   (i) If A has a complement in H, then A has a complement in G;

   (ii) if all the complements of A in H are conjugate in H, then all the complements of A in G are conjugate in G. (This is proved by embedding G in the wreath product $W = A \operatorname{Wr} F$ where $F = G/A$ (see Theorem 3.5 of [4]), and then applying (1) and (2) above.)

(4)   The theorem follows by transfinite induction on the members of the f-chain. The condition that $|G : C_G(A)|$ is finite is used when α is a limit ordinal, and (3) is applied when α is not a limit ordinal.

Notes.   1. Part (i) of the theorem still holds if A is nonabelian, and part (ii) remains valid provided A is solvable. The proofs of these assertions are similar to the proofs of Theorems 25 and 27 of [6].

2. The condition that $|G : C_G(A)|$ is finite may be relaxed when A is abelian, but some condition must be imposed (as is evident from the examples quoted earlier).

In a similar way we may prove the following generalization of a theorem of Gaschütz [1].

Theorem.   *Let G be a periodic group possessing a normal abelian subgroup of finite index (i.e., G is almost abelian). Let A be any normal abelian subgroup of G. Then*

(i) *if, for each prime p, there is a Sylow p-group S_p of G such that $S_p \cap A$ has a complement in S_p, then A has a complement in G;*

(ii) *if the complements of $S_p \cap A$ in S_p are all conjugate in S_p (for each p), then the complements of A in G are all conjugate in G.*

Note. Although the hypothesis on G is rather restrictive, it is satisfied in some important cases. Examples show that (ii) is no longer true if we only assume that G is a periodic almost nilpotent group.

Professor H. Wielandt has informed me that he has previously used similar methods to obtain results like those discussed here. Except for the survey [2], he has not published this work.

Added in proof (December 1966): Further results are given in the author's paper 'Complements of normal subgroups in infinite groups' to appear in *Proc. London Math. Soc.* (3) **17** (1967).

References

[1] WOLFGANG GASCHÜTZ, Zur Erweiterungstheorie der endlichen Gruppen, *J. Reine Angew. Math.* **190** (1952), 93–107.

[2] B. HUPPERT and H. WIELANDT, Arithmetical and normal structure of finite groups, *Proc. Sympos. Pure Math.*, vol. VI, pp. 17–38; Amer. Math. Soc., Providence, R.I., 1962.

[3] L. G. KOVÁCS, B. H. NEUMANN, and H. DE VRIES, Some Sylow subgroups, *Proc. Roy. Soc. London Ser. A.* **260** (1961), 304–316.

[4] B. H. NEUMANN, HANNA NEUMANN, and PETER M. NEUMANN, Wreath products and varieties of groups, *Math. Z.* **80** (1962), 44–62.

[5] PETER M. NEUMANN, On the structure of standard wreath products of groups, *Math. Z.* **84** (1964), 343–373.

[6] HANS J. ZASSENHAUS, *The theory of groups*, 2nd ed., Chelsea, New York, 1958 (pp. 162–163).

Proc. Internat. Conf. Theory of Groups, Austral. Nat. Univ. Canberra, August 1965, pp. 85–88. © Gordon and Breach Science Publishers, Inc. 1967

On groups with a cyclic Sylow subgroup

WALTER FEIT

Recently J. G. Thompson [11] has simplified and generalized some classical results of R. Brauer [1] concerning groups which have a Sylow group of prime order. His approach is based on results of J. A. Green [6], [7] and D. G. Higman's concept of relatively projective modules. Amongst other things he was able to simplify the proof of a critical special case of Theorem B of P. Hall and G. Higman [9]. We will first sketch another simplification of this special case and then state some results which can be proved by these methods. The details of the latter will appear elsewhere.

The following result includes [9, Theorem 2.5.1] and is the special case to which the proof of Theorem B is reduced in [9]. See also [11].

Theorem. *Let p, q be distinct primes. Let $\mathfrak{G} = \mathfrak{P}\mathfrak{Q}$ where $\mathfrak{Q} \lhd \mathfrak{G}$, $\mathfrak{Q}$ is a q-group and $\mathfrak{P} = \langle P_0 \rangle$ is a cyclic p-group such that $\mathbf{C}_{\mathfrak{G}}(P) = \mathfrak{P}\mathbf{Z}(\mathfrak{G})$ for all $P \in \mathfrak{P}^{\#}$. Let K be a field of characteristic p. Assume that there exists a faithful K-representation $\mathscr{L}$ of $\mathfrak{G}$ such that $\mathscr{L}|_{\mathfrak{Q}}$ is absolutely irreducible. Let $(x-1)^s$ be the minimum polynomial of $\mathscr{L}(P_0)$. Then $|\mathfrak{P}| - 1 \leq s \leq |\mathfrak{P}|$. Furthermore if $s = |\mathfrak{P}| - 1$ then $|\mathfrak{P}| - 1 = q^b$ for some b and $\mathfrak{Q}$ is nonabelian.*

Proof. Clearly $s \leq |\mathfrak{P}|$. Since $\mathscr{L}$ is faithful and irreducible $\mathfrak{G}$ has no normal p-subgroup. Thus it may be assumed that $\mathfrak{Q} \neq \langle 1 \rangle$. Hence $\mathfrak{N} = \mathbf{N}_{\mathfrak{G}}(\mathfrak{P}) = \mathfrak{P} \times \mathbf{Z}(\mathfrak{G})$. Let L be the $K[\mathfrak{G}]$ module corresponding to $\mathscr{L}$ and let $L_0(\mathfrak{G})$ be the trivial one dimensional $K[\mathfrak{G}]$ module. Set $d = \dim_K L$. Since $d > 1$, $\mathfrak{Q}$ is nonabelian and $d = q^b$ for some b.

Assume that $s < |\mathfrak{P}|$. Then $L|_{\mathfrak{N}}$ has no projective direct summand. Thus $L|_{\mathfrak{N}}$ is indecomposable by the Mackey decomposition and D. G. Higman's theorem since $\mathbf{C}_{\mathfrak{G}}(P) = \mathfrak{N}$ for all $P \in \mathfrak{P}^{\#}$. Hence $L|_{\mathfrak{P}}$ is indecomposable and so $d = s$. It remains to show that $d = |\mathfrak{P}| - 1$.

Let $\bar{\mathfrak{G}} = \mathfrak{G}/\mathbf{Z}(\mathfrak{G})$ and let $\bar{G}$ denote the image of G in $\bar{\mathfrak{G}}$. It is easily seen that $\mathbf{C}_{\overline{\mathfrak{G}}}(\overline{P}) = \overline{\mathfrak{P}}$ for all $\overline{P} \in \overline{\mathfrak{P}}^{\#}$. Thus every nonprincipal p-block of $\overline{\mathfrak{G}}$ has defect 0. Furthermore $\overline{\mathfrak{Q}}$ is in the kernel of every $K[\overline{\mathfrak{G}}]$ module in the principal p-block. Thus $L_0(\overline{\mathfrak{G}})$ is the only nonprojective irreducible $K[\overline{\mathfrak{G}}]$ module and so every nonprojective indecomposable $K[\overline{\mathfrak{G}}]$ module has invariant elements.

Let L^* denote the contragredient module of L. Since $L|_{\mathfrak{Q}}$ is absolutely irreducible, $p \nmid \dim_K L$. Hence $L_0(\mathfrak{G})$ is a direct summand of $L \otimes L^*$ (see for instance [4]). Furthermore $L \otimes L^*$ is a $K[\overline{\mathfrak{G}}]$ module. Let

$$L \otimes L^* \approx L_0(\mathfrak{G}) \oplus \sum L_i$$

where each L_i is indecomposable. By Schur's Lemma the space of invariant elements in $L \otimes L^*$ is one dimensional. Thus by the previous paragraph each L_i is projective. Consequently

$$L|_{\mathfrak{P}} \otimes L^*|_{\mathfrak{P}} \approx L_0(\mathfrak{P}) \oplus A$$

where A is projective. Let W denote the space of invariant elements in $L|_{\mathfrak{P}} \otimes L^*|_{\mathfrak{P}}$ and let $\mathfrak{C}$ be the commuting ring of $\mathscr{L}|_{\mathfrak{P}}$. Then

$$\dim_K \mathfrak{C} = \dim_K W = 1 + \frac{d^2 - 1}{|\mathfrak{P}|}.$$

However $\mathfrak{C}$ is the commuting ring of $\mathscr{L}(P_0)$ and so $\dim_K \mathfrak{C} = d$. Thus $d^2 - |\mathfrak{P}|d + |\mathfrak{P}| - 1 = 0$ and so $d = |\mathfrak{P}| - 1$ as required since $d > 1$.

The next two theorems generalize results of R. Brauer [2] and H-F. Tuan [12].

Theorem 1. *Let $\mathfrak{G}$ be a finite group with a cyclic S_p-subgroup $\mathfrak{P}$ for some prime p. Assume that $PSL_2(p)$ does not occur as a composition factor of $\mathfrak{G}$ and $\mathfrak{G}$ is not p-solvable. Assume further that $\mathfrak{G}$ has a faithful indecomposable representation $\mathscr{L}$ of degree $d \le p$ in a field of characteristic p. Then $p \ne 2$, $|\mathfrak{P}| = p$ and $d \ge \frac{2}{3}(p-1)$. Furthermore $d \ge \frac{7}{10}p - \frac{1}{2}$ in case $p \ge 13$.*

Theorem 2. *Suppose the S_p-subgroup $\mathfrak{P}$ of $\mathfrak{G}$ is not normal in $\mathfrak{G}$ and $\mathbf{Z}(\mathfrak{G}) = \langle 1 \rangle$. Assume that $\mathfrak{G}$ has a complex faithful irreducible representation of degree $d < p - 1$. Then either $\mathfrak{G} \approx PSL_2(p)$ or $p - 1 = 2^a$ and $\mathfrak{G} \approx PSL_2(p-1)$.*

The proof of Theorem 1 is a straightforward consequence of combining D. G. Higman's theorem with a detailed knowledge of the decomposition of the tensor product of a $K[N_{\mathfrak{G}}(\mathfrak{P})]$ module with its contragredient where K is a field of characteristic p. In case $p \leq 11$ the estimate is best possible. The extension of $\mathfrak{A}_5$, $\mathfrak{A}_6$, $\mathfrak{A}_7$ by its multiplier has a representation of degree 2, 3, 4 respectively in any algebraically closed field. In case $p = 11$ Janko's group [10] has a 7-dimensional representation in the field of 11 elements.

The proof of Theorem 2 is more difficult and depends on a recent paper of R. Brauer [3] as well as some results of Green [8] and Brauer and Tuan [5]. In case the hypothesis $\mathbf{Z}(\mathfrak{G}) = \langle 1 \rangle$ is dropped it is possible to obtain some partial results in this direction by the same methods.

In view of Janko's group the case $p = 11$ in Theorem 1 is of especial interest. G. Seligman and I have proved the following results for this case.

Theorem 3. *Let $\mathfrak{G}$ be a simple group which has an indecomposable 7-dimensional representation in a field of characteristic 11. Assume that $|\mathfrak{P}| = 11$ where $\mathfrak{P}$ is a Sylow 11-group of $\mathfrak{G}$. Then $\mathfrak{G}$ is isomorphic to a subgroup of $G_2(11)$.*

This in particular provides an alternative proof of a result of W. A. Coppel which asserts that Janko's group $\mathfrak{J}$ is isomorphic to a subgroup of $G_2(11)$.

Using some results of Janko it can further be shown that if $P = \exp D$ and

$$H = \begin{bmatrix} -1 & & & & & & \\ & 1 & & & & & \\ & & -1 & & & & \\ & & & 1 & & & \\ & & & & -1 & & \\ 3 & 0 & & & & 1 & \\ 0 & 3 & & & & & -1 \end{bmatrix},$$

$$J = \begin{bmatrix} & & & & & & 1 \\ & & & & & -4 & \\ & & & & 3 & & \\ & & & -1 & & & \\ & & 4 & & & & \\ & -3 & & & & & \\ 1 & & & & & & \end{bmatrix},$$

$$D = \begin{bmatrix} 0 & & & & & & \\ 1 & 0 & & & & & \\ & 1 & 0 & & & & \\ & & 2 & 0 & & & \\ & & & -1 & 0 & & \\ & & & & -1 & 0 & \\ & & & & & -1 & 0 \end{bmatrix},$$

then $\mathfrak{L} = \langle J, P \rangle$ is isomorphic to $PSL_2(11)$ and $\mathfrak{J} \approx \langle H, J, P \rangle$.

References

[1] RICHARD BRAUER, Investigations on group characters, *Ann. of Math.* (2) **42** (1941), 936–958.

[2] RICHARD BRAUER, On groups whose order contains a prime number to the first power. I., II., *Amer. J. Math.* **64** (1942), 401–420, 421–440.

[3] R. BRAUER, Some results on finite groups whose order contains a prime to the first power [to appear].

[4] R. BRAUER and W. FEIT, An analogue of Jordan's Theorem in characteristic *p, Ann of Math.* (2) **84** (1966), 119–131.

[5] RICHARD BRAUER and HSIO-FU TUAN, On simple groups of finite order, *Bull. Amer. Math. Soc.* **51** (1945), 756–766.

[6] J. A. GREEN, On the indecomposable representations of a finite group, *Math. Z.* **70** (1959), 430–445.

[7] J. A. GREEN, Blocks of modular representations, *Math. Z.* **79** (1962), 100–115.

[8] J. A. GREEN, A lifting theorem for modular representations, *Proc. Roy. Soc. London Ser. A.* **252** (1959), 135–142.

[9] P. HALL and GRAHAM HIGMAN, On the *p*-length of *p*-soluble groups and reduction theorems for Burnside's Problem, *Proc. London Math. Soc.* (3) **6** (1956), 1–42.

[10] ZVONIMIR JANKO, A new finite simple group with abelian 2-Sylow subgroups, *Proc. Nat. Acad. Sci. U.S.A.* **53** (1965), 657–658.

[11] J. G. THOMPSON, Vertices and sources [to appear].

[12] HSIO-FU TUAN, On groups whose orders contain a prime to the first power, *Ann. of Math.* (2) **45** (1944), 110–140.

Proc. Internat. Conf. Theory of Groups, Austral. Nat. Univ. Canberra,
August 1965, pp. 89–98. © Gordon and Breach Science Publishers, Inc. 1967

On orderable groups

L. FUCHS

The aim of this paper is to give a survey of some results on groups that admit a full order and to call attention to some open problems.

We shall adopt the multiplicative notation in groups; e will denote the neutral element, and $[a, b] = a^{-1}b^{-1}ab$ the commutator of a and b. For $a_1, \ldots, a_n$ in a group G, $S(a_1, \ldots, a_n)$ denotes the normal sub-semigroup generated by $a_1, \ldots, a_n$ in G. Standard reference will be made to the author's book [1]. The author is indebted to L. G. Kovács for an essential improvement in Theorem 5.

§1. O-groups and O*-groups

A group is called an *O-group* if it can be fully ordered and an *O*-group*[1] if every partial order of G can be extended to a full order of G. We introduce the corresponding local concepts: G is said to be an *LO-group* (*LO*-group*) if every finitely generated subgroup of G is an *O*-group (*O*-group*).

It is easy to see that *a group is an LO-group if and only if it is an O-group* (B. H. Neumann [11]). The nontrivial "only if" part follows, for instance, from the finite character of the following necessary and sufficient condition for G being an *O*-group (M. Ohnishi; see [1], p. 36):

(1) given $a_1, \ldots, a_n \in G$ with $a_i \neq e$, for at least one choice of the signs $\epsilon_i = \pm 1$ we have

$$e \notin S(a_1^{\epsilon_i}, \ldots, a_n^{\epsilon_n}).$$

Also, it follows readily that *every LO*-group is an O*-group*, this being a simple consequence of the fact that G is an *O*-group* if and only if it satisfies

---

[1] Thus an *O*-group* is characterized by the property that its maximal partial orders are full.

(2)   (i) if $b, c \in S(a)$ for some $a \in G$, then $S(b)$ and $S(c)$ intersect non-trivially; and

(ii) $e \in S(a)$ implies $a = e$ (M. Ohnishi; see [1], p. 39).

The following problem, however, is still open:

Problem 1. *Is an O*-group necessarily an LO*-group?*

This problem is obviously equivalent to the problem as to whether or not subgroups of O*-groups are again O*-groups.

Evidently, every O*-group is an O-group. However, *there exist O-groups that are not O*-groups* (see [2], [3], and [4]). This can be proved by means of the following group G_1.

Let A be the free nilpotent group of class 2 with two generators:

$$A = \mathrm{gp}\{a, b \mid [a, b] = z, az = za, bz = zb\};$$

this group has an automorphism of order 4 which maps a onto b and b onto a^{-1}. Hence we can define a group G_1 by

$$G_1 = \mathrm{gp}\{A, c \mid c^{-1}ac = b, c^{-1}bc = a^{-1}, c^4 = e\}.$$

Then a and c already generate G_1, and G_1 does not satisfy condition (2)(i). In fact, on the one hand, z is a central element of infinite order in G_1, and so $S(z^4) \cap S(z^{-4})$ is empty. On the other hand, as A is nilpotent of class 2, we have $z^{\pm 4} = [a^{\pm 2}, b^2]$, and so

$$
\begin{aligned}
z^{\pm 4} &= a^{\mp 2}b^{-2}a^{\pm 2}b^2 \\
&= a^{\mp 1}(c^{-2}a^{\pm 1}c^2)b^{-1}(c^{-2}bc^2)a^{\pm 1}(c^{-2}a^{\mp 1}c^2)b(c^{-2}b^{-1}c^2) \\
&= (a^{\mp 1}c^2a^{\pm 1})c^2(b^{-1}c^2b)c^2(a^{\pm 1}c^2a^{\mp 1})c^2(bc^2b^{-1})c^2 \in S(c^2)
\end{aligned}
$$

where we have used that $c^{-2}ac^2 = a^{-1}$ and $c^{-2}bc^2 = b^{-1}$. Since property (2)(i) is inherited by homomorphic images, no O-group can be an O*-group if it has G_1 for a homomorphic image. In particular, every noncyclic free group is an O-group which is not an O*-group.[2]

It is a trivial fact that an O-group G is always torsion-free, moreover, it is an *R-group* in the sense of P. G. Kontorovič (see [8], p. 242), i.e., if $a^n = b^n$ for some $a, b \in G$ and for some integer $n > 0$, then $a = b$. Call a group G an *R*-group* if

(3) $x_1^{-1}ax_1 \cdots x_n^{-1}ax_n = x_1^{-1}bx_1 \cdots x_n^{-1}bx_n$ implies $a = b$

[2] This is a slight improvement, due to L. G. Kovács, on the result presented at the Conference. The example then used in place of the group G_1 required three generators, so that this conclusion could only be obtained for free groups of rank ≥ 3.

where $x_i \in G$. Every R^*-group is manifestly an R-group, but the converse is not true, as shown, for example, by the group

$$G_2 = \text{gp}\{a, b \mid b^{-1}ab = a^{-2}\}.$$

If we agree to call a group satisfying (2)(ii) *generalized torsion-free*, then choosing $b = e$ in (3) we see that every R^*-group is generalized torsion-free. On the other hand, by making use of the equivalence of

$$\prod_{i=1}^{n} (x_n^{-1}b^{-1}x_n) \cdots (x_{i+1}^{-1}b^{-1}x_{i+1})x_i^{-1}(b^{-1}a)x_i(x_{i+1}^{-1}bx_{i+1}) \cdots (x_n^{-1}bx_n) = e$$

with the hypothesis of (3), we conclude that the converse also holds, and so *a group is an R^*-group if and only if it is generalized torsion-free*.

Since O-groups are necessarily generalized torsion-free, we are led to the result:

Theorem 1. *The following implications hold: LO^*-group $\Rightarrow$ O^*-group $\Rightarrow$ LO-group $=$ O-group $\Rightarrow$ generalized torsion-free group $=$ R^*-group $\Rightarrow$ R-group $\Rightarrow$ torsion-free group. All the implications except for the first and third are known to be irreversible.*

Problem 2. *Are all R^*-groups O-groups?*

§2. Locally nilpotent groups

We are going to show that all the classes of groups considered in Theorem 1 coincide under the hypothesis of local nilpotency (Mal'cev [9]).

Theorem 2. *A torsion-free locally nilpotent group is an LO^*-group.*

We have to show that a (finitely generated) torsion-free nilpotent group G is an O^*-group. Let

$$e = Z_0 \subset Z_1 \subset \cdots \subset Z_r = G$$

be the upper central chain of G, i.e., for $i = 1, \ldots, r$, Z_i/Z_{i-1} is the center of G/Z_{i-1}. Assume that P is a maximal partial order on G and, by way of contradiction, that $a \in G$ is incomparable with e. Then by the maximality of P both $S(a)$ and $S(a^{-1})$ meet P: say,

$$u = x_1^{-1}ax_1 \cdots x_m^{-1}ax_m \geq e \quad \text{and} \quad v = y_1^{-1}ay_1 \cdots y_n^{-1}ay_n \leq e$$

for some $x_i, y_j \in G$ and $m, n > 0$. We prove by induction on k that $a \in Z_k$ is impossible. As this is obvious for $k = 0$, assume that this has been proved for $k - 1$ $(k \geq 1)$, i.e., that P induces a full order on Z_{k-1}. For $a \in Z_k$ and for an arbitrary $x \in G$ we have $[x, a] \in Z_{k-1}$, and hence either $[x, a] \geq e$ or $[x, a] \leq e$. Thus any two conjugates of a are comparable. If we set

$$x^{-1}ax = \max_i (x_i^{-1}ax_i) \quad \text{and} \quad y^{-1}ay = \min_j (y_j^{-1}ay_j),$$

then we get

$$x^{-1}a^m x = (x^{-1}ax)^m \geq u \geq e \geq v \geq (y^{-1}ay)^n = y^{-1}a^n y.$$

Hence $a^m \geq e \geq a^n$, and so $e \leq (a^m)^n = (a^n)^m \leq e$, that is, $a^{mn} = e$. By torsion-freeness, we get $a = e$, a contradiction, establishing that Z_k too is fully ordered under P.   Q.E.D.

Since ZA-groups are locally nilpotent (Mal'cev, see [8], p. 223), we obtain the

Corollary.   *A torsion-free ZA-group is an LO^*-group.*

As a torsion-free nilpotent group can be obtained by successive central extensions of LO^*-groups by LO^*-groups (note that all G/Z_k are torsion-free), the following problem arises:

Problem 3.   *Is G an O^*-group $(LO^*$-group$)$ if it is a central extension of a torsion-free abelian group by an O^*-group $(LO^*$-group$)$?*

From the Corollary it is easy to derive

Theorem 3.   *The hypercenter of an arbitrary torsion-free group is an LO^*-group.*

In fact, a hypercenter has a well-ordered ascending central chain, and therefore it is a ZA-group.

Moreover, the proof of Theorem 2 implies that *every maximal partial order of a torsion-free group G induces a full order on the hypercenter of G.*

Let us observe that there exist LO^*-groups without center (cf. [2]), as shown by $G_3 = \text{gp}\{a, b \mid ba = ab^2\}$.

§3. The role of the center

Recently, A. I. Kokorin and V. M. Kopytov have pointed out the importance of the role played by the center in O-groups. Let us recapitulate here their most important results ([6], [7]).

The following lemmas will be made use of.

Lemma 1. *Every O-group G with center Z can be embedded in an O-group G^* with divisible center Z^* such that $G^*/Z^* \cong G/Z$. (If G is an O*-group, then G^* can be chosen as an O*-group.)*

The proof is straightforward. If Z^* is a minimal divisible abelian group containing Z, then let G^* be the group generated by the commutable subgroups G and Z^* subject to $G \cap Z^* = Z$. The elements of G^* are of the form au with $a \in G$ and $u^n \in Z$ for some $n > 0$, and $G/Z \cong G^*/Z^*$ holds under the correspondence $aZ \to aZ^*$ $(a \in G)$. If P is a full order on G and $au \in G^*$ is declared positive whenever $(au)^n \in P$ for some $n > 0$, then a full order P^* of G^* arises. (Since P^* turns out to be the only full order extending P, the second statement follows readily.)

Lemma 2. *If G is an O-group, if Z_0 lies in the center of G, and if Z_0 is isomorphic to the additive group of the rational numbers, then G admits a full order in which Z_0 is a convex subgroup.*

Let P be a full order on G and Σ the system of convex subgroups of G (cf. [1], p. 51). The elements $\neq e$ of Z_0 obviously generate the same convex subgroup C of G. If D denotes the union of all convex subgroups without elements $\neq e$ of Z_0, then $D \subset C$ and G has no convex subgroups between D and C. Every inner automorphism of G leaves Z_0 and hence C and D invariant, i.e., C and D are normal in G. Moreover, $[G, C] \subseteq D$ in view of $g^{-1}zg = z$ $(z \in Z_0, g \in G)$ and the archimedean character of C/D, thus the inner automorphisms of G induce the identity map on C/D. If we squeeze the subgroup $Z_0 D$ between D and C $(Z_0 D = C$ being possible), then we obtain a new system Σ' of subgroups from Σ. It is easy to check that G has a linear order P' with Σ' as the system of convex subgroups. Since in P' the product $Z_0 D = Z_0 \times D$ is ordered lexicographically [3] and the roles of Z_0 and D can be interchanged, we can define a new full order on G in which Z_0 will be a convex subgroup.

Lemma 3. *If N_λ $(\lambda \in \Lambda)$ is a chain of normal subgroups of a group G such that all the factor groups G/N_λ are O-groups, then G/N is an O-group where N is the union of all the N_λ.*

If G/N were not an O-group, then by (1) there would exist elements $a_1, \ldots, a_n \in G$ $(a_i \notin N)$ such that $S(a_1^{\epsilon_1}, \ldots, a_n^{\epsilon_n}) \cap N$ would be void for

[3] I.e., every element $> e$ of Z_0 is greater than every element of D.

no choice of the signs $\epsilon_i = \pm 1$. For each of the 2^n choices we select some N_λ intersecting $S(a_1^{\epsilon_1}, \ldots, a_n^{\epsilon_n})$. For the greatest N_μ of these 2^n subgroups N_λ, G/N_μ would not be an O-group.

The main result is the following theorem [6].

Theorem 4. *A group G is an O-group if and only if both its center Z and the factor group G/Z are O-groups.*

Embed the O-group G in an O-group G^* of Lemma 1. Then Z^*—as a divisible group—is a direct product of groups isomorphic to the additive group of the rational numbers. Thus Z^* is the union of a well-ordered ascending chain of subgroups N_α such that $N_0 = e$, $N_\alpha/N_{\alpha+1}$ is isomorphic to the group of rational numbers, $N_\alpha = \bigcup_{\beta < \alpha} N_\beta$ for limit ordinals α, with α ranging over all ordinals less than, say, α_0. The N_α are normal in G^* and $G^*/N_{\alpha+1} \cong (G^*/N_\alpha)/(N_{\alpha+1}/N_\alpha)$ is, by Lemma 2, an O-group if G^*/N_α is. By Lemma 3, G^*/N_α is an O-group for limit ordinals α. Thus $G/Z \cong G^*/Z^*$ is an O-group.—The converse follows immediately by taking arbitrary full orders on Z and G/Z, and then defining the lexicographic ordering on G.

Corollary 1. *A group G is an O-group if and only if both H and G/H are O-groups where H denotes the hypercenter of G.*

This follows at once from the preceding theorem by making use of Lemma 3 again. Moreover, taking Theorem 3 into account, we can conclude:

Corollary 2. *A torsion-free group G is an O-group if and only if G/H is an O-group.*

This reduces the problem of group-theoretical characterization of O-groups to the case of groups without center.

As a factor group of an O^*-group that is an O-group is likewise an O^*-group, we infer:

Corollary 3. *If G is an O^*-group (LO^*-group), then so are G/Z and G/H.*

The following problem is related to Problem 3.

Problem 4. *Is a torsion-free group G an O^*-group if G/H is an O^*-group (LO^*-group)?*

§4. Group-theoretical constructions

Finally, we are concerned with the problem: which group-theoretical constructions, applied to O-groups (O^*-groups), give rise to O-groups (O^*-groups)?

A. *Free products.* As shown by A. A. Vinogradov [13], *the free product of O-groups is again an O-group.* His proof was based on matrix-theoretical methods; so far no purely group-theoretical proof is available.

For O^*-groups the corresponding result does not hold. In fact, the group G_1 in Section 1 implies:

Theorem 5. *The free group F_n with $n \geq 2$ generators is not an O^*-group.*

B. *Direct products.* For O-groups it is easily shown: *the (discrete) direct product of O-groups is always an O-group.* Indeed, the lexicographic ordering of the factors furnished with arbitrary full orders defines a full order on the direct product. The corresponding result on O^*-groups lies much deeper ([4], [5]):

Theorem 6. *The direct product of O^*-groups (LO^*-groups) is again an O^*-group (LO^*-group).*

By (2) of Section 1, it suffices to consider the direct product $G = A \times B$ of *two* O^*-groups A and B. Let P be a maximal partial order on G. If $P \cap A$ were not a linear order on A, then there would be a linear extension Q of $P \cap A$ in A, and—as is readily checked—QP would be a larger order than P. Thus P induces full orders on A and B. Assume that $g = ab$ ($a \in A$, $b \in B$) is incomparable with e. Then

$$u = x_1^{-1}gx_1 \cdots x_m^{-1}gx_m \geq e \quad \text{and} \quad v = y_1^{-1}gy_1 \cdots y_n^{-1}gy_n \leq e$$

for some $x_i, y_j \in G$, $m, n > 0$. If $x^{-1}ax = \max_i (x_i^{-1}ax_i)$ in A and $x_0^{-1}bx_0 = \max_i (x_i^{-1}bx_i)$ in B, then $(x^{-1}ax)^m(x_0^{-1}bx_0)^m \geq u \geq e$. Since $x \in A$, $x_0 \in B$ may be assumed, we obtain $g^m \geq e$, and similarly, $g^n \leq e$. As in the proof of Theorem 2, we arrive at the contradiction $g = e$. Thus P must be a full order on G.—The statement on LO^*-groups is now obvious.[4]

---

[4] The proofs of Theorems 2 and 6 imply that in a torsion-free group G every maximal partial order $\geq$ satisfies: if $g \in G$ and if to any two conjugates g_1, g_2 of g there are conjugates g_3, g_4 of g such that g_3 is an upper bound and g_4 is a lower bound for g_1, g_2, then $g \geq e$ or $g \leq e$.

8—K.

C. *Cartesian products.* Again, the case of O-groups can be settled at once: *the cartesian (or complete direct) product of O-groups is an O-group.* In fact, the set of components can be well-ordered and then lexicographic ordering can be used. However, for O^*-groups we have [4]:

Theorem 7. *The cartesian product of O^*-groups is in general not an O^*-group.*

To each positive integer k, define a group J_k as the factor group of G_1 (of Section 1) over the normal closure of a^{2^k} in G_1; in other words, put

$$J_k = \mathrm{gp}\{a_k, c_k \mid a_k^{2^k} = c_k^4 = e, \; c_k^2 a_k c_k^2 = a_k^{-1},$$
$$[a_k, c_k^{-1} a_k c_k] = z_k, \; a_k z_k = z_k a_k\}.$$

As is readily seen, the map

(4) $a \to (a_1, \ldots, a_k, \ldots)$ and $c \to (c_1, \ldots, c_k, \ldots)$

induces an embedding of G_1 into the cartesian product J of the groups J_k, and in this embedding z becomes a central element, namely $(z_1, \ldots, z_k, \ldots)$, of J. Thus even in J we have that $S(z^4) \cap S(z^{-4})$ is empty, and of course also $z^{\pm 4} \in S(c^2)$ in J. Hence J does not possess property (2)(i). Note that each J_k is a finite 2-group and so a nilpotent group. Let N_k be a free nilpotent group of rank 2 and of the same class as J_k. The N_k are evidently torsion-free and so O^*-groups (by Theorem 2). On the other hand, since each N_k can be mapped homomorphically onto the corresponding J_k, the cartesian product G of the N_k has a homomorphism onto J. Thus G is not an O^*-group, for it does not have property (2)(i).

It also follows that the cartesian product of LO^*-groups need not be an O^*-group.

D. *Inverse limits.* If the groups G_λ form an inverse system with homomorphisms $\varphi_{\lambda\mu} : G_\lambda \to G_\mu$ for $\lambda \geq \mu$ (where $\varphi_{\lambda\lambda}$ is the identity map and $\varphi_{\lambda\nu} = \varphi_{\mu\nu}\varphi_{\lambda\mu}$ for $\lambda \geq \mu \geq \nu$), then inv lim G_λ is a subgroup of the cartesian product of the groups G_λ. Hence *the inverse limit of O-groups is always an O-group.* However,

Theorem 8. *The inverse limit of O^*-groups need not be an O^*-group.*

Choose the groups N_k and J_k as in the proof of the preceding theorem. Note that the class of nilpotency increases with k. Let $\varphi_{lk} : N_l \to N_k$

be the natural homomorphism $(l \geq k)$. This induces the natural map of J_l onto J_k, and it follows that inv lim $J_k = J'$ is a homomorphic image of inv lim $N_k = N$. The map induced by (4) embeds G_1 into J' in such a way that $S(z^4) \cap S(z^{-4})$ remains empty in J' while, of course, $z^{\pm 4} \in S(c^2)$ in J'. Hence N is not an O^*-group.

E. *Wreath products.* For the (restricted) wreath product[5] we obtain (see [2], [10]):

Theorem 9. *The wreath product of two O-groups is an O-group.*

Let A and B be O-groups with arbitrarily chosen full orders P and Q, respectively. Then P induces by conjugation a full order in each of $b^{-1}Ab = A_b$ $(b \in B)$. We order the direct product $\prod A_b$ lexicographically as indicated by Q; this order is invariant under conjugation by elements of B. Therefore $A \wr B$—as a semidirect product of $\prod A_b$ and B—can be lexicographically ordered, preserving the orderings of $\prod A_b$ and B. Thus $A \wr B$ is, in fact, an O-group.

Problem 5. *Prove or disprove that the wreath product of two O^*-groups is an O^*-group.*

It is to be noted that for the complete wreath product W of an O-group A by an O-group B the situation is different (B. H. Neumann [12]): W *is never an O-group* (unless A or B is trivial). Namely, in the cartesian product of the A_b $(b \in B)$ the element g whose components are a and a^{-1} in $A_{b^{2n}}$ and $A_{b^{2n+1}}$, respectively (for some $a \neq e$, $b \neq e$, and all $n = 0, \pm 1, \ldots$), and e elsewhere, satisfies $g(b^{-1}gb) = e$, and therefore W is not even an R^*-group.

References

[1] L. Fuchs, *Partially ordered algebraic systems*, Pergamon, Oxford, 1963.
[2] L. Fuchs and E. Sasiada, Note on orderable groups, *Annales Univ. Sci. Budapest Eötvös Sect. Math.* **7** (1965), 13–17.
[3] C. Holland, *Ph.D. Thesis*, New Orleans, 1964.
[4] M. I. Kargapolov, Orderable groups, *Algebra i Logika Sem.* **2:6** (1963), 5–14 [Russian].
[5] A. I. Kokorin, Ordering a direct product of ordered groups, *Ural. Gos. Univ. Mat. Zap.* **4:3** (1963), 95–96 [Russian].

---

[5] The wreath product $A \wr B$ of the group A by the group B is defined as the group generated by the direct product of the groups $b^{-1}Ab$ (for all $b \in B$) and by B where conjugation by elements of B is to be performed in the obvious way.

[6] A. I. KOKORIN, On linearly orderable groups, *Dokl. Akad. Nauk SSSR* **151** (1963), 31–33 [Russian].

[7] V. M. KOPYTOV, On the completion of the centre of an ordered group, *Ural. Gos. Univ. Mat. Zap.* **4**:**3** (1963), 20–24 [Russian].

[8] A. G. KUROSH, *The theory of groups*, vol. II, Chelsea, New York, 1956.

[9] A. I. MAL'CEV, On the completion of group order, *Trudy Mat. Inst. Steklov.*, vol. 38, pp. 173–175; Izdat. Akad. Nauk SSSR, Moscow, 1951 [Russian].

[10] B. H. NEUMANN, On ordered groups, *Amer. J. Math.* **71** (1949), 1–18.

[11] B. H. NEUMANN, An embedding theorem for algebraic systems, *Proc. London Math. Soc.* (3) **4** (1954), 138–153.

[12] B. H. NEUMANN, Embedding theorems for ordered groups, *J. London Math. Soc.* **35** (1960), 503–512.

[13] A. A. VINOGRADOV, On the free product of ordered groups, *Mat. Sb.* N.S. **25** (**67**) (1949), 163–168 [Russian].

Proc. Internat. Conf. Theory of Groups, Austral. Nat. Univ. Canberra, August 1965, pp. 99–100. © Gordon and Breach Science Publishers, Inc. 1967

On groups with abelian 2-Sylow subgroups

T. M. GAGEN

Definition. A finite group G is called an AZ-group if G has abelian 2-Sylow subgroups and cyclic Sylow subgroups for all odd primes.

Theorem. *A finite nonsoluble group G is an AZ-group if and only if*

(a) *G has a unique normal subgroup L isomorphic to $PSL(2, 2^n)$, $n > 1$, or to $PSL(2, p)$, $p > 3$, $p \equiv \pm 3$ (mod 8), p prime, or to the new Janko simple group J;*

(b) *the centralizer M of L in G is a soluble AZ-group;*

(c) *LM is complemented in G by a soluble AZ-group N;*

(d) *the orders of LM and N are co-prime, and the greatest common divisor of the orders of L and M is a power of 2.*

These groups have been studied by Sah [4]. The classification obtained by him did not include the case $L \cong J$, since his treatment of the nonsoluble case was based on a theorem of Thompson which has recently been corrected in a paper of Janko and Thompson [3]. The present theorem, based on the corrected result and on the characterization of J obtained in Janko [2], is proved by a suitable modification of Sah's argument.

The main properties of J used are:

(i) J is complete,

(ii) a 2-Sylow subgroup S of J is elementary abelian of order 8, and

(iii) a 2-Sylow normalizer of J has the form SH, where H is a non-cyclic group of order 21.

It follows quickly, using results of Suzuki [5] on groups with cyclic Sylow subgroups for all odd primes, that there is only one simple group of order $2^3 \cdot 3 \cdot 5 \cdot 7 \cdot 11 \cdot 19$, which is the order of J. In fact, any group of this order is either isomorphic to J or has a normal 19-Sylow subgroup.

For details, see [1].

References

[1] TERENCE M. GAGEN, On groups with abelian Sylow 2-groups, *Math. Z.* **90** (1965), 268–272.

[2] ZVONIMIR JANKO, A new finite simple group with abelian 2-Sylow subgroups and its characterization, *J. Algebra* **3** (1966), 147–186.

[3] ZVONIMIR JANKO and JOHN G. THOMPSON, On a class of finite simple groups of Ree, *J. Algebra* **4** (1966), 274–292.

[4] CHIH-HAN SAH, A class of finite groups with abelian 2-Sylow subgroups, *Math. Z.* **82** (1963), 335–346.

[5] MICHIO SUZUKI, On finite groups with cyclic Sylow subgroups for all odd primes, *Amer. J. Math.* **77** (1955), 657–691.

Proc. Internat. Conf. Theory of Groups, Austral. Nat. Univ. Canberra,
August 1965, p. 101. © Gordon and Breach Science Publishers, Inc. 1967

Nichtabelsche p-Gruppen
besitzen äussere p-Automorphismen

WOLFGANG GASCHÜTZ

Satz. *Ist S eine endliche nichtabelsche p-Gruppe, so besitzt S einen Automorphismus von p-Potenzordnung, der nicht innerer Automorphismus ist.*

Der Beweis dieses Satzes erfolgt mit dem in [1] bewiesenen Satz 1 über kohomologisch triviale G-Moduln und dem folgenden einfachen

Hilfssatz. *Ist unter den Voraussetzungen des voranstehenden Satzes N maximale Untergruppe von S und gilt $Z(N) \leq Z(S)$ für die Zentren von N und S, so gilt der Satz.*

Literatur

[1] WOLFGANG GASCHÜTZ, Kohomologische Trivialitäten und äussere Automorphismen von p-Gruppen, *Math. Z.* **88** (1965), 432–433.

Proc. Internat. Conf. Theory of Groups, Austral. Nat. Univ. Canberra, August 1965, p. 103. © Gordon and Breach Science Publishers, Inc. 1967

On stability groups of certain nilpotent groups

CHANDER KANTA GUPTA

Let G be a nilpotent group of class n and let

$$(1) \qquad G = G_1 > G_2 > \cdots > G_{n+1} = 1$$

be the lower central series of G. It is well-known [2] that the stability group A of (1) is nilpotent of class at most $n-1$.

Let I denote the group of inner automorphisms of G. Then $I \triangleleft A$ and A/I is nilpotent of class at most $n-1$. For every integer n (≥ 2) there is a nilpotent group G of class precisely n with A/I of class precisely $n-1$.

B. Chang [1] has proved that if G is a free group of rank 2, then $A = I$. We prove that if G is a free metabelian nilpotent-of-class-n group of rank 2, then A/I is abelian. If G is a free-nilpotent-of-class-5 group of rank 2, then A/I has class precisely 2.

References

[1] BOMSHIK CHANG, The automorphism groups of the free group with two generators, *Michigan Math. J.* **7** (1960), 79–81.
[2] P. HALL, Some sufficient conditions for a group to be nilpotent, *Illinois J. Math.* **2** (1958), 787–801.

Proc. Internat. Conf. Theory of Groups, Austral. Nat. Univ. Canberra, August 1965, pp. 105–109. © Gordon and Breach Science Publishers, Inc. 1967

Metabelian groups in the variety of certain two-variable laws

N. D. GUPTA

Let G be a group. For each element x of G, consider the mappings $\rho(x): g \to g^{-1}x^{-1}gx$, $\lambda(x): g \to x^{-1}g^{-1}xg$ of G into itself. With the usual product of mappings as operation, let $\Re(m, n)$ $[\Omega(m, n)]$ denote the variety of all groups G satisfying

$$\rho^m(x) = \rho^{m+n}(x) \quad [\lambda^m(x) = \lambda^{m+n}(x)]$$

for every x in G. It was shown in [2] that if $G \in \Re(m, 1)$, then the elements of odd order are mth left Engel elements of G and moreover if G has no element of order 3 then G satisfies the mth Engel condition. In [3] it was shown that a soluble group (and hence also any finite group) in $\Re(1, n)$ is abelian, and quite generally the variety $\Re(1, 2)$ is abelian. Other general results concerning the structure of the groups in terms of the invariants m, n shall appear in [4].

In this paper we restrict our attention to metabelian groups in $\Re(m, n)$ [in $\Omega(m, n)$] and prove the following theorems.

Theorem 1. *If G is a metabelian group in $\Re(m, n)$ [in $\Omega(m, n)$] then $G \in \Omega(m, \epsilon n)$ $[G \in \Re(m, \epsilon n)]$, where $\epsilon = 1$ or 2 according as n is even or odd.*

Let G_i denote the ith term of the lower central series of G and let G^j denote the subgroup of G generated by the jth powers of the elements of G.

Theorem 2. *If G is a metabelian group in $\Re(m, n)$ or in $\Omega(m, n)$, then*

$$[(G_r)^s, G^{\epsilon n}] = 1,$$

where $\epsilon = 1$ or 2 according as n is even or odd, $r = m^2 + m + 2$, and s divides $\prod_{i=1}^m \prod_{j=1}^{2i} j!$.

Theorem 3. *If G is a metabelian group in $\mathfrak{R}(m, n)$ or in $\mathfrak{L}(m, n)$ and if $(3, n) = 1$, then*

$$(G_u)^{t.f(\epsilon n)} = 1,$$

where $\epsilon = 1$ or 2 according as n is even or odd, $u = m+2$, t divides $(m-1)! \prod_{j=1}^{m-1} j!$ and $f(\epsilon n)$ is a certain function of ϵn associated to the circulant matrix $\left(1, \dbinom{\epsilon n}{1}, \ldots, \dbinom{\epsilon n}{n-1}\right)$.

Let G be a metabelian group. For each element x of G, let $\omega(x)$ denote the mapping $g \to x^{-1}gx$ of G into itself, and let $\mathfrak{M}(G)$ denote the ring generated by the $\rho(x)$, $\lambda(x)$, $\omega(x)$ as x ranges through G. [For θ_1, $\theta_2 \in \mathfrak{M}(G)$, addition is defined by $g(\theta_1 + \theta_2) = g\theta_1 \cdot g\theta_2$; the zero 0 of $\mathfrak{M}(G)$ is the mapping which takes every element of G to the identity of G.] We shall carry out our calculations in $\mathfrak{M}(G)$. The following results are easily verified:

$(\mathrm{M_1})$
$$\rho(x) = \lambda(x^{-1}) + \lambda(x^{-1})\rho(x) = \lambda(x^{-1})\omega(x) = \omega(x)\lambda(x^{-1})$$
$$= -\rho(x^{-1})\omega(x) = -\omega(x)\rho(x^{-1}) = -\lambda(x);$$

$(\mathrm{M_2})$
$$\rho(x^n) = \sum_{i=1}^{n} \binom{n}{i}\rho^i(x);$$

$(\mathrm{M_3})$ $\rho(x)\rho(y)\rho(z) = \rho(x)\rho(z)\rho(y)$ for all x, y, z in G.

Lemma 1. *Let G be a metabelian group in $\mathfrak{R}(m, n)$ or in $\mathfrak{L}(m, n)$ where n is even. Then, in $\mathfrak{M}(G)$, $\rho^m(x)\rho(x^n) = 0$.*

Proof. If $G \in \mathfrak{R}(m, n)$ then in $\mathfrak{M}(G)$ we have in turn

$$\rho^m(x^{-1}) = \rho^{m+n}(x^{-1});$$
$$(-1)^m\rho^m(x)\omega(x^{-m}) = (-1)^{m+n}\rho^{m+n}(x)\omega(x^{-m-n}) \quad \text{(by $(\mathrm{M_1})$)};$$
$$\rho^m(x)\omega(x^n) = \rho^{m+n}(x);$$
$$\rho^m(x) + \rho^m(x)\rho(x^n) = \rho^{m+n}(x);$$
$$\rho^m(x)\rho(x^n) = 0.$$

If $G \in \mathfrak{L}(m, n)$, then as before we get in turn

$$\lambda^m(x)\omega(x^n) = \lambda^{m+n}(x);$$
$$\lambda^m(x) + \lambda^m(x)\rho(x^n) = \lambda^{m+n}(x);$$
$$\lambda^m(x)\rho(x^n) = 0;$$
$$(-1)^m\rho^m(x)\rho(x^n) = 0;$$
$$\rho^m(x)\rho(x^n) = 0.$$

Proof of Theorem 1.   Let G be a metabelian group in $\mathfrak{R}(m, n)$; then $G \in \mathfrak{R}(m, \epsilon n)$ where $\epsilon = 1$ or 2 according as n is even or odd. In $\mathfrak{M}(G)$ we have in turn

$$\rho^m(x) = \rho^{m+\epsilon n}(x);$$

$$\lambda^m(x^{-1})\omega(x^m) = \lambda^{m+\epsilon n}(x^{-1})\omega(x^{m+\epsilon n});$$

$$\lambda^m(x^{-1})\omega(x^{-\epsilon n}) = \lambda^{m+\epsilon n}(x^{-1});$$

$$\lambda^m(x^{-1}) + \lambda^m(x^{-1})\rho(x^{-\epsilon n}) = \lambda^{m+\epsilon n}(x^{-1});$$

$$\lambda^m(x^{-1}) = \lambda^{m+\epsilon n}(x^{-1})$$

(by Lemma 1). Thus $G \in \mathfrak{L}(m, \epsilon n)$. The proof of the other part is analogous.

The following lemma is of independent interest:

Lemma 2.   *If G is a metabelian group and if $\rho^m(y)\rho(y^n) = 0$ for all y in G, then*

$$\rho(x) \prod_{i=1}^{m} \rho^{2i}(y_i)\rho(y_{m+1}^n) = 0$$

for all x, y_i in G.

Proof.   Since $\rho^m(y)\rho(y^n) = 0$, we have

(1) $$\rho(x)\rho^m(y)\rho(y^n) = 0 \qquad \text{for all } x, y \text{ in } G.$$

Replacing y by $y_1 z_1$ in (1) and then multiplying on the right by $\rho^m(y_1)\rho^{m-1}(z_1)$ gives, by using (M$_3$), that

$$\rho(x)\rho^m(y_1)\rho^m(y_1 z_1)\rho^{m-1}(z_1)\rho(y_1^n z_1^n) = 0,$$

since $(y_1 z_1)^n \equiv y_1^n z_1^n \bmod G_2$. This gives, on using (1) and (M$_3$), that

(2) $$\rho(x)\rho^{2m}(y_1)\rho^{m-1}(z_1)\rho(z_1^n) = 0.$$

Replacing z_1 by $y_2 z_2$ in (2), multiplying on the right by $\rho^{m-1}(y_2)\rho^{m-2}(z_2)$, and proceeding as before but using (2) instead of (1), we get

(3) $$\rho(x)\rho^{2m}(y_1)\rho^{2(m-1)}(y_2)\rho^{m-2}(z_2)\rho(z_2^n) = 0.$$

A repetition of this last step finally yields

(4) $$\rho(x)\rho^{2m}(y_1) \cdots \rho^{2 \cdot 2}(y_{m-1})\rho(z_{m-1})\rho(z_{m-1}^n) = 0.$$

In (4) replacing z_{m-1} by $y_m y_{m+1}$, multiplying on the right by $\rho(y_m)$, and using (4) and (M$_3$) gives the required result.

Proof of Theorem 2.   Let $G \in \Re(m, n)$ or $\mathfrak{L}(m, n)$; then, by Lemma 1, $\rho^m(x)\rho(x^n) = 0$ for all x in G. Hence by Lemma 2 we have

$$\rho(x) \prod_{i=1}^{m} \rho^{2i}(y_i)\rho(y_{m+1}^{\epsilon n}) = 0.$$

Now the proof of the theorem follows by using the main theorem of [5].

Proof of Theorem 3.   Suppose n is even and $(3, n) = 1$. Then, by Lemma 1, $\rho^m(x)\rho(x^n) = 0$ for all x in G. Now by (M_2) and by left distributivity this gives

$$\sum_{i=1}^{n} \binom{n}{i}\rho^{m+i}(x) = 0;$$

and since $\rho^m(x) = \rho^{m+n}(x)$, we have

$$\sum_{i=0}^{n-1} \binom{n}{i}\rho^{m+i}(x) = 0.$$

This gives for $j = 0, 1, \ldots, n-1$

$$\sum_{i=0}^{n-1} \binom{n}{i}\rho^{m+i+j}(x) = 0.$$

Corresponding to these n equations we have the $n \times n$ matrix

$$\begin{bmatrix} 1 & \binom{n}{1} & \cdots & \binom{n}{n-1} \\ \binom{n}{n-1} & 1 & \cdots & \binom{n}{n-2} \\ \vdots & & & \vdots \\ \binom{n}{1} & \binom{n}{2} & \cdots & 1 \end{bmatrix}.$$

The determinant of this matrix is $\prod_{i=1}^{n} ((1+\gamma_i)^n - 1)$, where γ_i are the nth roots of unity. Since $(3, n) = 1$, the determinant is nonzero. Thus the above matrix can be transformed (by elementary row operations) to a triangular form

$$\begin{bmatrix} \alpha_{11} & \alpha_{12} & \cdots & \alpha_{1n} \\ 0 & \alpha_{22} & \cdots & \alpha_{2n} \\ \vdots & & & \\ 0 & 0 & \cdots & \alpha_{nn} \end{bmatrix}$$

(see for instance [1], p. 136). Thus we have $\alpha_{nn}\rho^{m+n-1}(x) = 0$, which on multiplying by $\rho(x)$ gives $\alpha_{nn}\rho^{m}(x) = 0$. Now the proof of the theorem follows by using a particular case of the main theorem of [5].

References

[1] F. R. GANTMACHER, *The theory of matrices*, Vol. I, Chelsea, New York, 1960.
[2] N. D. GUPTA, Groups with Engel-like conditions, *Arch. Math.* **17** (1966), 193–199.
[3] N. D. GUPTA, Some group laws equivalent to the commutative law, *Arch. Math.* **17** (1966), 97–102.
[4] N. D. GUPTA and H. HEINEKEN, Groups with a two variable commutator identity, *Math Z.* [to appear].
[5] N. D. GUPTA and M. F. NEWMAN, On metabelian groups satisfying certain laws, *these Proc.*, pp. 111–113; *J. Austral. Math. Soc.* [to appear].

Proc. Internat. Conf. Theory of Groups, Austral. Nat. Univ. Canberra,
August 1965, pp. 111–113. © Gordon and Breach Science Publishers, Inc. 1967

On metabelian groups

N. D. GUPTA and M. F. NEWMAN

The investigation reported on in this note was motivated by the observation [3] that, if $\mathfrak{V}$ is a proper subvariety of $\mathfrak{A}_p\mathfrak{A}_p$ (the variety of extensions of elementary abelian p-groups by elementary abelian p-groups), then (in the notation explained below) there is an integer s greater than 1 such that

$$[a_1, (p-1)a_2, (p-1)a_1, (p-1)a_3, \ldots, (p-1)a_s] = e$$

(the identity element) whenever $a_1, \ldots, a_s$ are elements of a group G in $\mathfrak{V}$.

As usual $[x, y] = x^{-1}y^{-1}xy$ and $[x, y, z] = [[x, y], z]$; further $[x, 0y] = x$ and $[x, ky] = [x, (k-1)y, y]$ for positive integers k. A group G is metabelian if $[[x, y], [z, w]] = e$ for all x, y, z, w in G, and has finite exponent k if k is the least positive integer such that $x^k = e$ for all x in G. The sth term $\gamma_s(G)$ of the lower central series of a group G is the subgroup generated by $[a_1, \ldots, a_s]$ for all $a_1, \ldots, a_s$ in G; and G is nilpotent of class c if $\gamma_{c+1}(G)$ is the first term of the lower central series of G which is the identity subgroup.

Our main result is:

Theorem.[1] Let $n_1, \ldots, n_s$ be nonnegative integers and let $m = n_1 + \cdots + n_s + 2$. If G is a metabelian group such that

(A) $$[a_1, a_2, n_1a_1, n_2a_2, \ldots, n_sa_s] \in \gamma_{m+1}(G)$$

for all $a_1, \ldots, a_s$ in G, then $\gamma_m(G)/\gamma_{m+1}(G)$ has exponent dividing $(n_1+n_2+2)\prod_{i=1}^s n_i!$ and $\gamma_{m+1}(G)/\gamma_{m+2}(G)$ has exponent dividing

---

[1] This is a stronger result than that presented at the Conference. It has arisen out of suggestions made by the referee to a manuscript containing our earlier results which had been submitted to the Journal of the Australian Mathematical Society.

$d \prod_{i=1}^{s} n_i!$ *where* d *is the greatest common divisor of* $n_1 + 1$ *and* $n_2 + 1$. *If, moreover,*

(B) $\qquad\qquad [a_1, a_2, n_1 a_1, n_2 a_2, \ldots, n_s a_s] = e$

for all $a_1, \ldots, a_s$ *in* G, *then* $\gamma_{m+1}(G)$ *has exponent dividing*

$$(n_1 + 1)! \prod_{i=1}^{s} \prod_{j=1}^{n_i} j!.$$

The proof consists entirely of commutator calculations (see [2] for details).

The theorem is best possible in a number of senses:

(1) $\gamma_{m-1}(G)/\gamma_m(G)$ need not have finite exponent—every nilpotent group of class $m - 1$ satisfies (A);

(2) if p is a prime divisor of $d \prod_{i=1}^{s} n_i!$, then p may divide the exponent of $\gamma_{m+1}(G)/\gamma_{m+2}(G)$—the wreath product of a cyclic group of order p by a countably infinite elementary abelian p-group is a metabelian group satisfying (B) and having every factor of its lower central series of exponent p;

(3) if $n_1 + n_2 + 2$ is a prime, there is a metabelian group M which satisfies (B) and has $\gamma_m(M)/\gamma_{m+1}(M)$ of exponent $n_1 + n_2 + 2$;

(4) there is a group N which is both abelian-by-nilpotent-of-class-two and nilpotent-of-class-two-by-abelian which satisfies (B) and in which $\gamma_m(N)/\gamma_{m+1}(N)$ contains elements of infinite order.

We record separately the motivating case:

Corollary. *If* G *is a metabelian* p-*group such that*

$$[a_1, (p-1)a_2, (p-1)a_1, (p-1)a_3, \ldots, (p-1)a_s] = e$$

for all $a_1, \ldots, a_s$ *in* G, *then* G *is nilpotent of class at most* $s(p-1)$.

It follows that every proper subvariety of $\mathfrak{A}_p \mathfrak{A}_p$ is nilpotent.

The theorem includes as special cases all the related results we know: Theorem 1.10 of Gruenberg [1]; Weston [4]; and an unpublished result of Mrs U. Heineken (verbally communicated by Dr H. Heineken). Gruenberg's theorem gives a finite exponent for a term of the lower central series of a soluble Engel group—the term depending on the soluble length and the Engel condition. Our result improves the term of the lower central series involved for the metabelian case; using his techniques (embodied in his Lemma 4.4), we can improve his result also in the general soluble case—we will not, however, write these results down explicitly here as they are most unlikely to be best possible.

References

[1] K. W. Gruenberg, The upper central series in soluble groups, *Illinois J. Math.* **5** (1961), 436–466.

[2] N. D. Gupta and M. F. Newman, On metabelian groups, *J. Austral. Math. Soc.* [to appear].

[3] L. G. Kovács and M. F. Newman, On non-Cross varieties of groups, [in preparation].

[4] K. W. Weston, The lower central series of metabelian Engel groups, *Notices Amer. Math. Soc.* **12** (1965), 81.

Proc. Internat. Conf. Theory of Groups, Austral. Nat. Univ. Canberra, August 1965, pp. 115–144. © Gordon and Breach Science Publishers, Inc. 1967

Group theory and block designs

MARSHALL HALL, Jr.

1. Introduction and basic terminology

A block design D is an arrangement of v elements into b blocks, where each block contains k distinct elements and every element is in r different blocks, and such that every pair of distinct elements occurs together in exactly λ different blocks. Except in the trivial case when every block contains every element we have $r > \lambda$. The parameters satisfy two elementary relations

$$(1.1) \qquad bk = rv, \qquad r(k-1) = \lambda(v-1).$$

An automorphism α of a design D is a one to one mapping of elements onto elements and blocks onto blocks such that if $x \in B$, x an element and B a block then $(x)\alpha \in (B)\alpha$. Clearly the automorphisms of D form a group G.

If we are given a group of automorphisms G this can be very helpful in constructing a design D with such automorphisms. This is the subject of Section 2. The most striking result is that in certain instances if we have a group G of automorphisms of a design D, then D must necessarily have further automorphisms arising from the automorphisms of G. This involves what are called difference sets and their multipliers. Many of these difference sets are related to the theory of cyclotomy. Bose [2] has found many designs with prescribed cyclic groups of automorphisms. Other groups, such as linear fractional groups and unitary groups may be used to construct designs.

Section 3 deals with cases in which a group can be defined as the automorphism group of a certain block design or combinatorial configuration. Specifically the quintuply transitive Mathieu group M_{12} can be defined as the automorphism group of an Hadamard matrix H_{12} of order 12. Also the configuration of an inversive plane appropriately

115

defined from a field $GF(2^{2m+1})$ has an automorphism group which is the simple Suzuki group of order $q^2(q^2+1)(q-1)$ with $q=2^{2m+1}$.

Section 4 deals with designs that arise naturally in the theory of permutation groups. A primitive permutation group G with a subgroup H transitive on the letters it moves is always doubly transitive by a theorem of Jordan. If G is doubly but not triply transitive and H fixes at least 3 letters, then we call G a Jordan group and G is necessarily an automorphism group of a design D, whose blocks may be taken as the fixed letters of H (taken maximal) and its conjugates. If H fixes exactly 3 letters the design is shown to be the triple system of the lines either in a projective space over $GF(2)$ or an affine space over $GF(3)$. Relations of block designs to multiply transitive groups are discussed.

If the elements of a block design D are numbered $1, \ldots, v$ and the blocks $1, \ldots, b$ we define the incidence matrix $A = [a_{ij}]$, $i = 1, \ldots, v$, $j = 1, \ldots, b$ putting $a_{ij} = +1$ if the ith element is in the jth block and $a_{ij} = 0$ if not. The properties of the design D give

(1.2) $$AA^T = (r-\lambda)I_v + \lambda J_v = B$$

where I_v is the identity matrix of order v and J_v is the v by v matrix consisting entirely of ones. We readily find $\det B = (r-\lambda)^{v-1}(r+(v-1)\lambda)$. From the nonsingularity of B it follows that $b \geq v$. If $b = v$ then $r = k$ and we call D a symmetric design.

2. Designs with given automorphism groups

Block designs with $k = 3$ and $\lambda = 1$ are called Steiner triple systems, though in fact it was Kirkman [14] who first studied them in 1847, and proved the existence of these systems whenever $v \equiv 1, 3 \pmod 6$. These systems for $v = 15$ provide an excellent illustration as to what we should expect to find in general in the nature and existence of groups of automorphisms of designs. The systems were completely enumerated and their automorphism groups determined by White, Cole, and Cummings [25] in 1925. There are 80 nonisomorphic systems and the following table lists the orders of the groups and the number of systems whose automorphism group has the given order.

There are relatively few of these with large groups of automorphisms but more than half have nontrivial automorphisms. In general it appears to be true that whenever a design exists with given parameters, there also exists a design with these parameters possessing nontrivial automorphisms.

(2.1)

Group Order	Number of Systems	Group Order	Number of Systems
20160	1	21	1
288	1	12	3
192	1	8	2
168	1	6	1
96	1	5	1
60	1	4	8
36	1	3	12
32	1	2	6
24	2	1	36
Totals	10		70

In many cases the assumption that there is a cyclic group of automorphisms permuting the elements of the design in a single cycle enables us to construct the design easily. Here the v elements, without loss of generality, may be identified with the residues modulo v, and the mapping $\alpha: i \to i+1 \pmod{v}$ taken as a generator of the automorphism group on the elements. If the design is symmetric, i.e., $v=b$, $r=k$, the blocks will also be permuted in a single cycle. Here if a block B_0 is given:

(2.2) $$B_0: a_1, a_2, \ldots, a_k \mod v,$$

then the ith block, B_i, taking $i = 0, \ldots, v-1 \pmod{v}$ will be

(2.3) $$B_i: a_1+i, a_2+i, \ldots, a_k+i \mod v, \qquad i = 0, 1, \ldots, v-1.$$

For example with $v=b=11$, $r=k=5$, $\lambda=2$ we have a solution

(2.4)

$$
\begin{array}{llllll}
B_0: & 1, & 3, & 4, & 5, & 9, \\
B_1: & 2, & 4, & 5, & 6, & 10, \\
B_2: & 3, & 5, & 6, & 7, & 0, \\
B_3: & 4, & 6, & 7, & 8, & 1, \\
B_4: & 5, & 7, & 8, & 9, & 2, \\
B_5: & 6, & 8, & 9, & 10, & 3, \\
\end{array}
\qquad
\begin{array}{llllll}
B_6: & 7, & 9, & 10, & 0, & 4, \\
B_7: & 8, & 10, & 0, & 1, & 5, \\
B_8: & 9, & 0, & 1, & 2, & 6, \\
B_9: & 10, & 1, & 2, & 3, & 7, \\
B_{10}: & 0, & 2, & 3, & 4, & 8. \\
\end{array}
$$

Thus the residues $a_1, a_2, \ldots, a_k$ (mod v) in a single block determine the entire design. It is not difficult to show that residues

$$a_1, \ldots, a_k \quad (\text{mod } v)$$

determine a symmetric block design as above if they are a difference set, where we define:

Difference Set. *A set of k distinct residues modulo v:*

$$a_1, a_2, \ldots, a_k \quad (\text{mod } v)$$

where $k(k-1) = \lambda(v-1)$, *is called a difference set S if every nonzero residue* $d \not\equiv 0$ *can be expressed in exactly* λ *different ways as*

$$d \equiv a_i - a_j \quad (\text{mod } v)$$

with $a_i, a_j \in S$.

R. H. Bruck [3] has generalized the idea of these cyclic difference sets to consider an arbitrary automorphism group G of order v which is transitive and regular on the v elements of a symmetric block design D. Here we write G in multiplicative form and again identify the elements of D with the elements of G. If $a_1, \ldots, a_k$ are the elements of a block B we have

(2.5) $$B_x: a_1x, a_2x, \ldots, a_kx, \qquad x \in G.$$

We call $a_1, \ldots, a_k$ a group difference set and the corresponding property is:

Group Difference Set. *A set of k elements* $a_1, \ldots, a_k$ *in a group G of order v is a group difference set S where* $k(k-1) = \lambda(v-1)$ *if and only if:* (i) *for every* $d \neq 1$, $d \in G$ *there are exactly* λ *representations of the form* $d = a_i a_j^{-1}$, $a_i, a_j \in S$ *or* (ii) *for every* $d \neq 1$, $d \in G$ *there are exactly* λ *representations of the form* $d = a_i^{-1} a_j$, $a_i, a_j \in S$.

Each of the properties (i) and (ii) implies the other. As an example with $v = b = 16$, $r = k = 6$, $\lambda = 2$ let us take G as the abelian group of order 16 with generators a, b, c, d where $a^2 = b^2 = c^2 = d^2 = 1$. Then the following elements form a group difference set

(2.6) $$B_1: a, b, c, d, ab, cd.$$

There is no cyclic difference set with parameters $v = b = 16$, $r = k = 6$, $\lambda = 2$.

Let $G(Z)$ be the group ring of the group G over the integers, G being of order v. We define

(2.7) $$\theta(a) = a_1 + a_2 + \cdots + a_k, \qquad a_i \in D$$

and also write

(2.8) $$\theta(a^t) = a_1^t + a_2^t + \cdots + a_k^t.$$

In addition we define

(2.9) $$T = \sum_{x \in G} x.$$

Then the defining property of a difference set takes the form

(2.10) $$\theta(a)\theta(a^{-1}) = (k - \lambda) \cdot 1 + \lambda T.$$

In the case of cyclic difference sets it is convenient to consider the group ring $G(Z)$ as corresponding to the ring of polynomials in x taken modulo $x^v - 1$. In this case we write

(2.11) $$\theta(x) \equiv x^{a_1} + x^{a_2} + \cdots + x^{a_k} \pmod{x^v - 1}$$

and with $T(x) = 1 + x + x^2 + \cdots + x^{v-1}$ our defining relation becomes

(2.12) $$\theta(x)\theta(x^{-1}) = k - \lambda + \lambda T(x) \pmod{x^v - 1}.$$

In this form if $\epsilon \neq 1$ is a vth root of 1 we have $\theta(\epsilon)\theta(\epsilon^{-1}) = k - \lambda$. Thus cyclic difference sets are related to factorizations in cyclotomic fields. This viewpoint has been investigated in considerable detail by a number of writers, in particular by Yamamoto [27]. In the same way if G is an abelian group H. B. Mann [15] has used its characters to study (2.10). For finite groups G in general, the representations and characters of G are certainly relevant to an analysis of (2.10) but so far no significant results apply to this general case.

The remarkable fact about cyclic difference sets is that the corresponding block design usually has a larger group of automorphisms than the cyclic group assumed. The same holds true for abelian group difference sets. For this we need the concept of a "multiplier."

Definition. A *integer* t *is a multiplier of the cyclic difference set* $a_1, \ldots, a_k \pmod{v}$ *if* $x \to xt \pmod{v}$ *is an automorphism of the cyclic block design.*

It is easily checked in our example (2.4) that 3 is a multiplier. The chief theorem on the existence of multipliers is due to H. J. Ryser and the author [10].

Theorem 2.1. *If* $a_1, \ldots, a_k$ (mod v) *are a cyclic difference set, where* $k(k-1) = \lambda(v-1)$ *and if* p *is a prime dividing* $n = k - \lambda$, *such that* $(p, v) = 1$ *and* $p > \lambda$, *then* p *is a multiplier of the difference set.*

It has been shown by H. B. Mann [15], using a method originated by Parker [19] that every multiplier t fixes at least one block.

Theorem 2.2.   *A multiplier* t *fixes at least one block of a cyclic design.*

Frequently more is true:

Theorem 2.3. *If* t *is a multiplier of a cyclic block design modulo* v, *and if* $(t-1, v) = 1$, *then there is exactly one block fixed by* t. *If* $(k, v) = 1$ *there is a block fixed by every multiplier.*

These theorems greatly facilitate the construction of the designs when they exist. Note that without loss of generality we may multiply every residue modulo v by a number w, prime to v, in these designs. This amounts to a change in the choice of the generator of the cyclic group of order v. As an example take $v = b = 73$, $r = k = 9$, $\lambda = 1$. Here $p = 2$ is a multiplier. Let B_0 be a block fixed by 2 as multiplier and of the 9 residues in B_0 there will be an $a_i \not\equiv 0$ (mod 73). Multiplying by a_i^{-1} throughout we have 1 (mod 73) in B_0. But as B_0 is fixed by the multiplier 2 we find that 2, 4, 8, ... must be in B_0 whence

(2.13) $B_0 = \{1, 2, 4, 8, 16, 32, 37, 55, 64\}$ mod 73 .

This is a difference set and the argument above proves that it is essentially unique. Similarly if $v = b = 37$, $r = k = 9$, $\lambda = 2$, 7 is a multiplier and we immediately find the difference set

(2.14)   1, 7, 9, 10, 12, 16, 26, 33, 34   (mod 37) .

The condition $p > \lambda$ is necessary for the known proof of Theorem 2.1, but it is a quantitative condition whereas the other conditions are arithmetical. No case is known in which this condition seems to be required and so the following conjecture has been advanced:

Conjecture.   *Theorem 2.1 is true if the condition* $p > \lambda$ *is eliminated.*

There are a number of generalizations of Theorem 2.1 known, all of which would follow from the truth of the conjecture.

For group difference sets in general, Bruck defines a multiplier as an automorphism of the group G which is also an automorphism of the design D. Thus the multipliers are a subgroup of the group of auto-morphisms of G. For example the difference set of (2.6) has a multiplier group of order 72 fixing the given block. Theorems 2.1, 2.2, and 2.3 generalize readily to abelian groups where the multiplier t now takes the form of the automorphism $x \to x^t$ of an abelian group of order v, where we must have $(t, v) = 1$ for this mapping to be an automorphism. Such an automorphism is of course the identity for the difference set of (2.6).

With one exception all cyclic difference sets known to the writer fall into a number of special classes. The exception is

$$v = b = 133, \qquad r = k = 33, \qquad \lambda = 8;$$

(2.15) 1, 4, 5, 14, 16, 19, 20, 21, 25, 38, 54, 56, 57, 64, 66, 70, 76, 80, 83, 84, 91, 93, 95, 98, 100, 101, 105, 106, 114, 123, 125, 126, 131 (mod 133).

The first class is derived from finite geometries by a theorem of Singer [22]. Let $PG(n, p^r)$, $q = p^r$, p prime, be the projective space over the finite field $GF(q)$.

I. Type S: Singer difference sets. The blocks are hyperplanes in $PG(n, q)$, the elements points. Here

$$v = (q^{n+1} - 1)/(q - 1), \qquad k = (q^n - 1)/(q - 1), \qquad \lambda = (q^{n-1} - 1)/(q - 1).$$

The example (2.13) is of this type for $PG(2, 8)$.

II. Type Q: Quadratic residues in $GF(p)$, $p \equiv 3 \pmod 4$;

$$v = p = 4t - 1, \qquad k = 2t - 1, \qquad \lambda = t - 1.$$

This type also exists for $GF(p^r)$, $p^r \equiv 3 \pmod 4$ if we take our group G to be the additive group of $GF(p^r)$ which is elementary abelian, and so not cyclic for $r > 1$. The example of (2.4) is of this type.

III. Type H: Let p be a prime of the form $p = 4x^2 + 27$. The cubic residues modulo p and the class of sextic residues modulo p including the residue 3 form a cyclic difference set with

$$v = p = 4t - 1, \qquad k = 2t - 1, \qquad \lambda = t - 1.$$

Such a set would also exist in a finite field $GF(q)$, $q = p^r$ if we had $q \equiv 7 \pmod{12}$ and $q = 4x^2 + 27$. But arithmetic considerations show that these conditions force r to be one.

IV. Type T: Twin primes. Suppose that p and $q = p + 2$ are both primes. Of the $(p-1)(q-1)$ residues modulo pq prime to pq let $a_1, \ldots, a_m$, $m = (p-1)(q-1)/2$ be those for which the Legendre symbols have the property $(a_i/p) = (a_i/q)$. Also let $a_{m+1}, \ldots, a_{m+p}$ be $0, q, 2q, \ldots,$ $(p-1)q$. Here $m + p = (pq-1)/2 = k$. Here $a_1, \ldots, a_k$ are a difference set mod v and

$$v = pq, \qquad k = (pq-1)/2, \qquad \lambda = (pq-3)/4.$$

This type generalizes readily using finite fields $GF(p^r)$ and $GF(q^s)$ if $q^s = p^r + 2 \equiv 1 \pmod{2}$. Thus $3^3 = 5^2 + 2$ yields a design with $v = 25.27$, and this is the only case known to the writer with both $s > 1$ and $r > 1$.

Difference sets of types Q, H, and T are of Hadamard type. By this we mean that they may be used to construct an Hadamard matrix. This is a matrix H of order $4t$ where $v = 4t - 1$. We number our rows and columns ∞ and $0, 1, \ldots, v-1 \pmod{v}$; taking row and column ∞ as the first. Here row and column ∞ consist entirely of $+1$'s. In row i we place a $+1$ in column j if j is an element of the block B_i, and place a -1 if j is not an element of B_i. Thus the design of (2.4) gives the following matrix

(2.16)

	∞	0	1	2	3	4	5	6	7	8	9	10
∞	1	1	1	1	1	1	1	1	1	1	1	1
0	1	−	1	−	1	1	1	−	−	−	1	−
1	1	−	−	1	−	1	1	1	−	−	−	1
2	1	1	−	−	1	−	1	1	1	−	−	−
3	1	−	1	−	−	1	−	1	1	1	−	−
4	1	−	−	1	−	−	1	−	1	1	1	−
5	1	−	−	−	1	−	−	1	−	1	1	1
6	1	1	−	−	−	1	−	−	1	−	1	1
7	1	1	1	−	−	−	1	−	−	1	−	1
8	1	1	1	1	−	−	−	1	−	−	1	−
9	1	−	1	1	1	−	−	−	1	−	−	1
10	1	1	−	1	1	1	−	−	−	1	−	−

Here for simplicity of notation we have written $-$ for -1. An Hadamard matrix H_n of order n is a matrix consisting entirely of $+1$'s and -1's such that

(2.17) $$H_n H_n^T = nI_n.$$

With signs of rows and columns changed appropriately so that the first row and column consist entirely of $+1$'s, then (if $n > 2$) deleting the first row and column, taking rows as blocks, and the column positions of the $+1$'s as elements we have a symmetric block design with $v = 4t - 1$, $k = 2t - 1$, $\lambda = t - 1$. Conversely such a design may be used as above with a bordering of $+1$'s to construct an Hadamard matrix of order $n = 4t$.

V. Type B. Biquadratic residues of primes $p = 4x^2 + 1$, x odd. Here

$$v = p = 4x^2 + 1, \qquad k = x^2, \qquad \lambda = (x^2 - 1)/4.$$

The example (2.14) is of this type.

VI. Type B_0. Biquadratic residues and zero modulo a prime $p = 4x^2 + 9$, x odd. Here

$$v = p = 4x^2 + 9, \qquad k = x^2 + 3, \qquad \lambda = (x^2 + 3)/4.$$

VII. Type O. Octic residues of primes $p = 8a^2 + 1 = 64b^2 + 9$, a, b odd. Here

$$v = p, \qquad k = a^2, \qquad \lambda = b^2.$$

VIII. Type O_0. Octic residues and zero for primes $p = 8a^2 + 49 = 64b^2 + 441$, a odd, b even. Here

$$v = p, \qquad k = a^2 + 6, \qquad \lambda = b^2 + 7.$$

IX. Type W_4. This is a generalization of type T, developed by Whiteman [26], but uses biquadratic rather than quadratic residues. We have primes $p \equiv 1 \pmod 4$ and $q = 3p + 2$, and $pq - 1 = a^2$, a odd. Let g be a number which is a primitive root of both p and q. The difference set consists of

$$1, g, g^2, \ldots, g^{d-1}, 0, q, 2q, \ldots, (p-1)q \pmod{pq}$$

where $d = (p-1)(q-1)/4$. Here

$$v = pq, \qquad k = (v-1)/4, \qquad \lambda = (v-5)/16.$$

The theory of constructing block designs with given automorphisms for nonsymmetric designs is even less systematic than that for symmetric designs. The first major paper dealing with this subject is by R. C. Bose [2]. A later paper by Rao [20] extends this work.

Consider the problem of constructing a triple system with $v = 15$, $b = 35$, $r = 7$, $k = 3$, $\lambda = 1$ assuming an automorphism group G of order 15 cyclic and regular on the 15 elements. G must move the blocks in several orbits but the size of the orbits cannot be 15 blocks in every case. It is not hard to show that G must move the blocks in two orbits of 15 blocks and one of 5 blocks. A choice of blocks, one from each orbit, "base blocks" we may call them, determines the entire design. There are two essentially different solutions:

(2.18)   First solution: (0, 1, 4) and (0, 2, 8) in 15 orbits,
                              (0, 5, 10) in a 5 orbit.

(2.19)   Second solution: (0, 1, 4) and (0, 7, 13) in 15 orbits,
                              (0, 5, 10) in a 5 orbit.

The first solution consists of the lines in $PG(3, 2)$. This solution has the multiplier 2. Its full group of automorphisms is of order 20,160. By Singer's theorem the planes in $PG(3, 2)$ form a cyclic block design with $v = 15$, $k = 7$, $\lambda = 3$ given by the difference set

$$0, 1, 2, 4, 5, 8, 10 \pmod{15}.$$

The lines may of course be found as intersections of the planes. Incidentally $p = 2$ is a multiplier in this case but the condition $p > \lambda$ of Theorem 2.1 is not satisfied. Similarly Singer's theorem yields a triple system with a cyclic group of order $v = 2^m - 1$ for every m.

The second solution has the multiplier -2. It is the unique triple system whose automorphism group is of order 60 in the list of (2.1). It is a special case of the following construction:

Construction. Suppose $v = 6t + 3 = 3m$ *where* $m = 2t + 1 \not\equiv 0 \pmod{3}$. *Let us determine unordered pairs* (r, s) mod $3m$ *by the conditions* $r + s \equiv 0 \pmod{m}$, $r \equiv s \equiv 1 \pmod{3}$, $r, s \not\equiv 0 \pmod{m}$. *Then the base blocks modulo* $3m$

(2.20)
$$(0, r, s) \pmod{3m} \quad \textit{of orbit length } 3m$$
$$\textit{and } (0, m, 2m) \pmod{3m} \quad \textit{of orbit length } m$$

yield a design with $v = 6t + 3$, $b = (2t + 1)(3t + 1)$, $r = 3t + 1$, $k = 3$, $\lambda = 1$.

There is no known analogue to the multiplier theorem for the group constructions for the nonsymmetric designs. But the known constructions quite often involve some further automorphisms. In the most general case there will be several orbits of elements as well as several orbits of blocks, and of course orbits can be of varying lengths. Bose uses subscripts to distinguish orbits. A fixed element is represented by ∞ and if there are several fixed elements they are written ∞_1, ∞_2, etc. Let us consider an example with G the cyclic group of order 5, where $v = 16$, $b = 20$, $r = 5$, $k = 4$, $\lambda = 1$. Here there is one fixed element ∞ and three orbits of length 5, m_i (mod 5), $i = 1, 2, 3$; m (mod 5) and 4 orbits of blocks. We write the solution at length:

$$v = 16, \qquad b = 20, \qquad r = 5, \qquad k = 4, \qquad \lambda = 1;$$

$$
\begin{array}{llll}
(1_1, 4_1, 2_2, 3_2), & (1_2, 4_2, 2_3, 3_3), & (1_3, 4_3, 2_1, 3_1), & (\infty, 0_1, 0_2, 0_3), \\
(2_1, 0_1, 3_2, 4_2), & (2_2, 0_2, 3_3, 4_3), & (2_3, 0_3, 3_1, 4_1), & (\infty, 1_1, 1_2, 1_3), \\
(3_1, 1_1, 4_2, 0_2), & (3_2, 1_2, 4_3, 0_3), & (3_3, 1_3, 4_1, 0_1), & (\infty, 2_1, 2_2, 2_3), \\
(4_1, 2_1, 0_2, 1_2), & (4_2, 2_2, 0_3, 1_3), & (4_3, 2_3, 0_1, 1_1), & (\infty, 3_1, 3_2, 3_3), \\
(0_1, 3_1, 1_2, 2_2), & (0_2, 3_2, 1_3, 2_3), & (0_3, 3_3, 1_1, 2_1), & (\infty, 4_1, 4_2, 4_3).
\end{array}
$$

(2.21)

The four blocks at the top of the columns are base blocks. They can be tested by differences to see whether or not they do determine a block design. Bose distinguishes differences in the same orbit, calling them pure differences, from differences in different orbits, calling them mixed differences. For the pure differences in the first orbit we have $1_1 - 4_1 \equiv 2_1$, $4_1 - 1_1 \equiv 3_1$, $2_1 - 3_1 \equiv 4_1$, and $3_1 - 2_1 \equiv 1_1$. Thus every nonzero pure difference must occur λ times in the base blocks. The mixed differences of the first and second orbits are $1_1 - 2_2 \equiv 4_{12}$, $1_1 - 3_2 \equiv 3_{12}$, $4_1 - 2_2 \equiv 2_{12}$, $4_1 - 3_2 \equiv 1_{12}$ and $0_1 - 0_2 \equiv 0_{12}$. Thus every $i - j$ orbit mixed difference must occur λ times in the base orbits. If we are dealing mod m a fixed ∞_i and any a_j of the ith orbit gives the mixed ij difference m times. In case the orbit is of length less than m, say m/w, we must divide the count of the differences in a base block by w. Thus in (2.18) and (2.19) in the block (0, 5, 10) of period $5 = 15/3$ modulo 15 each difference 5 and 10 occurs 3 times but we divide by 3 to count each difference once. If we made an orbit of length 15 with base

$$(0, 5, 10) \quad (\text{mod } 15)$$

every difference $i - j \equiv 5$ or $i - j \equiv 10$ would occur three times. In general it is clear that given any group G of automorphisms base blocks determine the design and they may be tested by "differencing" for properties of the design desired. No attempt will be made here to give precise but unrevealing definitions for this differencing.

The example (2.21) is an instance of a more general construction, due to Bose.

Construction. Let $4t + 1 = p^n$, p a prime, and let x be a primitive root of $GF(p^n)$. Then there exists a pair of odd integers c, d such that $(x^c + 1)/(x^c - 1) = x^d$. Then the blocks

$$
\begin{aligned}
&(x_1^{2i},\ x_1^{2t+2i},\ x_2^{2i+c},\ x_2^{2t+2i+c}),\\
(2.22)\quad &(x_2^{2i},\ x_2^{2t+2i},\ x_3^{2i+c},\ x_3^{2t+2i+c}),\qquad i = 0,\ldots,t-1,\\
&(x_3^{2i},\ x_3^{2t+2i},\ x_1^{2i+c},\ x_1^{2t+2i+c}),\\
&(\infty,\ 0_1,\qquad 0_2,\qquad 0_3),
\end{aligned}
$$

form a base with respect to A, the additive group of $GF(p^n)$, of a design with

$$v = 12t + 4, \qquad b = (3t+1)(4t+1), \qquad r = 4t+1, \qquad k = 4, \qquad \lambda = 1.$$

A somewhat more complicated construction has been attempted by the writer and a former student, John Wilkinson, but so far only one case has been found in which this particular method has succeeded. If D is a symmetric block design with $\lambda = 2$, then we have $v = 1 + k(k-1)/2$. Following Hussain [12] and Atiqullah [1] we take one element as a fixed element X and if we number the blocks containing X as $B_1, \ldots, B_n$, where of course $k = n$, there will be a unique element besides X common to B_i and B_j and as $i \neq j$, $i, j = 1, \ldots, n$ we obtain every element except X exactly once. Represent this element by the unordered pair (i, j). Further blocks $B_{n+1}, \ldots, B_v$ will contain n elements $(i_1, j_1), (i_2, j_2), \ldots, (i_n, j_n)$ with each of $1, 2, \ldots, n$ occurring exactly twice in these unordered pairs. We may construct cycles from the unordered pairs taking a cycle $(a_1, a_2, \ldots, a_r)$ if $(a_1, a_2), (a_2, a_3), \ldots,$ (a_r, a_1) are among the unordered pairs. Of course a cycle and its inverse correspond to the same unordered pairs, but otherwise the correspondence between blocks and permutations in cycle form is unique. For the permutations corresponding to the remaining blocks it is easily seen that every cycle shall have length at least three, that no consecutive triple $\ldots a, b, c \ldots$ (or its reverse) shall occur in more than one permutation, and that any two permutations must have exactly two unordered pairs ...a, b... and c, d... in common. We seek a solution in which the permutations (identified with their inverses) form the elements of a conjugate class in an appropriate permutation group G. If we take the class of elements of order 3 in the group $LF(2, 8)$ we have a solution for $v = b = 37$, $r = h = 9$, $\lambda = 2$. The representative of this class is the permutation

$$(2.23) \qquad (8, 0, 1)(2, 6, 4)(3, 5, 7)$$

where the group is generated by

$$(2.24) \qquad \begin{aligned} &(0, 1)(2, 3)(4, 5)(6, 7)(8), \\ &(0)(8)(1, 2, 4, 3, 6, 7, 5), \\ &(0, 8)(1)(2, 5)(3, 6)(4, 7). \end{aligned}$$

The obvious generalization of this example is to consider a class of elements moving all letters in $LF(2, 2^m)$ represented as a permutation group on the $2^m + 1$ points of the projective line. The number of permutations and the total number of letters moved in the same way agrees with requirements for finding permutations yielding a symmetric design with $\lambda = 2$, but no further solutions have been found. No symmetric design with $\lambda = 2$ and k greater than 9 has been found by any method. With a little hesitance, the writer advances the conjecture that no such design exists.

Other groups may lead to designs in various ways. If we take the projective plane $PG(2, 9)$ its points are $z(x_1, x_2, x_3)$, $z, x_i \in GF(9)$ in homogeneous form. The mapping $x \to \bar{x} = x^3$ is an involutory automorphism of $GF(9)$. An *isotropic* point (or vector) is a point (x_1, x_2, x_3) such that

$$(2.25) \qquad x_1\bar{x}_1 + x_2\bar{x}_2 + x_3\bar{x}_3 = 0.$$

Clearly the isotropic points are permuted by the unitary group (or "hyperorthogonal group" in Dickson's terminology), this being the group of linear transformations leaving $x_1\bar{x}_1 + x_2\bar{x}_2 + x_3\bar{x}_3$ invariant. There are 28 isotropic points which lie in 63 sets of 4 on lines. This gives a design D with parameters $v = 28$, $b = 63$, $r = 9$, $k = 4$, $\lambda = 1$ which has the group $G = HO(3, 3^2)$ as a group of automorphisms. If we take our points as $1, 2, \ldots, 28$, the blocks containing 1 are given by

10—K.

$$
\begin{array}{rrrr}
1, & 2, & 3, & 4, \\
1, & 5, & 19, & 26, \\
1, & 6, & 15, & 24, \\
1, & 7, & 20, & 25, \\
1, & 8, & 14, & 21, \\
1, & 9, & 17, & 23, \\
1, & 10, & 22, & 28, \\
1, & 11, & 13, & 27, \\
1, & 12, & 16, & 18.
\end{array}
$$

(2.26)

Here $G = HO(3, 3^2)$ is a simple group of order 6048 and as an automorphism group of D is generated by the permutations

$$
\begin{aligned}
A = \ & (1, 16, 18)(2, 13, 19)(3, 14, 20)(4, 15, 17)(5, 21, 28)(6, 22, 25) \\
& (7, 23, 26)(8, 24, 27)(9, 11, 10)(12),
\end{aligned}
$$

(2.27)
$$
\begin{aligned}
B = \ & (1, 4)(2, 3)(5, 9)(6, 10)(7, 11)(8, 12)(13)(14, 17)(15, 21) \\
& (16, 25)(18)(19, 22)(20, 26)(23)(24, 27)(28),
\end{aligned}
$$

$$
\begin{aligned}
C = \ & (1, 5)(2, 6)(3, 7)(4, 8)(9, 12)(10, 11)(13)(14)(15)(16) \\
& (17, 28)(18, 25)(19, 26)(20, 27)(21, 23)(22, 24).
\end{aligned}
$$

The entire group of automorphisms of D is of order 12096 and in addition to the above group contains a further element.

(2.28)
$$
\begin{aligned}
E = \ & (1)(2)(3)(4)(5, 9, 7, 11)(6, 12, 8, 10)(13, 26, 23, 20) \\
& (14, 22, 24, 16)(15, 18, 21, 28)(17, 25, 27, 19).
\end{aligned}
$$

Here E^2 is in $HO(3, 3^2) = G$ and E induces an outer automorphism on G by conjugation.

3. Automorphism groups of given designs

Every block design has some group of automorphisms, but a glance at the list (2.1) shows that it may be the identity or a group of very small order, in which case the design does not contribute much to our knowledge of groups. But there are some cases in which the definition of a group as the automorphism group of a design is a simple and elegant method of defining a complicated group.

If H_n is an Hadamard matrix of order n, then from (2.17) it is clear that permuting rows or columns, or changing the sign of rows or columns leaves the property of being an Hadamard matrix unchanged. Let us call such Hadamard matrices equivalent. In terms of matrices this means that H_n and PH_nQ are equivalent where P and Q are monomial permutations with monomial factors ± 1. Hence if

(3.1)
$$PH_nQ = H_n,$$

we say that we have an automorphism of H_n. The automorphisms form a group isomorphic to the group of the Q's and anti-isomorphic to the group of the P's. This group always has a central element u for which $P = -I_n$, $Q = -I_n$. Thus the center is always of order at least 2.

For the Hadamard matrix H_{12} of (2.16) we have the following automorphisms.

$$\alpha:\ P_1^{-1} = Q_1 = \begin{pmatrix} \infty, 0, 1, 2, 3, 4, 5, 6, 7, 8, 9, 10 \\ \infty, 1, 2, 3, 4, 5, 6, 7, 8, 9, 10, 0 \end{pmatrix},$$

$$\beta:\ P_2^{-1} = Q_2 = \begin{pmatrix} \infty, 0, 1, 2, 3, 4, 5, 6, \ 7, 8, 9, 10 \\ \infty, 0, 3, 6, 9, 1, 4, 7, 10, 2, 5, \ 8 \end{pmatrix},$$

(3.2) γ:
$$\begin{cases} P_3^{-1} = \begin{pmatrix} \infty, & 0, & 1, & 2, 3, 4, 5, & 6, & 7, & 8, 9, & 10 \\ -0, & \infty, & 10, & -5, 7, 8, 2, & -9, & -3, & -4, 6, & -1 \end{pmatrix}, \\[2mm] Q_3 \ = \begin{pmatrix} \infty, & 0, & 1, & 2, 3, 4, 5, & 6, & 7, & 8, 9, & 10 \\ 0, & -\infty, & 10, & -5, 7, 8, 2, & -9, & -3, & -4, 6, & -1 \end{pmatrix}, \end{cases}$$

δ:
$$\begin{cases} P_4^{-1} = \begin{pmatrix} \infty, & 0, & 1, 2, & 3, & 4, & 5, & 6, 7, 8, & 9, 10 \\ \infty, & -0, & -5, 2, & -3, & -1, & -9, 10, 6, 7, & -4, & 8 \end{pmatrix}, \\[2mm] Q_4 \ = \begin{pmatrix} \infty, & 0, & 1, 2, 3, 4, 5, 6, 7, 8, 9, 10 \\ 0, & \infty, & 10, 1, 6, 2, 7, 5, 3, 9, 8, & 4 \end{pmatrix}. \end{cases}$$

Here $\begin{pmatrix} i \\ -j \end{pmatrix}$ in the monomial permutation means that $u_{ij} = -1$ in the appropriate matrix. The automorphisms α, β, γ modulo u generate the group $LF(2, 11)$. The automorphisms $\alpha, \beta, \gamma, \delta$ modulo u generate the group M_{12} quadruply transitive on 12 letters. This representation of M_{12} was found by the writer [9] and seems to be the simplest definition of this group. The automorphisms α, β, γ above readily generalize to Hadamard matrices of type Q, generating the group $LF(2, q)$. But only for H_{12} has an additional automorphism such as δ been found.

A block design D with $v = n^2$, $b = n^2 + n$, $r = n + 1$, $k = n$, $\lambda = 1$ is an affine plane if we consider the elements as points and the blocks as lines. It is easy to show that the lines can be divided into $n+1$ families of parallel lines, having shown that the parallel postulate must hold. Thus the affine plane can easily be extended (and in a unique way) to a projective plane, this being a symmetric block design with $v=b=n^2+n+1$, $r=k=n+1$, $\lambda=1$. But there is another way of extending the affine plane, this being, in its most familiar form, the extension of the complex plane by adjoining a point at infinity which is then topologically a sphere. On this sphere the finite lines, together with the point at infinity are circles, and these along with ordinary circles form the configuration of a circle geometry. In combinatorial terms we extend D by adjoining a single element to form a "tactical configuration" of sets such that any three elements lie in a unique block, and the blocks containing the new element are precisely the new element and the blocks of our original design D.

We may define our new system as an "inversive plane" or "Möbius plane." An axiomatic definition is the following: An *inversive plane I* is a set of *points* and a system of nonempty subsets called *circles* satisfying these axioms:

(1) *Any three distinct points are contained in exactly one circle.*

(2) *If P and Q are points and if c is a circle containing P but not Q, then there is a unique circle b such that $P, Q \in b$ and $b \cap c=\{P\}$.*

(3) *There are four points not on a common circle.*

For a fixed point P, let $I(P)$ be the system of elements of $\pi-P$ with "lines" as the circles of π containing P, deleting P from these circles. Here $I(P)$ is readily seen to be an affine plane, where (2) becomes the parallel postulate.

For a finite inversive plane I there is an integer $n \geq 2$ such that I contains n^2+1 points, and I is said to be of order n. The following properties are easily proved:

(a) I has n^2+1 points.

(b) Each circle of I consists of $n+1$ points.

(c) Each point of I is on $n(n+1)$ circles.

(d) I has $n(n^2+1)$ circles.

(e) Each circle of I is tangent to n^2-1 other circles.

(f) Each circle of I is disjoint from $n(n-1)(n-2)/2$ other circles.

One way of constructing a finite inversive plane is to take a non-ruled quadric Q in the projective space $PG(3, n)$, $n = p^r$, p a prime. Here the points of I are the points of Q and the circles of I are the intersections of Q by planes that cut Q in more than one point. The properties of Q that are relevant for these purposes are those we use to characterize an "ovoid."

Definition. In three-dimensional space $PG(3, n)$, $n = p^r$, an ovoid $\mathcal{O}$ is a set of $n^2 + 1$ points such that (i) any line of $PG(3, n)$ intersects $\mathcal{O}$ in at most two points, (ii) for any point P of $\mathcal{O}$, the union of all lines intersecting $\mathcal{O}$ in the point P alone is a plane of $PG(3, n)$.

Thus given an ovoid $\mathcal{O}$ in $PG(3, n)$, $n = p^r$, we may construct an inversive plane I of order n by taking as our circles the intersections of $\mathcal{O}$ with planes which cut $\mathcal{O}$ in more than one point. It has been shown by Segre [21] that for p odd an ovoid is necessarily a nonruled quadric surface, but that for $p = 2$ and $r > 1$ odd there exist ovoids which are not quadrics. It has been shown by Dembowski and Hughes [5] that for n even, an inversive plane I of order n is necessarily given as the intersection of an ovoid in $PG(3, n)$ by planes. In particular this means that if n is even, n is a power of 2, $n = 2^r$. Thus for n odd, there may be inversive planes not obtainable as the intersection of a quadric in $PG(3, n)$ by planes.

The simple groups of Suzuki can be regarded as automorphism groups of a class of inversive planes. Let $F = GF(2^{2m+1})$, and let θ be the automorphism of F such that $\theta(\alpha) = \alpha^r$, $\alpha \in F$ where $r = 2^m$. With $q = 2^{2m+1}$, there is an ovoid in $PG(3, q)$ whose points are in projective form

$$(3.3) \qquad (1, 0, 0, 0),$$
$$(\alpha^{2\theta+1} + \alpha^\theta\beta + \beta^{2\theta}, \ \alpha^{1+\theta} + \beta, \ \alpha^\theta, \ 1), \qquad \alpha, \beta \in F.$$

The Suzuki group is generated by matrices over F, $S(\alpha, \beta)$, $a, \beta \in F$, $M(\zeta)$, $\zeta \neq 0$, $\zeta \in F$, and J where

$$(3.4) \qquad S(\alpha, \beta) = \begin{bmatrix} 1 & , 0 & , 0, 0 \\ \alpha^\theta & , 1 & , 0, 0 \\ \beta & , \alpha & , 1, 0 \\ \alpha^{2\theta+1} + \alpha^\theta\beta + \beta^{2\theta} & , \alpha^{1+\theta} + \beta & , \alpha^\theta, 1 \end{bmatrix}.$$

(3.5) $$M(\zeta) = \begin{bmatrix} \zeta^\theta, & 0 & , 0 & , 0 \\ 0, & \zeta^{1-\theta}, & 0 & , 0 \\ 0, & 0 & , \zeta^{\theta-1}, & 0 \\ 0, & 0 & , 0 & , \zeta^{-\theta} \end{bmatrix}.$$

(3.6) $$J = \begin{bmatrix} 0, & 0, & 0, & 1 \\ 0, & 0, & 1, & 0 \\ 0, & 1, & 0, & 0 \\ 1, & 0, & 0, & 0 \end{bmatrix}.$$

Suzuki [23] first found these groups on the basis of a study of permutation groups. J. Tits [24] showed how the Suzuki groups could be defined in terms of ovoids over certain fields F of characteristic 2 in which the mapping $x \to x^2$ is an automorphism. Here the field F must have an automorphism σ such that $x^{\sigma^2} = x^2$. In Suzuki's notation $\sigma = 2\theta$. Tits obtains the ovoids by defining a polarity τ over the three-dimensional projective space $K = PG(3, F)$ so that $K \to K^{\tau^2}$ is the collineation determined by the automorphism $x \to x^2$ of F, and the ovoid consists of the absolute points of the polarity.

4. Designs arising in groups. Applications

In many cases there is a natural relation between a permutation group G on a set S and certain subsets of S which are permuted among themselves by G. If the subsets form a block design D (or a tactical configuration T), then the combinatorial properties of the design D (or T) are closely related to the properties of G.

A theorem due to Jordan [13] is the following:

Theorem 4.1. *Let G be a permutation group on n letters which is primitive, and let H be a subgroup of G, transitive on $m < n$ letters, fixing the remaining $n - m$ letters. Then (1) if H is primitive, G is $n - m + 1$-fold transitive; and (2) in any event G is doubly transitive.*

This theorem appears as Theorem 5.6.2 in the writer's book [7]. If G is not $n - m + 1$ fold transitive, then G may be r-fold but not $r + 1$-fold transitive where $r \geq 2$. Hence G contains a subgroup G_0 fixing $r - 2$ of the letters moved by G, such that (a) G_0 is doubly but not triply transitive, and (b) G_0 contains a subgroup H fixing $k \geq 3$ letters and transitive on the remaining letters. Let us call a group G_0 satisfying these two properties a Jordan group. Then the following theorem is almost immediate:

Theorem 4.2. *Let G be a Jordan group on n letters. Then G is an automorphism group of a block design D on $v = n$ elements with $\lambda = 1$.*

Proof. Choose H as the largest subgroup fixing $k \geq 3$ letters and transitive on the letters it moves. It easily follows that H moves more than half of the n letters of the set S permuted by G, whence for any element x of G, either $x^{-1}Hx = H$ or $K = x^{-1}Hx \cup H$ fixes at most one letter, since K is clearly transitive on the letters it moves. Let B_1, $B_2, \ldots, B_b$ be the sets of letters fixed by the distinct conjugates of H, letting B_i be the set fixed by $x_i^{-1}Hx_i = H_i$. Then $B_i \cap B_j$ is the set of letters fixed by $H_i \cup H_j$ and so is void or a single letter. This shows that a pair u, v of the letters of S occurs together in at most one of the sets B_i. But by the double transitivity of G every pair of letters occurs together the same number of times in the B's and so this number must be one, whence, if the B's are a block design we have $\lambda = 1$. But the transitivity of G ensures that every letter of S occurs the same number of times, say r in the B's. Hence the B's do form a block design D with b blocks on the $v = n$ letters of S and we have k the number of letters in each B, and $\lambda = 1$. Thus our theorem is proved.

Conversely given a block design D with $\lambda = 1$, a group G of automorphisms of D doubly transitive on the v elements of D will be a Jordan group if the subgroup H fixing all the elements of a block B_1 of D is transitive on the remaining elements of D. In particular the full automorphism groups of affine or projective geometries are Jordan groups. We write $EG(m, q)$ for the affine m dimensional geometry over $GF(q)$ and $PG(m, q)$ for the m-dimensional projective geometry over $GF(q)$.

Theorem 4.3. *A Jordan group G with $k = 3$ is the automorphism group of a Steiner triple system S. Either* (i) *for some m, S contains $2^m - 1 = v$ elements and S is isomorphic to the system whose elements are the points in $PG(m-1, 2)$ and whose triples are the lines of $PG(m-1, 2)$, or* (ii) *for some m S contains 3^m elements and S is isomorphic to the system whose elements are the points of $EG(m, 3)$ and whose triples are the lines of $EG(m, 3)$.*

Proof. Most of the proof of this theorem has already been given in paper [6] by the writer. Specifically it was shown in this paper that under the above hypothesis either (i) every triangle (three points not in a triple) generates a Steiner system $S(7)$ with 7 points, or (ii) every triangle generates a Steiner system $S(9)$ with 9 points. Alternative (i) leads directly to the isomorphism with $PG(m-1, 2)$ since the axioms for projective geometry are readily checked. The writer also proved the second alternative (ii) on the stronger hypothesis that G is transitive on sets of 4 independent points. It was brought to the writer's attention by R. H. Bruck that taking G as a Jordan group is sufficient for the conclusion. His proof will be given here:

Lemma 4.1. *If every triangle of a triple system S generates an $S(9)$, then S may be given the structure of a commutative Moufang loop.*

Proof. Designate an arbitrary element of S as the identity 1. Put $1x = x1 = x$ for every x of S. If $1a_1a_2$ is a triple put $a_1^2 = a_2$, $a_2^2 = a_1$, and $a_1a_2 = a_2a_1 = 1$. Thus the elements of a triple form a cyclic group of order 3. If 1, a_1, b_1 form a triangle we define a_1b_1 in the following way: Let triples of S be

$$1a_1a_2,$$
(4.1)
$$1b_1b_2,$$
$$a_2b_2c_2.$$

From this we define

(4.2) $a_1b_1 = c_2.$

The rule (4.2) clearly makes S a commutative loop. The triangle 1, a_1, b_1 generates an $S(9)$ whose triples can be put in the form:

$$1a_1a_2, \quad a_1b_1c_1, \quad a_2b_1d_2, \quad b_1c_2d_1,$$
(4.3)
$$1b_1b_2, \quad a_1b_2d_1, \quad a_2b_2c_2, \quad b_2c_1d_2,$$
$$1c_1c_2, \quad a_1c_2d_2, \quad a_2c_1d_1,$$
$$1d_1d_2.$$

Using the rule (4.2) the elements of the $S(9)$ form an elementary group of order 9. It is clear that this product rule is commutative and disassociative. Let e_1 be a further element of S not in the $S(9)$. We adjoin the triples of the $S(9)$'s generated by the triangles 1, a_1, e_1; 1, b_1, e_1 and if h_1 is given by the triple $b_1e_1h_1$ then also the $S(9)$ generated by 1, a_1, h_1. These triples take the form

$$
\begin{array}{llll}
1e_1e_2, & a_1e_1f_1, & a_2e_2f_2, & e_1f_2g_1, \\
1f_1f_2, & a_1e_2g_1, & a_2e_1g_1, & e_2f_1g_2, \\
1g_1g_2, & a_1f_2g_2, & a_2f_1g_1, & \\
1h_1h_2, & b_1e_1h_1, & b_2e_2h_2, & e_1h_2j_1, \\
1j_1j_2, & b_1e_2j_1, & b_2e_1j_2, & e_2h_1j_2, \\
& b_1h_2j_2, & b_2h_1j_1, & \\
1k_1k_2, & a_1h_1k_1, & a_2h_2k_2, & h_1k_2m_1, \\
1m_1m_2, & a_1h_2m_1, & a_2h_1m_2, & h_2k_1m_2, \\
& a_1k_2m_2, & a_2k_1m_1. &
\end{array}
$$

(4.4)

As has been shown by the writer, the property that every triangle generates an $S(9)$ is equivalent to the property that there is an automorphism α_x fixing an arbitrary x and interchanging the pairs of elements in triples with x. Since we have the triples

$$
\begin{array}{ll}
a_1b_1c_1, & b_1e_1h_1, \\
a_1e_1f_1, & \\
a_1h_1k_1, &
\end{array}
$$

(4.5)

taking $x = a_1$ and applying the automorphism α_x to the triple $b_1e_1h_1$ we have

(4.6)
$$
(b_1e_1h_1)\alpha_{a_1} = c_1f_1k_1
$$

whence $c_1f_1k_1$ is a triple of S.

Now let us take $x = a_1$, $y = b_1$, $z = e_1$.

The product rule (4.2) gives the following products

(4.7)
$$
\begin{array}{lll}
xy = c_2, & zx = f_2, & yz = h_2, \\
x(yz) = m_2, & (x(yz))x = k_1. &
\end{array}
$$

Also since $c_1f_1k_1$ is a triple by (4.6) we have

(4.8)
$$
(xy)(zx) = k_1.
$$

Combining the results of (4.7) and (4.8) we have the Moufang identity

(4.9)
$$
(xy)(zx) = [x(yz)]x.
$$

Thus our lemma is proved and we have given S the structure of a commutative Moufang loop in which every element not the identity is of order 3. Here we turn to the general theory of commutative Moufang loops as done by Bruck and Slaby and given in Bruck's monograph [4] "A survey of binary systems." Their main theorem is:

Theorem 4.4. (Bruck–Slaby). *Every commutative Moufang loop which can be generated by n elements $(n > 1)$ is centrally nilpotent of class at most $n - 1$.*

Our triple system S has a doubly transitive automorphism group G and regarding S as a commutative Moufang loop, the subgroup of G fixing the identity 1 is an automorphism group of the loop transitive on the nonidentity elements. But in a nilpotent loop there must be a nonidentity element x with the properties

$$(4.10)\quad (xy)z = x(yz), \qquad (yx)z = y(xz), \qquad (yz)x = y(zx) \quad \text{for all } y, z.$$

Such an x is in the *nucleus* of the loop. But by the transitivity of the loop automorphisms on nonidentity elements, it follows that every element of S as a loop element is in the nucleus. Thus the structure of S as a loop is in fact associative and so is a group, and indeed an elementary abelian group A of exponent 3, as determined by the product rule of (4.1) and (4.2). It now follows by a relatively easy argument which we leave to the reader, that S has the structure of $EG(m, 3)$ for some m where the elements of S are the points and the triples the lines of $EG(m, 3)$. The preceding argument completes the proof of Theorem 4.3 and even proves a slightly stronger statement:

Theorem 4.5. *If every triangle of a triple system S generates an $S(9)$ and if S has a group of automorphisms doubly transitive on its elements, then for some m, S is isomorphic to the triple system of the points and lines of $EG(m, 3)$.*

It may be that even more is true and we advance the following conjecture.

Conjecture. *A Steiner triple system with a group of automorphisms doubly transitive on its elements is isomorphic to the system of points and lines in either (i) a projective geometry $PG(m, 2)$ or (ii) an affine geometry $EG(m, 3)$.*

An automorphism α of a design D is a mapping of its elements onto themselves and its blocks onto themselves preserving incidences. Thus if A is the incidence matrix of D there will exist permutation matrices $P(\alpha)$ of order v and $Q(\alpha)$ of order b such that

$$(4.11) \qquad P(\alpha)AQ(\alpha) = A,$$

where $P(\alpha)^{-1}$ is the mapping of elements onto themselves and $Q(\alpha)$ is the mapping of blocks onto themselves. Clearly for the product $\alpha\beta$ of automorphisms

$$(4.12) \qquad P(\alpha\beta) = P(\beta)P(\alpha), \qquad Q(\alpha\beta) = Q(\alpha)Q(\beta).$$

Conversely permutation matrices P and Q such that $PAQ = A$ determine an automorphism of D. For a symmetric design with $v = b$, $r = k$, the incidence matrix A is nonsingular and so $Q(\alpha)$ and $P(\alpha)^{-1}$ are conjugate, and in particular have the same trace. Parker [19] treating the trace as a permutation character has shown that a group G of automorphisms of a design D has the same number of orbits on its blocks as on its elements. In particular if G is transitive on elements it is also transitive on blocks.

We shall give here a generalization of a theorem of Zassenhaus' [28] due to D. R. Hughes [11]. This uses the Parker result. The theorem in question deals with transitive extensions of groups. Let G be a permutation group on a set Ω of letters and let u be a further letter. Then a permutation group G^* is called a transitive extension of G if (a) G^* is transitive on $\Omega^* = \Omega + u$ and (b) G_u^*, the subgroup of G^* fixing u is G on Ω.

The theorem of Zassenhaus is

Theorem 4.6. (Zassenhaus). *Let G be a collineation group of $PG(n, q)$, $q = p^r$, $n \geq 2$ containing the unimodular projective group. Regarding G as a permutation group on the points of $PG(n, q)$, G cannot have a transitive extension unless $q = 2$ or $q = 4$, $n = 2$.*

For the generalization we begin with a lemma.

Lemma 4.2. *Let H be a collineation group of $PG(n, q)$, $q = p^r$, p prime, $n \geq 2$ fixing at least $q^{n-1} + q^{n-2} + \cdots + q + 1$ points. Then the fixed points of H are either (i) a hyperplane, (ii) a hyperplane and a point, or (iii) the whole space.*

Proof. We proceed by induction on n, beginning with $n = 2$. Here we are dealing with a projective plane of prime power order $q = p^r$. If $q = 2$ the result is true since a collineation fixing two points fixes the third point of the line. For $q \geq 3$ there are at least four fixed points. If there exist four fixed points no three on a line then all the fixed points generate a subplane all of whose points and lines are fixed. Let the subplane of fixed elements be of order s, so that there are $s^2 + s + 1$ fixed points. If this is a proper subplane it is known that either $q = s^2$ or $q \geq s^2 + s$. With $q = p^r$ we cannot have $p^r = s^2 + s = s(s+1)$ and so $q \geq s^2 + s + 1$, $q + 1 \geq s^2 + s + 2$. But we assumed that there were at least $q + 1$ fixed points or $s^2 + s + 1 \geq q + 1$, and this gives a conflict. Hence the fixed points do not form a proper subplane and there must be a line containing all but at most one of the points. Thus this line L contains at least q fixed points and so all $q + 1$ points of L are fixed. This completes the proof when $n = 2$.

Now suppose $n > 2$ and the lemma true for smaller values of n. Let s be the maximum number of independent fixed points. Then all fixed points lie within the space $PG(s, q)$ spanned by these points. But $PG(s, q)$ has only $q^s + q^{s-1} + \cdots + q + 1$ points and so we must have $s \geq n - 1$. If $s = n - 1$ all points of the hyperplane $PG(n-1, q)$ are fixed and our conclusion is true. If $s = n$, let independent fixed points be $P_0, P_1, \ldots, P_{n-1}$. Here $P_1, \ldots, P_{n-1}$ span a fixed hyperplane $K = PG(n-1, q)$. The intersection of the line L joining P_0 to any other fixed point is a fixed point of K and at most the remaining q points of L determine the same fixed point of K. Hence K contains at least $[(q^{n-1} + q^{n-2} + \cdots + 1) - 1]/q = q^{n-2} + q^{n-1} + \cdots + q + 1$ fixed points. If all of K is fixed our conclusion holds. Hence by induction on n, K contains a subspace T of dimension $n - 2$ all of whose $q^{n-2} + \cdots + q + 1$ points are fixed and possibly one further fixed point R. Hence all fixed points are on lines $P_0 X$, $X \in T$, and $P_0 R$. There must be at least $(q^{n-1} + \cdots + q^2 + q + 1) - 1 - q$ fixed points different from P_0 on lines $P_0 X$, $X \in T$, and so at least one of these contains a third fixed point Q. But then P, Q, and T generate a subspace $PG(n-1, q)$ all of whose points are fixed, and so our lemma is proved in all cases, since if $PG(n, q)$ had as many as two fixed points not on the hyperplane of fixed points, all points would be fixed.

Let us with Hughes call a collineation group H of $PG(n, q)$ a "translation group" if the fixed points of H are exactly the $q^{n-1} + q^{n-2} + \cdots$

$+q+1$ points of a hyperplane K. Note that H may very well contain homologies which fix K and a point not on K. The generalization of the Zassenhaus theorem is the following:

Theorem 4.7. (Hughes). *Let G be a collineation group of $PG(n, q)$, $q=p^r$, p prime, $n \geq 2$, containing a translation group H, and transitive on the points of $PG(n, q)$ as a permutation group. Then G cannot have a transitive extension unless $q=2$ or $q=4$ and $n=2$.*

Proof. Taking the points as elements and the hyperplanes as blocks the hyperplanes of $PG(n, q)$ form a symmetric block design D with

$$
\begin{aligned}
v = b &= q^n+q^{n-1}+\cdots+q+1, \\
r = k &= q^{n-1}+\cdots+q+1, \\
\lambda &= \phantom{q^n+q^{n-1}+} q^{n-2}+\cdots+q+1.
\end{aligned}
$$

(4.13)

Our given group G has a translation subgroup H whose fixed points are exactly the points of a hyperplane K. Without loss of generality we may take H to be the subgroup of G of all elements which fix the points of K. Thus the fixed points of H are a block of D regarding G as an automorphism group of D. As G is transitive on the points of D, by Parker's theorem G is transitive on the blocks of D, whence it follows that H has b conjugates and each has a block (hyperplane) as the set of its fixed points.

Consider G^*, a transitive extension of G, where a new element u has been adjoined to the set Ω of the v points of $PG(n, q)$. Let $B_1 = \{a_1, \ldots, a_k\}$ be one of the blocks of D (hyperplanes of $PG(n, q)$). We define a new tactical configuration T on $\Omega^* = \Omega + u$, taking a block $B_1^* = \{u, a_1, \ldots, a_k\}$ and all the images of B_1^* under G^* as blocks of T. Since G is the subgroup of G^* fixing u, T contains blocks $B_1^*, \ldots, B_b^*$ where B_i^* contains u and the elements of the ith block B_i of D. If H^* is the subgroup of G^* fixing $u, a_1, \ldots, a_k$ then $H^* \supseteq H$ where H is the subgroup of G fixing $a_1, \ldots, a_k$. As H fixes no further letters, it follows that H^* fixes no further letters. Thus if T contains a total of b' blocks then b' is the number of conjugates of H^* in G^*, or $b' = [G^* : N_{G^*}(H^*)]$. Since G^* is a doubly transitive group T must be a block design on $v' = v+1$ elements, b' blocks, with elements occurring r' times and each block containing $k' = k+1$ elements and pairs occurring λ' times. A block B_j^* of T containing the letter u corresponds to a conjugate H_j^* of H^* fixing the letter u. But then H_j^* restricted to Ω is a permutation group of G fixing exactly $k = q^{n-1} + \cdots + q + 1$ letters. By Lemma 4.2 these fixed letters must be the points of a hyperplane. But then B_g^* must be one of the b blocks $B_1^*, \ldots, B_b^*$ of T consisting of u and the points of the b hyperplanes of $PG(n, q)$. Hence u occurs in these b blocks and no others of T. Thus for T, the parameter $r' = b = q^n + q^{n-1} + \cdots + q + 1$. But now we have for the block design T

$$(4.14) \qquad\qquad\qquad b'k' = r'v'$$

or, using the known values of k', r', v'

$$(4.15) \qquad b'(q^{n-1} + \cdots + q + 2)$$
$$= (q^n + q^{n-1} + \cdots + q + 1)(q^n + q^{n-1} + \cdots + q + 2).$$

With $k' = q^{n-1} + \cdots + q + 2$, $r' = q^n + q^{n-1} + \cdots + q + 1$, $v' = q^n + q^{n-1} + \cdots + q + 2$ we have $v' - k' = q^n$ and $qk' - r' = q - 1$, and by (4.14) must have $k' | r'v'$. Suppose first that q is odd. A prime dividing v' and k' divides q^n and so must divide q, but then as it divides $k' = q^{n-1} + \cdots + q + 2$ it must also divide 2, a conflict when q is odd. Hence $(k', v') = 1$ and so k' divides r' and so also k' divides $qk' - r' = q - 1$. But this is a conflict since $k > q - 1$. Thus (4.15) is impossible for q odd. Next suppose $q = 2^s$ with $s \geq 2$. Then $k' \equiv v' \equiv 2 \pmod 4$. Here a prime dividing k' and v' must divide $v' - k' = q^n$ and so must be the prime 2, whence $(k', v') = 2$. Thus $k'/2$ divides r' and in turn $k'/2$ divides

$$qk' - r' = q - 1 = 2^s - 1, \qquad k' = (q^{n-1} + \cdots + q + 2)2(q - 1) = 2^{s+1} - 2.$$

This is impossible if $n \geq 3$ since $q^2 + q + 2 > 2q - 2$ for positive q. There remains only the possibility $n = 2$ giving $q + 2 | 2q - 2$ whence

$$q + 2 | (2q - 2) - 2(q + 2) = -6.$$

This gives $q = 4$, and the exception $q = 4$, $n = 2$ of the theorem. If $q = 2$, equation (4.15) becomes $b'(2^n) = (2^{n+1} - 1)(2^{n+1})$ giving $b' = 2(2^{n+1} - 1)$, a case also permitted by the theorem. The exceptions do arise, those for $q = 2$ including at least the holomorph of the elementary abelian group. For $q = 4$, $n = 2$ we have a group on $4^2 + 4 + 1 = 21$ letters and the transitive extension is the Mathieu group M_{22}, which possesses further transitive extensions to the groups M_{23} and M_{24}.

In almost the same way Hughes proves analogous results for affine spaces.

Lemma 4.3. *If a collineation group H of $EG(n, q)$, $q = p^r$, p prime, $n \geq 2$ fixes exactly q^{n-1} points, then the set of fixed points is an affine hyperplane if $n \geq 3$ and if $n = 2$ is a line or a "Baer subplane" of order $\sqrt{q}$.*

Theorem 4.8. *Let G be a collineation group of $EG(n, q)$, $q = p^r$, p prime, $n \geq 2$, transitive on its points and on its hyperplanes, containing a subgroup H whose fixed points are the points of a hyperplane. Then if $n \geq 3$, G has no transitive extension.*

Here $n = 2$ is an exception. There are always extensions and these are the groups of inversive planes discussed in Section 3 and the groups include the simple Suzuki groups.

As a final point let us consider the relation between multiply transitive permutation groups and block designs. Let G be a doubly transitive group on v letters Ω. For any k, $3 \leq k < v$ choose k distinct letters $\{a_1, a_2, \ldots, a_k\}$ from Ω. The b distinct images of this set under G will be the blocks $B_1, \ldots, B_b$ of a block design, for transitivity assures us that each of the v letters occurs equally often, say r times, and double transitivity that every pair of elements $a_i a_j$ occurs equally often in the B's, say λ times.

Thus every multiply transitive group G can be considered as an automorphism group of some block design D. But clearly some designs are more intimately related to the group than others.

Let us turn briefly to the subject of quadruply transitive groups. Jordan [13] showed that a quadruply transitive group in which only the identity fixes four letters is necessarily one of: S_4, S_5, A_6, or M_{11}, the last the Mathieu group on 11 letters. The writer [6] generalized this to show that if G is a quadruply transitive group and if $H = G_{1234}$, the subgroup fixing the four letters 1, 2, 3, 4 is of finite odd order, then necessarily G is one of the following groups: S_4, S_5, A_6, A_7, or M_{11}. In an arbitrary quadruply transitive group G on a finite set $\Omega = \{1, 2, \ldots, n\}$ let $H = G_{1234}$, and let P be a Sylow 2-subgroup of H. Let H fix a set Δ of letters and P a set Δ'. Then of course $\Delta \subseteq \Delta'$ and the results above show that $|\Delta| = 4$, 5, 6, or 11 and $|\Delta'| = 4$, 5, 6, 7, or 11. Let $N = N_G(H)$ and $N' = N_G(P)$ and write N^Δ, $(N')^{\Delta'}$ for the restriction of these groups to the sets Δ, Δ' respectively. Then the above theorems show that $N^\Delta = S_4$, S_5, A_6, or M_{11} and $(N')^{\Delta'} = S_4$, S_5, A_6, A_7, or M_{11}. In the Mathieu group M_{23} we find $N^\Delta = S_4$, and $(N')^{\Delta'} = A_7$. The letters fixed by conjugates of H form a tactical configuration of blocks each containing $k = |\Delta|$ elements with the property that every quadruple occurs in exactly one of these sets. If $|\Delta| = 4$ this is trivially the set of all quadruples chosen from $\{1, 2, \ldots, n\}$ but if $|\Delta| = 5$, 6, or 11 this is a nontrivial configuration. In particular for $|\Delta| = 5$ if we take all the quintuples of letters fixed by conjugates of H, and consider those containing the pair 1, 2, the remaining three letters in the quintuples form a Steiner triple system on 3, 4, $\ldots$, n which has the doubly transitive group G_{12} as a group of automorphisms. If the conjecture of the preceding section is correct, then $n - 2 = 2^m - 1$ or $n - 2 = 3^m$. But Steiner triple systems arise in another way in quadruply transitive groups with $N^\Delta = M_{11}$. Here we have a tactical configuration with $k = 11$ and every set of 4 letters occurring in exactly one block. The involutions of M_{11} form a single class and each fixes exactly 3 letters. Thus in N^Δ if Δ contains 1 and 2 we will have a transposition

$$(4.16) \qquad\qquad a = (12)(i_1)(i_2)(i_3)(jk)(mn)(pq)\cdots,$$

where 1, 2, $\ldots$, p, q are the letters of Δ. Here 1, 2 and any two others of Δ determine Δ completely. This fact has been used by Nagao and Oyama [16], [17], [18] to show that if r is the number of letters fixed by an arbitrary involution, then $n = r^2 + 2$, when $|\Delta| = 11$ and also by a similar argument when $|\Delta| = 6$. In particular in (4.16) any two of i_1, i_2, i_3 uniquely determine the third. Thus i_1, i_2, i_3 and their images under the doubly transitive group G_{12} form a Steiner triple system. The triples formed from i_1, i_2, i_3, $\ldots$, p, q form an $S(9)$ and so if the conjecture of the preceding section is true then $n - 2 = 3^m$, where by the Nagao–Oyama result m must be even. Since we must have $r \geq 11$ by our assumption the smallest value of n satisfying these conditions is $n = 731$. But, of course, if the translation subgroup H of Theorem 4.8 should exist, then G_{12} could have no transitive extension and G would not exist. If $|\Delta| = |\Delta'| = 4$, no combinatorial design would be directly related to these sets, but similar considerations would apply to the fixed letters of a Sylow subgroup fixing more than 4 letters if such groups exist.

References

[1] M. ATIQULLAH, Some new solutions of symmetrical balanced incomplete block design with $\lambda = 2$ and $k = 9$, *Bull. Calcutta Math. Soc.* **50** (1958), 23–28.

[2] RAJ CHANDRA BOSE, On the construction of balanced incomplete block designs, *Ann. Eugenics* **9** (1942), 353–399.

[3] R. H. BRUCK, Difference sets in a finite group, *Trans. Amer. Math. Soc.* **78** (1955), 464–481.

[4] RICHARD HUBERT BRUCK, *A survey of binary systems*, Springer, Berlin-Göttingen-Heidelberg, 1958.

[5] P. DEMBOWSKI and D. R. HUGHES, On finite inversive planes, *J. London Math. Soc.* **40** (1965), 171–182.

[6] MARSHALL HALL, JR., On a theorem of Jordan, *Pacific J. Math.* **4** (1954), 219–226.

[7] MARSHALL HALL, JR., *The theory of groups*, Macmillan, New York, 1959.

[8] MARSHALL HALL, JR., Automorphisms of Steiner triple systems, *Proc. Sympos. Pure Math.*, vol. VI, pp. 47–66; Amer. Math. Soc., Providence, R.I., 1962.

[9] MARSHALL HALL, JR., Note on the Mathieu group M_{12}, *Arch. Math.* **13** (1962), 334–340.

[10] MARSHALL HALL, JR. and H. J. RYSER, Cyclic incidence matrices, *Canad. J. Math.* **3** (1951), 495–502.

[11] D. R. HUGHES, Extensions of designs and groups: projective, symplectic, and certain affine groups [to appear].

[12] Q. M. HUSSAIN, Symmetrical incomplete block designs with $\lambda = 2$, $k = 8$ or 9, *Bull. Calcutta Math. Soc.* **37** (1945), 115–123.

[13] C. JORDAN, Recherches sur les substitutions, *J. Math. Pures Appl.* (2) **17** (1872), 351–363.

[14] T. P. KIRKMAN, On a problem in combinations, *Cambridge and Dublin Math. J.* **2** (1847), 191–204.

[15] H. B. MANN, Balanced incomplete block designs and abelian difference sets, *Illinois J. Math.* **8** (1964), 252–261.

[16] H. NAGAO, On multiply transitive groups I, *Nagoya Math. J.* **27** (1966) 15–19.

[17] H. NAGAO and T. OYAMA, On multiply transitive groups II, *Osaka Math. J.* [to appear].

[18] H. NAGAO and T. OYAMA, On multiply transitive groups III, *Osaka Math. J.* [to appear].

[19] E. T. PARKER, On collineations of symmetric designs, *Proc. Amer. Math. Soc.* **8** (1957), 350–351.

[20] C. RADHAKRISHNA RAO, A study of BIB designs with replications 11 to 15, *Sankhyā Ser. A.* **23** (1961), 117–127.

[21] B. SEGRE, On complete caps and ovaloids in three-dimensional Galois spaces of characteristic 2, *Acta Arith.* **5** (1959), 315–332.

[22] JAMES SINGER, A theorem in finite projective geometry and some applications to number theory, *Trans. Amer. Math. Soc.* **43** (1938), 377–385.

[23] MICHIO SUZUKI, A new type of simple groups of finite order, *Proc. Nat. Acad. Sci. U.S.A.* **46** (1960), 868–870.

[24] J. TITS, Ovoides et groupes de Suzuki, *Arch. Math.* **13** (1962), 187–198.

[25] A. S. WHITE, F. N. COLE, and LOUISE D. CUMMINGS, Complete classification of triad systems on fifteen elements, *Mem. Nat. Acad. Sci. U.S.A.* **14** (1925), Second memoir, 1–89.

[26] ALBERT LEON WHITEMAN, A family of difference sets, *Illinois J. Math.* **6** (1962), 107–121.

[27] KOICHI YAMAMOTO, Decomposition fields of difference sets, *Pacific J. Math.* **13** (1963), 337–352.

[28] HANS ZASSENHAUS, Über transitive Erweiterungen gewisser Gruppen aus Automorphismen endlicher mehrdimensionaler Geometrien, *Math. Ann.* **111** (1935), 748–759.

Proc. Internat. Conf. Theory of Groups, Austral. Nat. Univ. Canberra, August 1965, pp. 145–150. © Gordon and Breach Science Publishers, Inc. 1967

Analogues of Prefrattini subgroups

TREVOR HAWKES

§1. Introduction

In [2] W. Gaschütz investigates a characteristic conjugacy class of subgroups of a finite soluble group called Prefrattini subgroups. If W is a Prefrattini subgroup of G and α is a homomorphism of G onto G^*, then $\alpha(W)$ is a Prefrattini subgroup of G^*. Moreover, the largest normal subgroup of G contained in W is $\varphi(G)$, the Frattini subgroup of G, and consequently W avoids all complemented chief factors of G and covers the rest. Here we shall describe a way of constructing classes of subgroups with properties closely analogous to those of the Prefrattini subgroups. These classes are defined in terms of the Sylow systems of G and the method yields a simple characterization of Prefrattini subgroups which now appear as a special case. Our results lean heavily on the elegant theory of formations expounded by W. Gaschütz in [3]. All groups considered are finite and soluble.

§2. Definitions and preliminary results

For each prime p let $f(p)$ denote a formation, and let $\mathscr{F}$ be the local formation defined by $f(p)$ (see [3] for details of these concepts). A maximal subgroup will be called *p-maximal* if its index is a power of the prime p. Thus in a finite soluble group every maximal subgroup is p-maximal for some prime p. If M is a subgroup of G, *Core M* will denote the largest normal subgroup of G contained in M. A p-maximal subgroup M of G will be called *f-normal* if $M/(\text{Core } M) \in f(p)$ and *f-abnormal* otherwise. $\text{Aut}_G (H/K)$ will denote the group of automorphisms induced by G in its chief factor H/K. Thus $\text{Aut}_G (H/K) \cong G/C_G(H/K)$ where $C_G(H/K)$ is the subgroup of G consisting of elements which centralize H/K. A p-chief factor H/K of G is said to be *f-central* if

145

$\mathrm{Aut}_G(H/K) \in f(p)$ and *f-eccentric* otherwise. When $f(p)$ is the identity formation these concepts take on their usual meanings (i.e., "*f*-normal" becomes simply "normal," "*f*-central" becomes simply "central," etc.). We shall use the convention that if $f(p) = \emptyset$, the empty set, every *p*-chief factor is *f*-eccentric and every *p*-maximal subgroup is *f*-abnormal.

Lemma 2.1. *If* M *is a maximal subgroup which complements the chief factor* H/K *of* G *then* $M/(\mathrm{Core}\ M) \cong \mathrm{Aut}_G(H/K)$.

Let $C = C_G(H/K)$. Then $M \cap C = \mathrm{Core}\ M$ and $G/(\mathrm{Core}\ M)$ has a unique self-centralizing minimal normal subgroup $C/(\mathrm{Core}\ M)$. Hence $G/C = MC/C \cong M/M \cap C = M/(\mathrm{Core}\ M)$ and the result follows at once.

Corollary 2.2. M *is an f-normal maximal subgroup of* G *if, and only if, it complements an f-central chief factor of* G.

Lemma 2.3. *If* M *and* M^g, *two conjugate maximal subgroups of* G, *both contain a Sylow p-complement* S^p *of* G, *then* $M = M^g$.

This well-known result follows from the fact that if $S^p \leq M$ and M is not normal in G then $N_G(S^p) \leq M$ (see P. Hall [5], Theorem 3.5); since $N_G(S^p)$ is an abnormal subgroup of G and cannot therefore be contained in two distinct conjugates, the result follows.

Lemma 2.4. *Let* N *be a minimal normal subgroup of* G *and let* M_1 *and* M_2 *be two nonconjugate maximal subgroups complementing* N *in* G. *Then* $M = (M_1 \cap M_2)N$ *is a maximal subgroup of* G *such that* $\mathrm{Aut}_G(N) \cong M/(\mathrm{Core}\ M)$.

Let $D_i = \mathrm{Core}\ M_i$ $(i = 1, 2)$ so that $D_i = M_i \cap C$ where $C = C_G(N)$. Thus $ND_i = N(M_i \cap C) = C$. G/D_i has a unique minimal normal subgroup C/D_i and all complements of C/D_i are conjugate. Since by hypothesis M_1 and M_2 are not conjugate, we must have $D_1 \neq D_2$ and therefore $D_1D_2 = C$. It is readily seen that the following are G-isomorphisms:

$$C/N(D_1 \cap D_2) \cong D_i/D_1 \cap D_2 \cong C/D_i \cong N \qquad (i = 1, 2).$$

Hence $C/N(D_1 \cap D_2)$ is a chief factor of G complemented by M; for $N(D_1 \cap D_2) \leq M$ and

$$CM = D_1D_2(M_1 \cap M_2) = D_1[D_2(M_1 \cap M_2)] = D_1M_2 = G.$$

The result now follows from Lemma 2.1. Using the same notation we therefore have

Corollary 2.5. *If N is f-eccentric, then M complements some f-eccentric chief factor of G/N.*

§3. The f-Prefrattini subgroups

Let $\mathfrak{S}$ be a Sylow system of G and S^p the Sylow p-complement of G contained in $\mathfrak{S}$. Denote by $\mathfrak{M}$ the set of f-abnormal maximal subgroups of G containing S^p. Let $M^p = \bigcap \{M \mid M \in \mathfrak{M}\}$; then we have

Theorem 3.1. *If $\mathfrak{M}^*$ is a subset of $\mathfrak{M}$ comprising exactly one complement of each complemented f-eccentric p-chief factor in a given chief series of G*

$$1 = G_0 < G_1 < \cdots < G_k = G,$$

then $M^p = \bigcap \{M \mid M \in \mathfrak{M}^*\}$.

Assume the theorem is false for some $\mathfrak{M}^*$. Of all subsets of $\mathfrak{M}$ containing $\mathfrak{M}^*$ whose intersections are equal to M^p consider just those with the smallest number of members, r say; then by assumption $r > |\mathfrak{M}^*|$, and each of these subsets has at least two distinct members complementing the same chief factor G_i/G_{i-1} in the above chief series. Let $\mathfrak{M}^{**}$ be one of these subsets with i as large as possible; clearly $i < k$. Then $\mathfrak{M}^* \subset \mathfrak{M}^{**} \subseteq \mathfrak{M}$ and $\mathfrak{M}^{**}$ contains M_1 and M_2 which complement the f-eccentric chief factor G_i/G_{i-1}. By Corollary 2.5, $(M_1 \cap M_2)G_i = M$ is a maximal subgroup of G complementing an f-eccentric chief factor G_j/G_{j-1} such that $j > i$. Without loss of generality assume $M_2 \notin \mathfrak{M}^*$ and consider the set $\overline{\mathfrak{M}} = (\mathfrak{M}^{**} - \{M_2\}) \cup \{M\}$. $\mathfrak{M}^* \subseteq \overline{\mathfrak{M}}$ and since $M_1 \cap M = M_1 \cap M_2$ the intersection of the members of $\overline{\mathfrak{M}}$ is M^p. Either we have $M \in \mathfrak{M}^{**} - \{M_2\}$ and therefore $|\overline{\mathfrak{M}}| < r$, or $M \notin \mathfrak{M}^{**} - \{M_2\}$ and $\overline{\mathfrak{M}}$ contains two distinct complements of G_j/G_{j-1} with $j > i$. Either conclusion contradicts the definition of $\mathfrak{M}^{**}$ and the result follows.

Definition. If M^p is defined as above for each prime $p \mid |G|$, we define $W^f(\mathfrak{S}) = \bigcap_{p \mid |G|} M^p$ and call it the f-Prefrattini subgroup of G corresponding to $\mathfrak{S}$.

Since the set of Sylow systems of G is transitively permuted by the inner automorphisms of G (see P. Hall [4]), we have

Corollary 3.2. *The set of all f-Prefrattini subgroups of G forms a characteristic conjugacy class.*

Corollary 3.3. *If $\mathfrak{N}$ is a set of f-abnormal maximal subgroups of G, each containing a Sylow p-complement of $\mathfrak{S}$ for some p, and if $\mathfrak{N}$ contains at least one complement of each complemented f-eccentric chief factor of a given chief series of G, then $W^f(\mathfrak{S}) = \bigcap \{M \mid M \in \mathfrak{N}\}$.*

Corollary 3.4. *An f-eccentric maximal subgroup M of G contains $W^f(\mathfrak{S})$ whenever the Sylow system $\mathfrak{S}$ reduces into M.*

Corollary 3.5. *If $\alpha: G \twoheadrightarrow G^*$ is an epimorphism and $W = W^f(\mathfrak{S})$ then $W^* = \alpha(W)$ is an f-Prefrattini subgroup of G^* corresponding to $\mathfrak{S}^* = \alpha(\mathfrak{S})$.*

This is perhaps most readily seen by considering a chief series of G passing through $N = \mathrm{kernel}(\alpha)$ and taking W as the intersection of a set of complements comprising exactly one for each complemented f-eccentric chief factor of this chief series.

Corollary 3.6 *An f-Prefrattini subgroup avoids every complemented f-eccentric chief factor of G and covers the rest.*

To conclude this section we note that if $f(p) = \emptyset$ for all primes p then $W^f(\mathfrak{S})$ is a Prefrattini subgroup of G; for Corollaries 3.4 and 3.6 together yield the characteristic property of Prefrattini subgroups given in Satz 6.3 of [2].

§4. Prefrattini subgroups and f-normalizers

In [1] new conjugacy classes of subgroups are investigated which generalize system normalizers in much the same way as Gaschütz extends the concept of a Carter subgroup in [3]. These subgroups may be defined as follows: Let $\mathfrak{S}$ be a Sylow system of G and let K be the smallest normal subgroup of G such that $G/K \in f(p)$. Let $T^p = K \cap S^p$ where S^p is the Sylow p-complement of $\mathfrak{S}$. The set

$$\mathfrak{T} = \{T^p \mid p \text{ divides } |G|\}$$

is called an *f-system* of G and the normalizer of $\mathfrak{T}$ is called the *f-normalizer* of G corresponding to $\mathfrak{S}$; we shall denote it by $D^f(\mathfrak{S})$. For a fixed f the set of all $D^f(\mathfrak{S})$ obtained as $\mathfrak{S}$ runs through the Sylow systems of G forms a homomorphism-invariant characteristic conjugacy class of G. $D^f(\mathfrak{S})$ covers the f-central and avoids the f-eccentric chief factors of G. Another property of f-normalizers inherited from system normalizers is the following: every f-abnormal maximal subgroup of G into which $\mathfrak{S}$ reduces contains $D^f(\mathfrak{S})$. This enables us to prove

Theorem 4.1. *If D is an f-normalizer and W a Prefrattini subgroup of G, both corresponding to the same Sylow system $\mathfrak{S}$, then D and W permute and $DW = W^f(\mathfrak{S})$.*

For both D and W are contained in every f-abnormal maximal subgroup into which $\mathfrak{S}$ reduces and therefore

$$DW \leq \langle D, W \rangle \leq W^f(\mathfrak{S}),$$

where $\langle D, W \rangle$ denotes the group generated by D and W. But $D \cap W$ avoids all f-eccentric chief factors and all complemented chief factors of G and therefore $|DW| = |D| \cdot |W|/|D \cap W| \geq |W^f(\mathfrak{S})|$ since $|W^f(\mathfrak{S})|$ is equal to the product of the orders of those chief factors of G which are not simultaneously f-eccentric and complemented. Hence $|DW| = |W^f(\mathfrak{S})|$ and the result follows at once.

Since D and W permute, the order of the subgroup $D \cap W$ is equal to the product of the orders of the noncomplemented f-central chief factors of G. If we denote this subgroup by $V^f(\mathfrak{S})$ the set of all such subgroups obtained as $\mathfrak{S}$ runs through the Sylow systems of G forms a homomorphism-invariant characteristic conjugacy class of subgroups which cover the noncomplemented f-central chief factors of G and avoid the rest.

The case when $f(p) = 1$ for all primes p perhaps deserves special mention. For then $D = D^f(\mathfrak{S})$ is the normalizer of $\mathfrak{S}$ and therefore, as D clearly normalizes the intersection M^p of all maximal subgroups of G containing $S^p \in \mathfrak{S}$, D normalizes $\bigcap_{p \mid |G|} M^p = W$, the Prefrattini subgroup of G corresponding to $\mathfrak{S}$. Since the highest chief factor of a chief series of G is central and complemented W can never be self-normalizing. If $\Delta(G)$ denotes the intersection of all the nonnormal maximal subgroups of G then $\Delta(G)$ is the largest normal subgroup of G contained in $W^1(\mathfrak{S})$ ($= DW$), and so W^1 bears the same relation to $\Delta(G)$ as W to $\varphi(G)$. It may be conjectured that $\varphi(D) = W \cap D$ but a counterexample is provided by the split extension G of a quarternion group of order 8 by its group of outer automorphisms which has order 6. In this case D is elementary abelian of order 4, while W has order 2 and is contained in D. However, we always have $\varphi(D) \leq W \cap D$ and equality occurs when G is an A-group. More generally, if the $f(p)$ are arbitrary subject to the condition $f(p) \leq \mathscr{F}$, and if U is a Prefrattini subgroup of $D = D^f(\mathfrak{S})$ corresponding to the Sylow system $\mathfrak{S} \cap D$ of D, then $U \leq W^f(\mathfrak{S})$.

This account is by no means exhaustive and it is hoped to give elsewhere a more detailed description of the consequences of these observations and to investigate other special conjugacy classes which arise from related constructions.

References

[1] R. W. CARTER and T. O. HAWKES, The $\mathscr{F}$-normalizers of a finite soluble group, *J. Algebra* [to appear].

[2] WOLFGANG GASCHÜTZ, Praefrattinigruppen, *Arch. Math.* **13** (1962), 418–426.

[3] WOLFGANG GASCHÜTZ, Zur Theorie der endlichen auflösbaren Gruppen, *Math. Z.* **80** (1963), 300–305.

[4] P. HALL, On the Sylow systems of a soluble group, *Proc. London Math. Soc.* (2) **43** (1937), 316–323.

[5] P. HALL, On the system normalizers of a soluble group, *Proc. London Math. Soc.* (2) **43** (1937), 507–528.

Proc. Internat. Conf. Theory of Groups, Austral. Nat. Univ. Canberra, August 1965, pp. 151–152. © Gordon and Breach Science Publishers, Inc. 1967

Groups with an existence property with respect to commutators

HERMANN HEINEKEN

Let $x \circ y = x^{-1}y^{-1}xy$. Groups satisfying the following condition have been considered earlier ([1], [2]):

(C)   To each pair of elements x, y of G there is an element z in G such that, for all g in G, $x \circ (y \circ g) = z \circ g$.

Groups satisfying (C) are metabelian; furthermore, a group G satisfying (C) is nilpotent if $G/C(G')$ is finitely generated.

Obviously each such group satisfies the condition:

(Q)   For each element x of G there is an element y in G such that $x \circ (x \circ g) = y \circ g$ for all g in G.

One would hope that all groups satisfying (Q) have properties similar to those mentioned above. However, if G is a group with an elementary abelian normal 2-subgroup N such that G/N is abelian and acts regularly on N, then G satisfies (Q) and is nonnilpotent (unless $G = N$). On the other hand, groups G satisfying (Q) satisfy also the condition that $(a \circ b) \circ (a \circ c) = 1$ for all triplets a, b, c of elements of G; hence $G/Z(G)$ is metabelian and G'' is of exponent 2 (see Macdonald [3]). So if G' contains no elements of even order, G is again metabelian; and if furthermore $G/C(G')$ is finitely generated, G is nilpotent. The proof of this fact rests on the following lemma:

If R is a ring and $pR = 0$ for some prime $p \neq 2$, then the following statements on the subgroup U of the unit group of R are equivalent:

(1)   if $x \in U$, then $1 - (-x+1)^2 \in U$;

(2)   if $x \in U$, then $1 + k(-x+1) \in U$ for every integer k;

(3)   if $x \in U$, then $1 + (-x+1)g(x) \in U$ for every polynomial $g(x)$ in x.

If we consider the reversed condition,

(Q') for each element x of G there is an element y in G such that $(g \circ x) \circ x = g \circ y$ for all g in G,

we have a quite different situation. Again $(a \circ b) \circ (a \circ c) = 1$, but for every prime $p \equiv 3 \bmod 4$ the holomorph of Z_p satisfies (Q').

References

[1] HERMANN HEINEKEN, Commutator closed groups, *Illinois J. Math.* **9** (1965), 242–255.
[2] HERMANN HEINEKEN, Linkskommutatorgeschlossene Gruppen, *Math. Z.* **87** (1965), 37–41.
[3] I. D. MACDONALD, On certain varieties of groups, *Math. Z.* **76** (1961), 270–282.

Proc. Internat. Conf. Theory of Groups, Austral. Nat. Univ. Canberra, August 1965, pp. 153–165. © Gordon and Breach Science Publishers, Inc. 1967

The orders of relatively free groups

GRAHAM HIGMAN

If $\mathfrak{B}$ is a locally finite variety of groups then, in particular, the free group in $\mathfrak{B}$ with n generators is finite, of order $f(n)$ say, for each positive integer n. The object of this paper is to suggest directions in which the sequence $f(1), f(2), f(3), \ldots$ might profitably be studied. There are two classes of questions which seem to arise.

First we may ask approximative questions. How fast does $f(n)$ tend to infinity with n, and how smoothly? Some answers to these questions can be obtained fairly easily, but many problems remain. Secondly, there is a fairly obvious conjecture, which we shall make later, about the analytic form of $f(n)$. In this direction, we have not done more than consider a number of special cases.

1. Approximative questions

1.1. *Convexity.* Let G_n be the free group in $\mathfrak{B}$ freely generated by $a_1, a_2, \ldots, a_n$, for each integer n, and let π_{n+1} be the projection of G_{n+1} onto G_n, in which $a_i \pi_{n+1} = a_i$, $i = 1, \ldots, n$ and $a_{n+1} \pi_{n+1} = 1$. If we put for the moment $g(n) = f(n+1)/f(n)$, we have

$$g(n) = |\ker(\pi_{n+1})| = |\langle a_{n+1} \rangle^{G_{n+1}}| = |\langle a_1 \rangle^{G_{n+1}}|.$$

Next, π_{n+2} plainly maps $\langle a_1 \rangle^{G_{n+2}}$ onto $\langle a_1 \rangle^{G_{n+1}}$, so that if we put, again for the moment, $h(n) = g(n+1)/g(n)$, we have

$$h(n) = |\langle a_1 \rangle^{G_{n+2}} \cap \ker(\pi_{n+2})| = |\langle a_1 \rangle^{G_{n+2}} \cap \langle a_{n+2} \rangle^{G_{n+2}}|$$
$$= |\langle a_1 \rangle^{G_{n+2}} \cap \langle a_2 \rangle^{G_{n+2}}|.$$

This process can be continued indefinitely, and the result expressed in the formula

(1.1) $$\Delta^r \log f(n) = \log \left| \bigcap_{i=1}^{r} \langle a_i \rangle^{G_{n+r}} \right|.$$

153

The sequence on the right is nonnegative for all r, so that the sequence $\log f(n)$ is totally convex. This is one way of saying that $f(n)$ tends to infinity very smoothly, and it is natural to inquire whether more can be said in the same direction. In particular, it follows from (1.1) that

$$(1.2) \qquad \log f(n) = k_1 n + k_2 \binom{n}{2} + \cdots + k_r \binom{n}{r} + \cdots + k_n \binom{n}{n},$$

where

$$(1.3) \qquad k_r = \log \left| \bigcap_{i=1}^{r} \langle a_i \rangle^{G_r} \right|.$$

It is plausible that the k_r should in turn exhibit some sort of smoothness, but I know of nothing that has been proved along these lines.

1.2. *Rates of growth.* The most obvious question here is whether there is a bound, independent of $\mathfrak{B}$, to the rate of growth of $f(n)$. This, however, is probably not very suitable for direct attack. Such a bound could be found if one could show that a locally finite locally soluble variety is soluble and also that a locally finite variety can contain only a finite number of finite simple groups. This second statement would follow in turn either from the conjecture that there are only a finite number of finite simple groups of given exponent, or from the conjecture that there is a bound to the number of elements necessary to generate a finite simple group. The existence of a bound to the rate of growth would also follow if it could be proved that every locally finite variety has its identical relations finitely based. Thus, the discovery of a bound to the rate of growth of $f(n)$, if it exists, is likely to be connected with the solution of problems which do not need the present context to point up their interest.

1.3. *Gap theorems.* There are, however, other questions which it might be profitable to attack. For instance, by [4, Lemma 3.2], we have, in the notation of Section 1.1,

$$[a_1, a_2, \ldots, a_r] \in \bigcap_{i=1}^{r} \langle a_i \rangle^{G_r} \subset \gamma_r(G_r),$$

where $\gamma_r(G)$ is the rth term of the lower central series of G, so that $k_r = 0$ if and only if $\mathfrak{B}$ is nilpotent of class less than r. Since also $k_r \geq \log 2$ whenever $k_r \neq 0$, it follows from (1.2) that $\log f(n)$ is a polynomial if $\mathfrak{B}$ is nilpotent, and $\log f(n) \geq 2^n \log 2$ otherwise. That is, there is a

gap in the set of possible rates of growth of $f(n)$, and this gap is corre-
lated with a change in the group-theoretically interesting properties of
$\mathfrak{B}$. It is natural to ask if there are any more instances of this
phenomenon.

In particular, it is easy to see that if groups in $\mathfrak{B}$ have nilpotent
derived groups, and therefore derived groups of bounded class, then
there is a bound for $n^{-1} \log \log f(n)$. It is tempting to conjecture that
if $n^{-1} \log \log f(n)$ is unbounded then $\log \log f(n) \geq kn^2$ for some positive
k and n large enough, and, perhaps less confidently, that for locally
nilpotent varieties $\mathfrak{B}$, $n^{-1} \log \log f(n)$ is bounded only if groups in $\mathfrak{B}$
have nilpotent derived groups. It is necessary to restrict oneself to
locally nilpotent varieties in the second part of the conjecture, since the
variety $\mathfrak{B}$ generated by any finite group has $n^{-1} \log \log f(n)$ bounded.

The first half of the conjecture would follow from the conjecture
made below on the analytic form of $f(n)$. If the second half could be
proved, it would have one perhaps unexpected consequence. For it
follows from Wright's result [7] on the class of an n-generator group of
exponent 4 that, for $\mathfrak{B}$ the variety of groups of exponent 4, $f(n) \leq 2^{n^3 n}$,
n large. Thus the conjecture would imply that there is a bound for the
classes of the derived groups of groups of exponent 4. (The result of
Tobin [6] makes this less improbable than it might seem at first sight.)

2. The form of $f(n)$

2.1. *A conjecture.* If $p_1, p_2, \ldots, p_r$ are the distinct primes dividing
$f(1)$, then $f(n)$ can be written $p_1^{g_1(n)} \cdots p_r^{g_r(n)}$, so that a conjecture about
the form of $f(n)$ is a conjecture about the $g_1(n), \ldots, g_r(n)$, and vice
versa. To see what is reasonable, recall first that for nilpotent varieties
each $g_i(n)$ is a polynomial, and second that the order of the free n-
generator group in the product variety $\mathfrak{A}_s\mathfrak{B}$ is $f(n)s^{1+(n-1)f(n)}$, where
$\mathfrak{A}_s$ is the variety of abelian groups of exponent s. This suggests that we
consider the least class Σ of functions such that (i) the function $i(n)$
which is always equal to n belongs to Σ, (ii) if $g_1(n), \ldots, g_r(n)$ belong
to Σ, and $h(x_1, \ldots, x_r)$ is a polynomial with rational coefficients, then
$h(g_1(n), \ldots, g_r(n))$ belongs to Σ, and (iii) if $f(n)$ belongs to Σ and b is a
positive rational number then $b^{f(n)}$ belongs to Σ. We shall describe a
function in Σ as CREAM (on the grounds that it Comes by Repeated
Exponentiations, Additions, and Multiplications).

Our conjecture is then that for all locally finite varieties $\mathfrak{B}$, $f(n)$ is CREAM. We shall not get anywhere near proving this in this paper: we shall have to content ourselves with exhibiting some varieties, and classes of varieties, for which it is true.

2.2. *Generalities.* As we have seen the conjecture is true for nilpotent varieties, and also for the product $\mathfrak{A}_s\mathfrak{B}$ if it is true for $\mathfrak{B}$. More generally, if it is true for varieties $\mathfrak{B}$ and $\mathfrak{W}$, it is true for $\mathfrak{B}\mathfrak{W}$.

If $\mathfrak{B}$, $\mathfrak{W}$ are locally finite varieties so are their intersection $\mathfrak{B} \cap \mathfrak{W}$ and join $\langle \mathfrak{B}, \mathfrak{W} \rangle$, and if $f(n)$, $g(n)$, $h(n)$, $k(n)$ are the orders of the n-generator free groups in these varieties, then

$$f(n)g(n) = h(n)k(n).$$

Thus if the conjecture is true for three of the varieties it is true for the fourth. In particular if the conjecture is true for $\mathfrak{B}$ and $\mathfrak{W}$, and for all subvarieties of $\mathfrak{W}$, it is true for $\langle \mathfrak{B}, \mathfrak{W} \rangle$. To be still more particular, if the conjecture is true for $\mathfrak{B}$, it is true for the join of $\mathfrak{B}$ with any nilpotent variety.

2.3. *Adjoining critical groups to varieties.* One can also adjoin a simple group to a variety without disturbing the truth of the conjecture. To be precise, let $\mathfrak{B}$ be a variety for which the conjecture is true, and which contains all proper subgroups of the finite simple group X, but not X itself. Then the conjecture is true for $\langle \mathfrak{B}, X \rangle$.

To see this, recall first that if $k_X(n)$ is the number of ways in which n elements can be chosen to generate X, then (Hall [2])

$$k_X(n) = \sum_{Y \leq X} m_Y |Y|^n,$$

where the coefficients m_Y are integers independent of n. Thus $k_X(n)$ is CREAM. But the n-generator free group in $\langle \mathfrak{B}, X \rangle$ is the direct product of $k_X(n)/|\mathrm{Aut}(X)|$ copies of X and the n-generator free group in $\mathfrak{B}$. This gives the result.

For instance, the variety generated by the proper subgroups of the simple group of order 60 is $\langle \mathfrak{A}_3\mathfrak{A}_2, \mathfrak{A}_5\mathfrak{A}_2, \mathfrak{A}_2\mathfrak{A}_3 \rangle = \mathfrak{B}$ say. But each of $\mathfrak{A}_3\mathfrak{A}_2$, $\mathfrak{A}_5\mathfrak{A}_2$, $\mathfrak{A}_2\mathfrak{A}_3$ is a product of nilpotent varieties whose proper subvarieties are all nilpotent (cf. [4, Example 4.9]), so that the conjecture is true for $\mathfrak{B}$. Hence it is true also for the variety generated by the simple group of order 60 itself.

A similar result is that if X is a (critical) group of the form NY, where N is an abelian minimal normal subgroup on which Y acts faithfully, and $\mathfrak{V}$ is a variety containing all proper subgroups and proper homomorphic images of X but not X itself, then the conjecture is true for $\langle \mathfrak{V}, X \rangle$ if it is true for $\mathfrak{V}$.

Here, if F_n is the n-generator free group in $\langle \mathfrak{V}, X \rangle$, the kernel of the natural homomorphism of F_n onto G_n, the n-generator free group in $\mathfrak{V}$, is the direct product of a number, $r(n)$ say, of copies N^* of N, each normal in F_n and satisfying $F_n/C(N^*) \simeq Y$. Then it is enough to prove that $r(n)$ is CREAM. But the number of choices of $C(N^*)$ is the number of kernels of homomorphisms of F_n onto Y, which is $k_Y(n)/|\mathrm{Aut}(Y)|$. For each such kernel K, the number of direct factors with $C(N^*) = K$ is $b(n-1)$, where b is the dimension of N over the centralizer algebra of the induced representation of Y. Thus $r(n) = b(n-1)k_Y(n)/|\mathrm{Aut}(Y)|$ is CREAM. We leave the verification of these things to the reader.

Of course, if we could prove the analogous results for a general critical group X, it would follow that the conjecture is true for any variety generated by a single finite group, by the Oates–Powell theorem.

2.4. *Subvarieties of varieties* $\mathfrak{A}_p\mathfrak{W}$. For the remainder of the paper, we shall be concerned mainly with varieties $\mathfrak{V}$ satisfying $\mathfrak{W} \subset \mathfrak{V} \subset \mathfrak{A}_p\mathfrak{W}$, where $\mathfrak{W}$ is a variety for which the conjecture is true, and p is a prime not dividing the exponent of $\mathfrak{W}$. These varieties were determined in [4], and we recall the facts briefly. These varieties are in one to one correspondence with the closed classes of irreducible linear groups in $\mathfrak{W}$ over $GF(p)$. Here, if X, Y are linear groups, we write $Y \leq X$ if Y is linearly isomorphic to a component of a subgroup of X, and a class is called closed if it contains with X every irreducible group Y such that $Y \leq X$. If $\mathfrak{X}$ is such a closed class, a group G in $\mathfrak{A}_p\mathfrak{W}$ belongs to the corresponding variety $\mathfrak{U}(\mathfrak{X})$ if, for every minimal normal p-subgroup P of G, the linear group induced in P by G belongs to $\mathfrak{X}$.

There is one case in which it is practically immediate that $\mathfrak{U}(\mathfrak{X})$ also satisfies the conjecture, namely in the case when $\mathfrak{X}$ consists of all linear groups over $GF(p)$ in some subvariety $\mathfrak{W}_0$ of $\mathfrak{W}$, and $\mathfrak{W}_0$ satisfies the conjecture. For then $\mathfrak{U}(\mathfrak{X}) = \langle \mathfrak{W}, \mathfrak{A}_p\mathfrak{W}_0 \rangle$, and $\mathfrak{W} \cap \mathfrak{A}_p\mathfrak{W}_0 = \mathfrak{W}_0$.

In most cases, however, we shall have to do more work. If G_n, of order $g(n)$, is the n-generator free group in $\mathfrak{W}$, the n-generator free group in $\mathfrak{A}_p\mathfrak{W}$ is AG_n, where A is an elementary abelian normal subgroup of order $p^{1+(n-1)g(n)}$, and it is important to recall that we know not only the order of A but also its structure as a G_n-module. By a theorem of Gaschütz [1], the representation of G_n induced on A is the sum of the trivial representation and $n-1$ copies of the regular representation. If $\mathfrak{X}$ is a class of irreducible linear groups over $GF(p)$, closed or not, we denote by $c_{\mathfrak{X}}(n)$ the sum of the dimensions of the components of the regular representation of G_n which belong, as linear groups, to $\mathfrak{X}$. Then if $\mathfrak{X}$ is a nonempty closed class, the order of the n-generator free group in $\mathfrak{U}(\mathfrak{X})$ is $g(n)p^{1+(n-1)c_{\mathfrak{X}}(n)}$, so that the conjecture is true for $\mathfrak{U}(\mathfrak{X})$ if and only if $c_{\mathfrak{X}}(n)$ is CREAM.

For a single irreducible linear group X, the number of homomorphisms of G_n onto X is just the number of ways that n elements can be chosen to generate X, that is, $k_X(n)$. Any such homomorphism gives a representation of G_n in which the induced linear group is isomorphic to X, and two such homomorphisms give equivalent representations if one is the other followed by a linear automorphism of X. If $d(X)$ is the dimension over $GF(p)$ of the space on which X operates, and $d_0(X)$ the dimension of the same space over the centralizer algebra of X, each representation of G_n in which the induced linear group is isomorphic to X occurs $d_0(X)$ times in the regular representation, and so contributes $d_0(X)d(X)$ to the dimension of the regular representation. Thus

$$c_X(n) = d_0(X)d(X)k_X(n)/|\text{lin Aut}(X)|,$$

so that $c_X(n)$ is CREAM. Hence so also is $c_{\mathfrak{X}}(n)$ for a finite closed class $\mathfrak{X}$, and so, for such a class, the conjecture is true for $\mathfrak{U}(\mathfrak{X})$.

If $\mathfrak{X}$, $\mathfrak{Y}$ are closed classes so are $\mathfrak{X} \cup \mathfrak{Y}$ and $\mathfrak{X} \cap \mathfrak{Y}$, and $\mathfrak{U}(\mathfrak{X} \cup \mathfrak{Y}) = \langle \mathfrak{U}(\mathfrak{X}), \mathfrak{U}(\mathfrak{Y}) \rangle$, and $\mathfrak{U}(\mathfrak{X} \cap \mathfrak{Y}) = \mathfrak{U}(\mathfrak{X}) \cup \mathfrak{U}(\mathfrak{Y})$. Thus we may combine the two cases so far dealt with; the conjecture is true for $\mathfrak{U}(\mathfrak{X})$ if $\mathfrak{X}$ is the union of the set of irreducible linear groups in a subvariety and a finite set.

There are some "small" varieties $\mathfrak{W}$ for which this is the complete answer. This is so, for instance if $\mathfrak{W}$ is abelian, or consists of p-groups of Φ-class 2 [4, Examples 4.9 and 4.10], and for $\mathfrak{W} = \mathfrak{A}_s\mathfrak{A}_t$ where s and t are distinct primes (cf. Section 2.6 below).

2.5. *The case when $\mathfrak{W}$ is a direct join.* Suppose next that $\mathfrak{W} = \langle \mathfrak{W}_1, \mathfrak{W}_2 \rangle$ where $\mathfrak{W}_1$ and $\mathfrak{W}_2$ have coprime exponents. Then a group in $\mathfrak{W}$ is a direct product of a group in $\mathfrak{W}_1$ and a group in $\mathfrak{W}_2$. An irreducible group X in $\mathfrak{W}$ over $GF(p)$ determines its direct factors X_1 and X_2 in $\mathfrak{W}_1$ and $\mathfrak{W}_2$ not only as abstract groups but also as linear groups over $GF(p)$; they are the irreducible components of the space on which X acts when considered as X_1-module and X_2-module, respectively. If a second irreducible group Y has direct factors Y_1 and Y_2, it is clear that $Y \leq X$ implies $Y_1 \leq X_1$ and $Y_2 \leq X_2$. Thus if $\mathfrak{X}_i$ are closed classes of irreducible groups over $GF(p)$ in $\mathfrak{W}_i$, $i = 1, 2$, the class $\mathfrak{X}_1 \otimes \mathfrak{X}_2$ of irreducible groups X in $\mathfrak{W}$ such that the direct factor X_i belongs to $\mathfrak{X}_i$, $i = 1, 2$ is also closed. If G_{1n}, G_{2n} are the n-generator free groups in $\mathfrak{W}_1$, $\mathfrak{W}_2$ respectively, the n-generator free group in $\mathfrak{W}$ is the direct product $G_{1n} \times G_{2n}$, and the component in its regular representation consisting of linear groups in $\mathfrak{X}_1 \otimes \mathfrak{X}_2$ is the tensor product of the components in the regular representations of G_{1n} and G_{2n} consisting of linear groups in $\mathfrak{X}_1$ and $\mathfrak{X}_2$ respectively. Thus

$$c_{\mathfrak{X}_1 \otimes \mathfrak{X}_2}(n) = c_{\mathfrak{X}_1}(n) c_{\mathfrak{X}_2}(n),$$

and the conjecture is true for $\mathfrak{u}(\mathfrak{X}_1 \otimes \mathfrak{X}_2)$ if it is true for $\mathfrak{u}(\mathfrak{X}_1)$ and $\mathfrak{u}(\mathfrak{X}_2)$.

Of course, we would like to be able to say that if the conjecture is true for all varieties between $\mathfrak{W}_1$ and $\mathfrak{A}_p\mathfrak{W}_1$, and for all between $\mathfrak{W}_2$ and $\mathfrak{A}_p\mathfrak{W}_2$, it is true for all varieties between $\mathfrak{W}$ and $\mathfrak{A}_p\mathfrak{W}$. There are two difficulties that stand in the way of this.

The first is that though X, above, determines X_1 and X_2, X_1 and X_2 do not determine X. Indeed, if X_1 and X_2 are given, X can be any irreducible component of their tensor product $X_1 \otimes X_2$ over $GF(p)$, and $X_1 \otimes X_2$ is not, in general, irreducible or even the direct sum of isomorphic irreducible linear groups. If it is not, there is no hope of building all closed classes in $\mathfrak{W}$ out of the classes $\mathfrak{X}_1 \otimes \mathfrak{X}_2$. I see no way out of this difficulty except to note that there are many cases in which it does not arise, and to assume that we are in one of them. For instance, if every irreducible group in $\mathfrak{W}_1$ over $GF(p)$ is absolutely irreducible then certainly $X_1 \otimes X_2$ will be irreducible; and there are always primes p for which this is so, namely the primes congruent to 1 modulo the exponent of $\mathfrak{W}_1$. Another condition sufficient to avert the difficulty is that irreducible groups in $\mathfrak{W}_1$ over the algebraic closure of $GF(p)$ which are isomorphic as abstract groups are isomorphic as linear groups. This does not necessarily make $X_1 \otimes X_2$ irreducible, but it does make it the direct sum of isomorphic linear groups. The condition is

12—K.

presumably quite restrictive, but it has the merit of being easy to verify when it is true. It holds, for instance, if $\mathfrak{W}_1$ is nilpotent and of Φ-class two, and if $\mathfrak{W}_1 = \mathfrak{A}_s \mathfrak{A}_t$ where s, t are coprime and t is square-free.

If we assume that $X_1 \otimes X_2$ is always the direct sum of isomorphic irreducible linear groups, then it is clear that for irreducible groups X, Y in $\mathfrak{W}$, $Y \leq X$ if and only if $Y_1 \leq X_1$ and $Y_2 \leq X_2$, and hence any closed class $\mathfrak{X}$ is the union of the maximal classes of the form $\mathfrak{X}_1 \otimes \mathfrak{X}_2$ contained in it. The second difficulty is that in general it is not obvious that this union is finite. However, we know how to deal with this sort of difficulty. The assumption necessary to cope with it is that both the set of varieties between $\mathfrak{W}_1$ and $\mathfrak{A}_p \mathfrak{W}_1$ and the set between $\mathfrak{W}_2$ and $\mathfrak{A}_p \mathfrak{W}_2$ are partially well-ordered under inclusion (cf. [3], where however a partially well-ordered set is said to have the finite basis property). For if $\mathfrak{X}_1^{(\lambda)} \otimes \mathfrak{X}_2^{(\lambda)}$, $\lambda \in \Lambda$, are the maximal classes of the form $\mathfrak{X}_1 \otimes \mathfrak{X}_2$ contained in $\mathfrak{X}$ it is evident that the correspondence $\mathfrak{X}_1^{(\lambda)} \leftrightarrow \mathfrak{X}_2^{(\lambda)}$ is an anti-isomorphism between the sets $\{\mathfrak{X}_1^{(\lambda)}, \lambda \in \Lambda\}$ and $\{\mathfrak{X}_2^{(\lambda)}, \lambda \in \Lambda\}$, each ordered by inclusion. Thus if these sets are partially well-ordered, Λ is finite. It then follows by induction on the cardinal of Λ that the conjecture is true for $\mathfrak{U}(\mathfrak{X})$. It can also be seen that the varieties between $\mathfrak{W}$ and $\mathfrak{A}_p \mathfrak{W}$ are partially well-ordered by inclusion, so that we are in a position to deal, by induction, with direct joins of more than two varieties. Our partial well-order condition is equivalent to the requirements that the sets of varieties in question satisfy the minimal condition, and that there is no infinite subset of them no member of which contains another. It is, of course, a well-known open question whether or not the set of all varieties satisfies the minimal condition. It is certainly not partially well-ordered, though I know no counterexample to the conjecture that the set of subvarieties of any locally finite variety is partially well-ordered.

The methods of this section are sufficient to prove the following. If $\mathfrak{W}_1$, $\mathfrak{W}_2, \ldots$, $\mathfrak{W}_r$ are varieties whose exponents are prime to one another and to p, each of which is either nilpotent of Φ-length two, or of the form $\mathfrak{A}_s \mathfrak{A}_t$ with s, t distinct primes, then the conjecture is true for every subvariety of $\mathfrak{A}_p \langle \mathfrak{W}_1, \mathfrak{W}_2, \ldots, \mathfrak{W}_r \rangle$, and every such subvariety has a finite basis for its identical relations.

If $\mathfrak{W}_1$, $\mathfrak{W}_2$ are varieties whose exponents are prime to one another and to r, then any group in $\langle \mathfrak{A}_r \mathfrak{W}_1, \mathfrak{A}_r \mathfrak{W}_2 \rangle$ is uniquely expressible as a direct product $X_0 X_1 X_2$, where X_0 belongs to $\mathfrak{A}_r$, and, for $i = 1$, 2, X_i

belongs to $\mathfrak{A}_r\mathfrak{W}_i$ and has factor derived group of order prime to r. This situation can be exploited in a similar way to prove the conjecture in simple cases. For instance, it holds for subvarieties of $\mathfrak{A}_p\langle\mathfrak{A}_s\mathfrak{A}_{t_1}, \mathfrak{A}_s\mathfrak{A}_{t_2}\rangle$ where p, s, t_1, t_2 are distinct primes.

2.6. *The case* $\mathfrak{W} = \mathfrak{A}_s\mathfrak{A}_t$, *s prime to t.* If X belongs to $\mathfrak{A}_s\mathfrak{A}_t$, where s is prime to t, then $X = X_sX_t$, where X_s is normal, abelian, and of exponent dividing s, and X_t is abelian of exponent dividing t. If u is a divisor of s, we write $\pi_u(X)$ for $\mathrm{Hol}_X(X_s^{s/u})$, that is, for the split extension of $X_s^{s/u}$ by $\mathrm{Aut}_X(X_s^{s/u})$. $\pi_u(X)$ belongs to $\mathfrak{A}_u\mathfrak{A}_t$ and has center of order prime to t. If T is any such group, we write $\mathfrak{P}_s^u(T)$ for the class of irreducible linear groups X in $\mathfrak{A}_s\mathfrak{A}_t$ over $GF(p)$ such that $\pi_u(X)$ is isomorphic to T. Our first aim is to show that for $\mathfrak{P} = \mathfrak{P}_s^u(T)$, $c_{\mathfrak{P}}(n)$ is CREAM. For better intuitive understanding, observe that X_s can be written as a direct product of groups $X_s^{(q)}$, one for each prime power q dividing s, where $X_s^{(q)}$ is normal in X and homocyclic of exponent q. The statement that X belongs to $\mathfrak{P}_s^u(T)$ prescribes the factor $X_s^{(q)}$ and the automorphisms induced by X_t in it, if q does not divide s/u, but leaves them free if it does.

Now if K is a normal subgroup of the free n-generator group G_n in $\mathfrak{A}_s\mathfrak{A}_t$, $\pi_u(G_n/K) = \mathrm{Hol}_{G_n}(G_{n,s}^{s/u}/(K \cap G_{n,s}^{s/u}))$. Thus G_n/K belongs to $\mathfrak{P}$ if and only if $\mathrm{Hol}_{G_n}(G_{n,s}^{s/u}/(KG_{n,s}^{s/u})) \simeq T$. For L contained in $G_{n,s}^{s/u}$ and normal in $G_{n,s}$, let d_L be the contribution to the dimension of the regular representation of G_n made by those irreducible representations whose kernels K satisfy $K \cap G_{n,s}^{s/u} = L$. We have then $c_{\mathfrak{P}}(n) = \sum d_L$, summed over those L such that $\mathrm{Hol}_{G_n}(G_{n,s}^{s/u}/L) \simeq T$. However, it is evident that $\sum_{M \geq L} d_M = |G_n|/|L|$, and so

$$d_L = \sum_{M \geq L} j_{M,L}\frac{|G_n|}{|M|} = \frac{|G_n|}{|G_{n,s}^{s/u}|}\sum_{M \geq L} j_{M,L}\frac{|G_{n,s}^{s/u}|}{|M|}.$$

Here the coefficient $j_{M,L}$ depends only on the lattice of normal subgroups between M and L, so that the sum in the last expression depends only on the structure of $G_{n,s}^{s/u}/L$ as G_n-module. But $|G_n|/|G_{n,s}^{s/u}| = t^n u^{1 + (n-1)t^n}$ so that, for L such that $\mathrm{Hol}_{G_n}(G_{n,s}^{s/u}/L) \simeq T$, d_L depends only on T and n, and as function of n is CREAM. But, for such L, if $L^{u/s}$ denotes the set of elements x of $G_{n,s}$ such that $x^{s/u}$ belongs to L, $J = L^{u/s}(C(G_{n,s}^{s/u}/L) \cap G_{n,t})$ is a normal subgroup of G_n such that $G_n/J \simeq T$, and since $L = (J \cap G_{n,s})^{s/u}$

the correspondence between subgroups L and subgroups J is one to one. Thus the number of subgroups L is $k_T(n)/|\mathrm{Aut}(T)|$, and is CREAM. This concludes the proof that $c_{\mathfrak{P}}(n)$ is CREAM.

It is easy to see that the intersection of a class $\mathfrak{P}_s^u$ and a class $\mathfrak{P}_s^v$ is the union of a finite number of classes $\mathfrak{P}_s^w$, where w is the least common multiple of u and v. Hence if $\mathfrak{X}$ is a closed class which is the union of a finite set of classes $\mathfrak{P}_s^u$, for various u's, and a finite set, then the conjecture is true for $\mathfrak{U}(\mathfrak{X})$. Our next aim is to show that if t is prime every closed class is of this form. We shall only sketch the proof of this, leaving many details to be filled in by the reader.

First, it is necessary, as it has not been hitherto, to say something precise about irreducible linear groups in $\mathfrak{A}_s\mathfrak{A}_t$. By a theorem of Kochendörffer [5], a metabelian group X has a faithful irreducible representation if and only if no two distinct minimal normal subgroups of X are isomorphic as X-modules. One can also see that two faithful irreducible representations of the same group in $\mathfrak{A}_s\mathfrak{A}_t$ give rise to isomorphic linear groups if t is prime. Thus the irreducible linear groups in $\mathfrak{A}_s\mathfrak{A}_t$ can, in this case, be identified with a certain set of abstract groups, and it is then easy to verify that the relation $X \leq Y$ becomes simply "X is isomorphic to a subgroup of Y." Note that, by the condition mentioned, if X is an irreducible linear group in $\mathfrak{A}_s\mathfrak{A}_t$, $\pi_u(X)$ is also an irreducible linear group.

Now let $s = p_1^{a_1} \cdots p_j^{a_j}$ where $p_1, \ldots, p_j$ are distinct primes, and let $(b) = (b_1, \ldots, b_j)$ be a sequence of integers such that $0 \leq b_\lambda \leq a_\lambda$, $\lambda = 1, \ldots, j$. The (b)-measure of a group X in $\mathfrak{A}_s\mathfrak{A}_t$ is defined as the largest integer k such that, whenever $b_\lambda \neq 0$, X_s has a direct factor which is a direct product of k cyclic groups of order $p_\lambda^{b_\lambda}$. A set of groups in $\mathfrak{A}_s\mathfrak{A}_t$ will be said to be unbounded for (b) if there is no upper bound to the (b)-measures of the groups contained in it. We shall write $u = u(b) = p_1^{a_1 - b_1} \cdots p_j^{a_j - b_j}$; and for sequences (b), (c) we write $(b) \leq (c)$ if $b_\lambda \leq c_\lambda$, $\lambda = 1, \ldots, j$.

We are home if we can prove the following three propositions for any closed class $\mathfrak{X}$, nonzero sequence (b), and $u = u(b)$:

(i) if $\mathfrak{T}$ is a closed class in $\mathfrak{A}_u\mathfrak{A}_t$, and $\mathfrak{X} \cap \mathfrak{P}_s^u(\mathfrak{T})$ is unbounded for (b), then there exists T not in $\mathfrak{T}$ such that $\mathfrak{X} \cap \mathfrak{P}_s^u(T)$ is unbounded for (b);

(ii) if $\mathfrak{X} \cap \mathfrak{P}_s^u(T)$ is unbounded for (b) for an infinity of T, $\mathfrak{X}$ is unbounded for (c), for some (c) with $(b) < (c)$;

(iii) if $\mathfrak{X} \cap \mathfrak{P}_s^u(T)$ is unbounded for (b), then $\mathfrak{P}_s^u(T) \subseteq \mathfrak{X}$.

For if the closed class $\mathfrak{X}$ is not unbounded for any $(b) \neq 0$, it is finite. Otherwise, let (b) be maximal among the sequences for which $\mathfrak{X}$ is unbounded. By (ii), there are only a finite number of T, say $T_1, \ldots, T_k$, such that $\mathfrak{X} \cap \mathfrak{P}_s^u(T)$ is unbounded for (b). By (iii) $\mathfrak{X}$ contains $\bigcup_\alpha P_s^u(T_\alpha)$, and by (i) the complement $\mathfrak{Y}$ of $\bigcup_\alpha P_s^u(T_\alpha)$ is not unbounded for (b), and so neither is $\mathrm{cl}(\mathfrak{Y})$. Thus the result follows by induction on the number of sequences for which $\mathfrak{X}$ is unbounded.

Of the three statements (ii) is a straightforward deduction from the definitions. To prove (i), we make the inductive hypothesis that the main result has been proved for closed classes in $\mathfrak{A}_u \mathfrak{A}_t$. This implies fairly easily that closed classes in $\mathfrak{A}_u \mathfrak{A}_t$ satisfy the minimum condition, and hence that irreducible groups in $\mathfrak{A}_u \mathfrak{A}_t$ are partially well-ordered by the relation $T_1 \leq T_2$. If the hypotheses of (i) hold, there exists, for each integer k, a group X_k in $\mathfrak{X}$ of (b)-measure greater than k, with $\pi_u(X_k) = T_k$, T_k not in $\mathfrak{T}$. By the partial well-order, there is a finite number, e say, such that for all k, there exists i in $1 \leq i \leq e$ with $T_i \leq T_k$. The fact that $\mathfrak{X}$ is closed implies that $\mathfrak{X} \cap \mathfrak{P}_s^u(T_i)$ is unbounded for any i which satisfies this relation for an infinity of k.

To prove (iii), notice first that there is a unique minimal element T^* in $\mathfrak{P}_s^u(T)$; as we have said, the fact that X belongs to $\mathfrak{P}_s^u(T)$ determines the direct factors of X_s which are homocyclic of exponent q for q not dividing s/u; T_s^* has these direct factors, and no others. Next, if $b_\lambda \neq 0$, we define B_λ to be a group with $B_{\lambda, s}$ homocyclic of exponent $p_\lambda^{b_\lambda}$, and $B_{\lambda, t}$ cyclic of order t, operating faithfully and irreducibly on the Frattini factor group of $B_{\lambda, s}$. We show that if Y is a group with the property that, for all X in $\mathfrak{P}_s^u(T)$ with large enough (b)-measure, $Y \leq X$, then the direct product $Y B_\lambda$ is another. This is because, if $Y \leq X$, $|X_t/C(Y) \cap X_t|$ is bounded, and hence the number of non-isomorphic homocyclic direct factors W of X_s of exponent $p_\lambda^{b_\lambda}$ with $C(W) \geq C(Y)$ is bounded. Since each of these can occur in X_s only once, if the (b)-measure of X is large enough, there must be such a W not centralized by $C(Y)$. From here it is easy to see that $Y B_\lambda \leq X$. In particular, if $\mathfrak{X}$ is a closed class such that $\mathfrak{X} \cap \mathfrak{P}_s^u(T)$ is unbounded for (b), $\mathfrak{X}$ contains every direct product $T^*(B_1 B_2 \cdots B_j)^k$ for some k, and so $\mathfrak{P}_s^u(T)$ is contained in $\mathfrak{X}$.

This completes our sketch of the proof that if p, t are primes not dividing s, every subvariety of $\mathfrak{A}_p \mathfrak{A}_s \mathfrak{A}_t$ satisfies the conjecture. As we have already had occasion to remark, it is a corollary of the proof that these subvarieties satisfy the minimal condition.

2.7. *Subvarieties of varieties* $\mathfrak{A}_r\mathfrak{W}$. We observe that the considerations in [4], and hence much of the last three sections, can be extended to deal with varieties $\mathfrak{V}$ such that $\mathfrak{W} \subset \mathfrak{V} \subset \mathfrak{A}_r\mathfrak{W}$, where r is 'not prime, but is coprime to the exponent of $\mathfrak{W}$.

First, if $r = r_1 r_2$ where r_1, r_2 are coprime, it is easy to see that the varieties in question are in one to one correspondence with the pairs $(\mathfrak{V}_1, \mathfrak{V}_2)$ of varieties such that $\mathfrak{W} \subset \mathfrak{V}_1 \subset \mathfrak{A}_{r_1}\mathfrak{W}$ and $\mathfrak{W} \subset \mathfrak{V}_2 \subset \mathfrak{A}_{r_2}\mathfrak{W}$. Here $\mathfrak{V}$ corresponds to the pair $(\mathfrak{V} \cap \mathfrak{A}_{r_1}\mathfrak{W}, \mathfrak{V} \cap \mathfrak{A}_{r_2}\mathfrak{W})$, and the pair $(\mathfrak{V}_1, \mathfrak{V}_2)$ corresponds to $\langle \mathfrak{V}_1, \mathfrak{V}_2 \rangle$. This reduces us to the case when r is a prime power.

Suppose then that $r = p^a$. If X is an irreducible linear group in $\mathfrak{W}$ over $GF(p)$, we define a critical group (X, b) in $\mathfrak{A}_r\mathfrak{W}$, but not in $\mathfrak{W}$, for each b in $1 \leq b \leq a$, namely the product PX, where P is a homocyclic normal subgroup of exponent p^b, and X induces the right linear group in $P/\Phi(P)$. These are all critical groups in $\mathfrak{A}_r\mathfrak{W}$ but not in $\mathfrak{W}$. Moreover, a variety containing (X, b) contains (Y, c) if $Y \leq X$ and $c \leq b$, and this is the only restriction on the critical groups in a variety (by an argument similar to that in [4]). Thus the varieties we are considering may be written $\mathfrak{U}(\mathfrak{X}_1, \ldots, \mathfrak{X}_a)$, where $\mathfrak{X}_1 \geq \mathfrak{X}_2 \geq \cdots \geq \mathfrak{X}_a$ is a chain of closed classes of irreducible groups, and $\mathfrak{X}_b$ is the set of X such that (X, b) belongs to the variety. If $\mathfrak{X}_a$ is nonempty, the order of the free n-generator group in $\mathfrak{U}(\mathfrak{X}_1, \ldots, \mathfrak{X}_a)$ is $f(n)p^{a+(n-1)c(n)}$ where $c(n) = c_{\mathfrak{X}_1}(n) + \cdots + c_{\mathfrak{X}_a}(n)$ so that the conjecture holds for $\mathfrak{U}(\mathfrak{X}_1, \ldots, \mathfrak{X}_a)$ if it holds for each subvariety $\mathfrak{U}(\mathfrak{X}_b)$ of $\mathfrak{A}_p\mathfrak{W}$.

In particular, if r, s are coprime, and t is a prime dividing neither, the conjecture holds for all subvarieties of $\mathfrak{A}_r\mathfrak{A}_s\mathfrak{A}_t$. Also, the minimal condition holds for these varieties.

References

[1] WOLFGANG GASCHÜTZ, Über modulare Darstellungen endlicher Gruppen, die von freien Gruppen induziert werden, *Math. Z.* **60** (1954), 274–286.
[2] P. HALL, The Eulerian functions of a group, *Quart. J. Math. Oxford Ser.* (1) **7** (1936), 134–151.
[3] GRAHAM HIGMAN, Ordering by divisibility in abstract algebras, *Proc. London Math. Soc.* (3) **2** (1952), 326–336.

[4] GRAHAM HIGMAN, Some remarks on varieties of groups, *Quart. J. Math. Oxford Ser.* (2) **10** (1959), 165–178.

[5] RUDOLF KOCHENDÖRFFER, Über treue irreduzible Darstellungen endlicher Gruppen, *Math. Nachr.* **1** (1948), 25–39.

[6] SEÁN TOBIN, On a theorem of Baer and Higman, *Canad. J. Math.* **8** (1956), 263–270.

[7] C. R. B. WRIGHT, On the nilpotency class of a group of exponent four, *Pacific J. Math.* **11** (1961), 387–394.

Proc. Internat. Conf. Theory of Groups, Austral. Nat. Univ. Canberra, August 1965, pp. 167–173. © Gordon and Breach Science Publishers, Inc. 1967

Representations of general linear groups and varieties of p-groups

GRAHAM HIGMAN

1. *Introduction.* It is one of my deeper mathematical convictions that the theory of the representations of the general linear groups is and must always be a topic of central importance to algebra, that it needs to be rewritten every generation or so, in the idiom of the day and to meet current needs, and in particular that it needs to be rewritten in this generation. The purpose of this lecture is to illustrate this. As the second part of the title indicates, I shall try to do this by indicating applications of the theory to a subject which previous contributions to this Conference have shown to be of interest to many of its members.

2. *The problem.* We are interested in representations, over the field K, of the groups $GL(n, K)$, that is, in homomorphisms $GL(n, K) \to GL(m, K)$. We shall, however, restrict ourselves, in two steps, to a subclass of these. First, the representations that occur naturally in algebra arise not singly but in series, one for each integer n. They arise, in fact, from functors from the category of vector spaces over K into itself. Such a functor is, of course, simply a rule which assigns to each vector space U over K a vector space U^t, and to each map $\alpha: U \to V$ of vector spaces a map $\alpha^t: U^t \to V^t$, in such a way that $(\alpha\beta)^t = \alpha^t\beta^t$ whenever $\alpha\beta$ is defined. If U is of dimension n and U^t is of dimension m, the map carrying α to α^t is in particular a homomorphism of $GL(n, K)$ into $GL(m, K)$, and we consider only representations of this kind.

It is clear how subfunctors and quotient functors can be defined, and the direct sum of two functors. We can also define equivalence between functors, and hence indecomposable and irreducible functors. Another very important notion is that of the tensor product of two functors s

167

and t. This assigns to U the vector space $U^{s \otimes t} = U^s \otimes U^t$ spanned in the usual way by expressions $u \otimes v$, u in U^s, v in U^t, and $(u \otimes v)\alpha^{s \otimes t} = u\alpha^s \otimes v\alpha^t$. The second restriction that we make is that we only consider functors built up from the identity functor (in which $U^i = U$ and $\alpha^i = \alpha$) by repeated formation of subfunctors, quotient functors, direct sums, and tensor products. The central problem is then the structure of the r-fold tensor power $U^{(r)}$ as $GL(n, K)$-module, or, more precisely, the subfunctor structure of the functor (r).

3.   The classical solution. In case K has characteristic 0, this problem has a classical solution, which depends on the fact that the symmetric group S_r on $(1, 2, \ldots, r)$ acts in an obvious way on $U^{(r)}$: $(u_1 \otimes \cdots \otimes u_r)\sigma = (u_{1_\sigma} {}^{-1} \otimes \cdots \otimes u_{r_\sigma} {}^{-1})$. It is clear that the action of S_r commutes with the action of $GL(n, K)$. It is less obvious, but true, that the algebras $\overline{KS_r}$ and $\overline{KGL(n, K)}$ of linear transformations spanned by these actions are centralizer algebras of one another. But $\overline{KS_r}$ is a homomorphic image of the group algebra KS_r, and so is semisimple. Hence $\overline{KGL(n, K)}$ is also semisimple and $U^{(r)}$ is completely reducible, as representation space for $GL(n, K)$. Moreover the irreducible representations of S_r and of $GL(n, K)$ occurring in $U^{(r)}$ are in one to one correspondence, the degree of a representation of S_r being the multiplicity of the corresponding representation of $GL(n, K)$ and vice versa (we are here using the fact that the rational field, and hence any field K of characteristic 0, is a splitting field for S_r). If the dimension n of U is at least r, all irreducible representations of S_r occur in $U^{(r)}$, and so the irreducible representations of $GL(n, K)$ occurring in $U^{(r)}$, like those of S_r, can be indexed by the partitions of r. If $n < r$, the representations of S_r that occur are those indexed by partitions of r into not more than n parts, so that these partitions index the representations of $GL(n, K)$.

All this is classical, and I add here only that S_r, of course, acts not only on $U^{(r)}$, but also on any $V^{(r)}$, and if $\alpha: U \to V$ is a map, $\alpha^{(r)}\sigma = \sigma\alpha^{(r)}$. Thus the decomposition is genuinely functorial. To any partition (λ) there corresponds an irreducible functor $[\lambda]$, which is a subfunctor of (r) if (λ) is a partition of r. If the number of parts in (λ) is s, $U^{[\lambda]} = 0$ if and only if the dimension of U is less than s. Any subfunctor t of (r) is determined completely by the $GL(n, K)$-module U^t, provided that the dimension n of U is at least r.

If K no longer has characteristic 0, but has characteristic p, all these considerations remain valid, provided that $r < p$, since all that matters is that KS_r is semisimple. It is this case that we apply below.

4. *An explicit construction.* We give next a construction which exhibits the functors [λ] explicitly. Let $R(U)$ be the coordinate ring of the manifold M of affine flags in the dual U^* of U. Thus a point of M is a sequence $(x_1, x_2, \ldots, x_n)$ of elements of U^*, where we regard $(x_1, x_2, \ldots, x_n)$ as equal to $(x_1', x_2', \ldots, x_n')$ if, for $i = 1, \ldots, n$, $x_i' = x_i + \sum_{j<i} \lambda_{ij} x_j$ for some λ_{ij} in K. As we have said, $R(U)$ is the ring of functions on M. Then $R(U)$ contains U; for we can identify u in U with the function given by $u(x_1, \ldots, x_n) = \langle ux_1 \rangle$. This is well-defined on M, since $(x_1, \ldots, x_n) = (x_1', \ldots, x_n')$ implies $x_1 = x_1'$. Next, $R(U)$ also contains the exterior square of U, for we can identify $u \wedge v$ with the function

$$(u \wedge v)(x_1, x_2, \ldots, x_n) = \begin{vmatrix} \langle ux_1 \rangle & \langle vx_1 \rangle \\ \langle ux_2 \rangle & \langle vx_2 \rangle \end{vmatrix}.$$

This, too, is well defined on M, for if $(x_1, x_2, \ldots, x_n) = (x_1', x_2', \ldots, x_n')$ we have $x_1' = x_1$, $x_2' = \lambda x_1 + x_2$, so that the expression is unaltered if x_1, x_2 are replaced by x_1', x_2'. Continuing so, we embed the whole exterior algebra $E(U)$, as vector space, in $R(U)$. In fact $R(U)$ is generated by $E(U)$, and so is a homomorphic image of the polynomial algebra $P(E(U))$. It is a proper homomorphic image, for instance

$$\{u(v \wedge w) + v(w \wedge u) + w(u \wedge v)\}(x_1, \ldots, x_n)$$

$$= \begin{vmatrix} \langle ux_1 \rangle & \langle vx_1 \rangle & \langle wx_1 \rangle \\ \langle ux_1 \rangle & \langle vx_1 \rangle & \langle wx_1 \rangle \\ \langle ux_2 \rangle & \langle vx_2 \rangle & \langle wx_2 \rangle \end{vmatrix} = 0,$$

so that $u(v \wedge w) + v(w \wedge u) + w(u \wedge v) = 0$ in $R(U)$, though, if u, v, w are linearly independent, $u(v \wedge w) + v(w \wedge u) + w(u \wedge v) \neq 0$ in $P(E(U))$. We shall not explain here precisely what the relations in $R(U)$ are. We are concerned only to say that, if the partition conjugate to [λ] is $(1^{a_1} 2^{a_2} 3^{a_3} \cdots)$ then the elements of $R(U)$ homogeneous of degree a_1 in U, of degree a_2 in $U \wedge U$, of degree a_3 in $U \wedge U \wedge U$, and so on, form an irreducible $GL(n, K)$ module of type [λ], so that $R(U)$ contains each representation once and once only.

For instance, [r] is the representation afforded by the terms of degree

r in $P(U)$, and $[1^r]$ the representation afforded by the terms of degree r in $E(U)$.

5. *The free group of exponent p and class $p-1$.* Let G_n be the free group of exponent p and class $p-1$ on n generators. Written additively, the Frattini factor group of G_n is a vector space U over the field $GF(p)$. So is the factor group $U^{L_c} = \gamma_c(G_n)/\gamma_{c+1}(G_n)$ of any two successive terms of the lower central series of G_n. Evidently L_c is a functor on the category of vector spaces over $GF(p)$. Moreover U^{L_c} is spanned by the elements corresponding to the commutators $[g_1, g_2, \ldots, g_c]$, and the element so corresponding is a multilinear function of the elements of U corresponding to $g_1, g_2, \ldots, g_c$. It follows that L_c is a quotient functor of (c) (where $U^{(c)}$ is the c-fold tensor power of U). Since $c < p$, the classical theory applies, L_c is completely reducible, and its irreducible components are functors $[\lambda]$.

The structure of L_c is in fact known completely. As the notation is intended to suggest, U^{L_c} can be identified with the component of degree c in the free Lie ring $L(U)$ generated by U. A well-known formula of Witt states that the dimension of U^{L_c} is

$$\frac{1}{c} \sum_{d|c} \mu(d) n^{c/d},$$

where n is the dimension of U. This is only a special case of a formula (perhaps also due to Witt, though I first met it in P. M. Cohn's Ph.D. thesis) for the character of $GL(n, K)$ afforded by L_c, namely

$$\chi^{L_c}(M) = \frac{1}{c} \sum_{d|c} \mu(d) s_d^{c/d},$$

where $s_d = \sum_i \alpha_i^d = \mathrm{tr}(M^d)$. From this it follows easily that the multiplicity of $[\lambda]$ in L_c is

$$\frac{1}{c} \sum_{d|c} \mu(d) \chi^{(\lambda)}(\sigma^{c/d}),$$

where $\chi^{(\lambda)}$ is the corresponding character of the symmetric group, and σ is a cyclic permutation of order c.

This can be worked out explicitly when c is small. Of the results, we note only that for $c \leq 3$, L_c is irreducible; for $c = 4, 5$, L_c is no longer irreducible but is multiplicity free; but for $c \geq 6$, L_c is no longer multiplicity free, for instance $[3, 2, 1]$ occurs three times in L_6.

6. Varieties between $\mathfrak{B}_{p,c-1}$ and $\mathfrak{B}_{p,c}$. Let $\mathfrak{B}$ be a variety of p-groups of exponent p and class at most c containing all groups of exponent p and class $c-1$, for some $c<p$. Then the kernel of the natural homomorphism of G_n onto $\overline{G_n}$, the free group in $\mathfrak{B}$ with n generators, contains $\gamma_{c+1}(G_n)$ and is contained in $\gamma_c(G_n)$, and so determines, and is determined by a subspace W of U^{L_c}, where U is the Frattini factor group of G_n, as vector space over $GF(p)$. The correspondence between U and W is obviously functorial, so that $W=U^t$, for some subfunctor t of L_c. Conversely, if t is a subfunctor of L_c, then the normal subgroup of G_n determined by U^t is fully invariant, so that its factor group is relatively free. Thus, if $\mathfrak{B}_{p,k}$ denotes the variety of groups of exponent p and class at most k, then, as long as $p>c$, the varieties between $\mathfrak{B}_{p,c-1}$ and $\mathfrak{B}_{p,c}$ are in one to one correspondence with the subfunctors of L_c. As long as $c\leq 5$, so that L_c is multiplicity free, the number of such varieties is independent of p, but for $c\geq 6$, when L_c is not multiplicity free, the number will tend to infinity with p, and will, in fact, be a nonconstant polynomial in p.

As we have seen, a subfunctor t of L_c is determined by U^t, if $\dim(U)\geq c$. Thus a variety $\mathfrak{B}$ between $\mathfrak{B}_{p,c-1}$ and $\mathfrak{B}_{p,c}$ is determined by its c-generator free group, a reflection, in this context, of the fact that a critical group of class c can be generated by c elements. The fact, proved by Paul Weichsel, that if $p>c>2$ a critical p-group of class c can be generated by $c-1$ elements corresponds to the fact that, for $c>2$, L_c does not contain a subfunctor of type $[1^c]$.

7. Join irreducible varieties. Let us denote by $\mathfrak{B}(t)$, where t is a subfunctor of L_c, the variety between $\mathfrak{B}_{p,c-1}$ and $\mathfrak{B}_{p,c}$ corresponding to it. If $\mathfrak{U}$ is any variety contained in $\mathfrak{B}_{p,c}$ but not in $\mathfrak{B}_{p,c-1}$, $\langle\mathfrak{U},\mathfrak{B}_{p,c-1}\rangle$ will be the variety $\mathfrak{B}(t)$ for some proper subfunctor t of L_c. If t is not a maximal subfunctor of L_c then $\mathfrak{U}$ cannot be join irreducible; if $t=t_1\cap t_2$, then $\mathfrak{U}=\langle\mathfrak{U}\cap\mathfrak{B}(t_1),\mathfrak{U}\cap\mathfrak{B}(t_2)\rangle$, but $\mathfrak{U}$ is not contained in $\mathfrak{B}(t_i)$, $i=1,2$. If, however, t is a maximal subfunctor of L_c, there will be at least one join irreducible variety $\mathfrak{U}$ such that $\langle\mathfrak{U},\mathfrak{B}_{p,c-1}\rangle=\mathfrak{B}(t)$; indeed any variety minimal subject to this condition is join irreducible. Thus the join irreducible varieties of class c and exponent p are at least as numerous as the maximal subfunctors of L_c. In particular, their number tends to infinity with p if L_c is not multiplicity free, that is, if $c\geq 6$.

Furthermore, if L_c is not multiplicity free, we can choose three maximal subfunctors t_1, t_2, t_3 such that $t_1 \cap t_2 = t_1 \cap t_3 = t_2 \cap t_3 = t$, say. If $\mathfrak{U}$ is a variety minimal with respect to the condition $\langle \mathfrak{U}, \mathfrak{B}_{p,c-1} \rangle = \mathfrak{B}(t)$, and, for $i = 1, 2, 3$, $\mathfrak{U}_i$ is a subvariety of $\mathfrak{U}$ minimal with respect to the condition $\langle \mathfrak{U}_i, \mathfrak{B}_{p,c-1} \rangle = \mathfrak{B}(t_i)$, then $\mathfrak{U}_1$, $\mathfrak{U}_2$, and $\mathfrak{U}_3$ will be three join irreducible varieties whose joins in pairs coincide (in $\mathfrak{U}$). This answers, in a strong form, a question raised orally at this Conference by Dr L. G. Kovács.

8. *Critical groups.* Let t be a proper subfunctor of L_c such that every partition (λ) such that $[\lambda]$ occurs in L_c/t is a partition into exactly s parts, and let $\mathfrak{U}$ be a variety of groups of exponent p and class c, minimal with respect to the condition that $\langle \mathfrak{U}, \mathfrak{B}_{p,c-1} \rangle = \mathfrak{B}(t)$.

If G is an n-generator group belonging to $\mathfrak{U}$, then $G = G_n/H$ for some normal subgroup H, and we write W for the subspace of U^{L_c} corresponding to $(H \cap \gamma_{c-1}(G_n))\gamma_c(G_n)/\gamma_c(G_n)$, where U, as usual is $G_n/\Phi(G_n)$ written additively. The fact that G belongs to $\mathfrak{U}$ implies that G belongs to $\mathfrak{B}(t)$, and so that $W \supset U^t$. The condition that we have imposed on t implies that, if $n < s$, $U^t = U^{L_c}$, and hence that G has class less than c. If G generates $\mathfrak{U}$, then $W \not\supset U^{t'}$ for any subfunctor t' of L_c not contained in t; conversely, by the minimality of $\mathfrak{U}$, this condition is sufficient for G to generate $\mathfrak{U}$. If, in particular t' is an irreducible subfunctor of L_c not contained in t, $U_0^{t'} \neq 0$, where U_0 is an s-dimensional subspace of U. Hence the sum of all such $U_0^{t'}$, being a nonzero submodule of $U^{t'}$, is $U^{t'}$ itself. Hence if $W \not\supset U^{t'}$, $W \not\supset U_0^{t'}$ for some s-dimensional subspace U_0 of U. This means that if G generates $\mathfrak{U}$, then the s-generator subgroups of G generate $\mathfrak{U}$.

In particular, if G is a critical group which generates $\mathfrak{U}$, then G has s generators precisely. Moreover, W is then of codimension 1, and so U^{L_c}/t has a subspace of codimension 1 not containing $U^{t'}$ for any nonzero subfunctor t' of L_c/t, where $\dim U = s$. Conversely, if such a subspace exists, we can choose W to be a subspace of U^{L_c} of codimension 1 containing U^t, but not $U^{t'}$ for any t' not contained in t. If H is a normal subgroup of G_s such that $G = G_s/H$ belongs to $\mathfrak{U}$ and

$$(H \cap \gamma_{c-1}(G_s))\gamma_c(G_s)/\gamma_c(G_s)$$

corresponds to W, then G generates $\mathfrak{U}$, and if H is maximal then G is critical. For the maximality of H ensures that proper factor groups of

G have class less than c, and the fact that $(s-1)$-generator groups in $\mathfrak{u}$ have class less than c ensures that proper subgroups of G do.

Thus $\mathfrak{u}$ is generated by a single critical group if and only if, for dim $U = s$, $U^{L_c/t}$ has a subspace of codimension 1 not containing any nonzero $U^{t'}$. This will be true if and only if, for each (λ), the multiplicity of $[\lambda]$ in L_c/t does not exceed the dimension of $U^{[\lambda]}$. For $c \geq 8$, there exist t violating this condition. For instance, the multiplicity of $[5, 3]$ in L_8 is 4, but if dim $U = 2$, $\dim(U^{[5,3]}) = 3$, so that we can take t to be the sum of all irreducible components of L_8 not of type $[5, 3]$. For such t, $\mathfrak{u}$ cannot be generated by a single critical group.

If W and W' are distinct subspaces of $U^{L_c/t}$ satisfying the given conditions, the critical groups constructed from them will certainly not be isomorphic unless W and W' are equivalent under the action of the general linear group on $U^{L_c/t}$. In particular, if L_c/t is irreducible, the number of pairwise nonisomorphic critical groups generating $\mathfrak{u}$ is at least equal to the number of orbits under the general linear group of the set of subspaces of codimension 1 in $U^{L_c/t}$. For instance, the number of distinct critical groups generating the variety of metabelian groups of exponent p and class c tends to infinity with p if $c \geq 6$.

Finally, if G is a critical group generating $\mathfrak{u}$, then, as we have seen, the variety $\mathfrak{u}_0$ generated by its proper factors has class less than c. It follows that the lattice of varieties between $\mathfrak{u}_0$ and $\mathfrak{u}$ is at least as complicated as the lattice of functors between t and L_c. If c has k distinct partitions into s parts, t can be chosen so that this lattice is a Boolean algebra of order 2^k.

Proc. Internat. Conf. Theory of Groups, Austral. Nat. Univ. Canberra, August 1965, pp. 175–183. © Gordon and Breach Science Publishers, Inc. 1967

Periodic linear groups[1]

K. A. HIRSCH

I shall present here, with his permission, some of the results obtained by a research student of mine, Mr[2] B. Wehrfritz. His work is still in progress and my report is sketchy and incomplete,[3] but in due course Mr Wehrfritz is going to publish a full account.[4]

We are concerned with the class $\mathfrak{P}$ of groups isomorphic to a periodic subgroup of the general linear group $GL(n, F)$ of an arbitrary finite degree n over an arbitrary field F. Occasionally it is advantageous to assume that F is algebraically closed.

It is a well-known phenomenon that linear groups are in many respects better behaved than abstract groups in general. I quote a few examples which are frequently used in Wehrfritz' work.

1. First and foremost there is *Schur's Theorem* of 1911 [6]: *periodic linear groups are locally finite.* Schur proved this for the field **C** of complex numbers and the result readily extends to char $F = 0$. For char $F = p$ it has been known in interested circles for some time and was first stated explicitly, as far as I am aware, by Kargapolov whose proof, however, contains a *non sequitur*; a correct proof was published in 1962 by Suprunenko and Platonov [7][5].

Schur's Theorem, of course, gives an affirmative solution to the Burnside problem in the case of linear groups. Burnside's previous result of 1909 plays a lesser role in our present context: *if a group,*

---

[1] *Added in proof* (December 1966): Owing to the delay in publication this report, written immediately after the Conference, need some minor modifications which I indicate in footnotes.

[2] Now Dr.

[3] It is also out of date, because a better proof of Theorem A1 is now available.

[4] In the Journal and the Proceedings of the London Mathematical Society.

[5] Irving Kaplansky has told me that, even before this date, a set of mimeographed notes of his lectures in Chicago contained a complete general proof; see [3a].

13—K.   175

$G \in \mathfrak{P}$ *is of finite exponent* e, *then its order* $|G|$ *divides* e^{n^3} (provided G is absolutely irreducible, or char $F = 0$, or char F does not divide e).

2. *Zassenhaus' Theorem* of 1938 [8]: *every linear group has a unique maximal soluble normal subgroup* (the soluble radical) *and there is a bound on the length of the derived series in terms of the degree only*. Consequently for linear groups the concepts of "locally soluble" and "soluble" coincide.

3. For all primes $p \neq$ char F linear p-groups are of a simple structure: they are hypercentral and satisfy the minimal condition for subgroups, hence are Černikov p-groups in the terminology of Kuroš. For char $F = 0$ this can be derived from Schur's other main result of 1911, an extension of Jordan's Theorem: a periodic linear group (over $\mathbf{C}$) has an abelian normal subgroup of finite index, bounded by a function of the degree n only. For char $F \neq 0$ the result above is not hard to establish with the help of a theorem of Mal'cev of 1951, [4], in conjunction with Zassenhaus' Theorem under **2**. Another perhaps more lucid approach is based on the existence of monomial representations for linear p-groups whenever $p \neq$ char F. [It was announced at the Conference that Brauer and Feit [1] have recently proved an extended form of Jordan's Theorem for finite linear groups and $p =$ char F: if the Sylow p-subgroups of such a group are of order p^m, then the group has an abelian normal subgroup whose index is bounded by a function of m, n, and p only.]

4. Recent results of K. W. Gruenberg [3] show that the Engel elements in linear groups, and indeed in automorphism groups of finitely generated modules over a noetherian ring, have the best possible structural behavior that can be expected:

The left Engel elements form a group, namely the locally nilpotent (Hirsch–Plotkin) radical $\eta(G)$, and $\eta(G)$ is hypercentral; the bounded left Engel elements form the Baer radical $\beta(G)$ consisting of all subnormal elements, and $\beta(G)$ is nilpotent of finite class; the right Engel elements form a group, namely the hypercenter $\alpha(G)$, and $\alpha(G)$ has central height less than $\omega 2$; the bounded right Engel elements coincide with the elements of $\alpha_\omega(G)$, the ωth term of the upper central series.

The general trend of Mr Wehrfritz' investigations is to show how closely the groups of the class $\mathfrak{P}$ are related to finite groups. Many structural theorems known for finite groups can be extended to groups in $\mathfrak{P}$, but I should like to emphasize at once that the proofs are by no means of a routine transfer nature, as I shall show at least in one example. The relevant definitions (π-subgroup, Sylow p-subgroup, Sylow basis, basis normalizer) carry over verbatim; except that a *Hall π-subgroup* H of $G \in \mathfrak{P}$ has to be defined as a maximal π-subgroup such that for each prime $p \in \pi$ every Sylow p-subgroup of H is also a Sylow p-subgroup of G, and a *Carter subgroup* of G as a *locally* nilpotent self-normalizing subgroup.

When no additional assumptions are made on the groups in $\mathfrak{P}$, the most obvious challenge comes from the Sylow theorems on maximal p-subgroups. These theorems are indeed valid. Once they have been established, one naturally enquires about the existence of normal p-complements, about p-factors, etc.

When the additional assumption is made that the groups under discussion are soluble, then the entire theories of P. Hall and Carter of finite soluble groups go over, with suitable modifications. An interesting feature here is that with induction proofs in mind we take as the appropriate object of study not just the class $\mathfrak{P} \cap \mathfrak{S}$, where $\mathfrak{S}$ is the class of soluble groups, but the wider class $\mathfrak{Q} = \mathsf{Q}(\mathfrak{P} \cap \mathfrak{S})$ of all homomorphic images of periodic soluble linear groups. Wehrfritz has constructed a class of examples of linear p-groups with homomorphic images having no linear representation.

I ought to mention that in contrast to the Hall and Carter theories the closely related Gaschütz theory of formations in finite soluble groups does not appear so far to have its counterpart in the class $\mathfrak{Q}$, mainly owing to the lack of a good and manageable substitute for the Frattini subgroup.

Here is a partial list of theorems that can be proved on the groups in the classes $\mathfrak{P}$ and $\mathfrak{Q}$. The names in square brackets, of course, refer to the finite analogues of these theorems.

A.   Let G be a periodic linear group over an arbitrary field.

1. For every prime p the Sylow p-subgroups of G are conjugate[6] and this property is preserved under homomorphisms.

2. [Wielandt] If G has a locally nilpotent Hall π-subgroup, then all maximal π-subgroups of G are conjugate.

---

[6] Platonov has recently given a sketch of another possible proof of the conjugacy of the Sylow subgroups [5].

3. [Hall] The hypercenter of G is the intersection (a) of the normalizers of all Sylow p-subgroups; (b) of all self-normalizing subgroups of G.

4. [Schur–Zassenhaus] Every normal Hall π-subgroup H of G is complemented and the complements of H are conjugate.

5. Theorems on the existence of normal complements. Let P be a Sylow p-subgroup of G. Normal complements of P exist if

(a) [Burnside] $N_G(P) = C_G(P)$, i.e., P is in the center of its normalizer;

(b) [Frobenius] $N_G(H)/C_G(H)$ is a p-group for every finite p-subgroup H of G;

(c) [Grün] G is p-normal and $N_G(Z(P))$ has a normal p-complement;

(d) [Wielandt] P is regular and $N_G(P)$ has a normal p-complement;

(e) [Thompson] $p \neq \mathrm{char}\ F$ and both $C_G(Z(P))$ and $N_G(J(P))$ have normal p-complements. (The definition of $J(P)$ has to be suitably modified.)

6. A Hall π-subgroup H of G has a normal complement if

(a) [D. G. Higman] H is locally hyperfocal in G;

(b) [Wielandt] H is not a Sylow subgroup of G and $N_G(P) = H$ for all Sylow subgroups P of H;

(c) [Carter] H is a Carter subgroup with regular Sylow subgroups.

B. Let G be a homomorphic image of a souble periodic linear group.

1. For an arbitrary set π of primes the maximal π-subgroups of G are conjugate and so are Hall subgroups. If H is a Hall π-subgroup and H' a Hall π'-subgroup, then $G = HH'$.

2. G has Sylow bases and all such bases are conjugate. The basis normalizers are locally nilpotent. If $\{Q_i\}$ is a set of commuting p-subgroups, one for each prime, then it can be extended to a Sylow basis $\{P_i\}$, $Q_i \leq P_i$.

3. The basis normalizers cover every hypercentral factor of G and avoid every eccentric chief factor of G.

4. No proper normal subgroup of G can contain a basis normalizer; but every maximal nonnormal subgroup contains at least one basis normalizer. The union of the basis normalizers is G, their intersection is the hypercenter.

5. G has Carter subgroups and they are all conjugate. The Carter subgroups are abnormal and each contains a basis normalizer of G.

A Carter subgroup H covers every H-central H-factor of G and avoids every H-irreducible H-eccentric H-factor of G.

6. Hall subgroups, Sylow bases, basis normalizers, and Carter subgroups map correctly under homomorphisms.

As soon as the condition of periodicity is dropped, things go very wrong: for example, the infinite dihedral group

$$D_\infty = \mathrm{gp}\{a, b \parallel b^2 = (ab)^2 = 1\}$$

is soluble and has an absolutely irreducible faithful linear representation, but has two conjugacy classes of Sylow 2-subgroups and Carter subgroups.

Tools in the proofs of many of the preceding theorems are:

(i) Some rudimentary facts from the theory of algebraic groups (such as decomposition into semisimple and unipotent parts, etc.);

(ii) Mal'cev's Theorem of 1951 [4]: a soluble linear group has a triangularizable subgroup of finite index (a result which is, of course, extremely easy to establish with the apparatus of algebraic groups);

(iii) induction on a quantity called the *level* of G.

Level of G. Let $G \subseteq GL(n, F)$. We embed F in a universal field Ω, algebraically closed and of countable transcendence degree over F. With any $G \subseteq GL(n, \Omega)$ we can now associate a pair of integers $\lambda(G) = (d(G), n(G))$. Here $d(G) = \dim \mathfrak{A}(G)_0 \geq 0$ is the dimension of the connected component of the identity in the algebraic hull of G, and $n(G) = |G : G_0| \geq 1$ is the index of G_0 in G. We call $\lambda(G)$ the level of G and well-order all the levels lexicographically.

If G is a finite group, then $\lambda(G) = (0, |G|)$. This gives us a starting point for induction proofs.

The following remark is important. If X is any subset of G not contained in the center $Z(G)$, and $C = C_G(X)$ the centralizer of X in G (or, more generally, if C is any proper subgroup of G that is closed in the induced Zariski topology), then $\lambda(C) < \lambda(G)$. For we always have $d(C) \leq d(G)$, and since C is closed, $d(C) = d(G)$ implies that $C_0 = G_0$, whereas $n(C) = |C : C_0| < |G : G_0| = n(G)$.

As a sample we now prove

Theorem A1. *Let G be a periodic linear group. Then for each prime p the Sylow p-subgroups of G are conjugate.*

Proof.

I.   We first dispose of the simpler case $p = \operatorname{char} F$. Here we can make use of the fact that a subgroup of $GL(n, F)$ is a p-group if and only if it is unitriangularizable. The idea is to find, for any two Sylow p-subgroups S_1 and S_2 of G, two larger subgroups L_1 and L_2 that are known to be conjugate and to contain each a unique Sylow p-subgroup, which must then coincide with S_1 and S_2, respectively. We choose finitely many elements $s_1^1, s_2^1, \ldots \in S_1$ and $s_1^2, s_2^2, \ldots \in S_2$ whose linear hulls contain S_1 and S_2. Then the group $\Gamma = \operatorname{gp}\{s_i^1, s_j^2\}$ is finitely generated, hence finite, and has its Sylow p-subgroups conjugate. So there exist Sylow p-subgroups of Γ, say Σ_1 and Σ_2, such that

$$\operatorname{gp}\{s_i^1\} \subseteq \Sigma_1, \qquad \operatorname{gp}\{s_j^2\} \subseteq \Sigma_2,$$

and an element $x \in \Gamma$ such that $\Sigma_1^x = \Sigma_2$.

We denote by L_1 and L_2 the groups generated by the intersections of G with the linear hulls of Σ_1 and Σ_2: $L_i = \operatorname{gp}\{G \cap L(\Sigma_i)\}$. Then $S_i \subseteq L_i \subseteq G$ and so S_1 and S_2 are Sylow p-subgroups of L_1 and L_2, respectively. But the Sylow p-subgroup of L_i is unique. For, as Σ_i is unitriangularizable, L_i is triangularizable; assuming, without a change of notation, that the triangularizing transformations have already been made, we see that L_i has a unique unitriangular Sylow p-subgroup, which must then be equal to S_i. Therefore $S_1^x = S_2$.

II.   We now come to the case $p \neq \operatorname{char} F$ and use as an inductive hypothesis that the conjugacy theorem holds for groups whose level is less than that of G.

Let S_1 and S_2 be any two Sylow p-subgroups of G. We shall show first of all that we may assume both S_1 and S_2 to be infinite. Suppose that S_1 is finite. If S_2 is infinite, then S_2 being locally finite has a finite subgroup T_2 with $|T_2| > |S_1|$. Now $\operatorname{gp}\{S_1, T_2\}$ is finite and has its Sylow p-subgroups conjugate. But T_2 is a p-group so that a conjugate of T_2 lies in S_1 which gives us the contradiction $|T_2| \leq |S_1|$. So S_2 is also finite and so is $\Gamma = \operatorname{gp}\{S_1, S_2\}$. But then S_1 and S_2 are Sylow p-subgroups of G, hence of Γ, and are therefore conjugate. Let us note that whenever a locally finite group has even one finite Sylow p-subgroup, the conjugacy theorem holds.

We may now assume that S_1 and S_2 are infinite. Let A_1 and A_2 be their smallest subgroups of finite index. Then A_1 and A_2 are abelian, divisible, and nontrivial.

Suppose next that A_1 or A_2 is contained in the center of G, say $A_1 \subseteq Z(G)$. Now $Z(G)$ has a unique Sylow p-subgroup P and $A_1 \subseteq P$. If a group H lies between P and G and H/P is a p-group, so is H itself. But $P \lhd G$, hence $P \subseteq S_1 \cap S_2 \subseteq G$ and $|S_1 : P_1|$ is finite, because $|S_1 : A_1|$ is. Therefore S_1/P is a finite Sylow p-subgroup of G/P and by the remark above the Sylow p-subgroups of G/P are conjugate. The same is then true for G itself.

We may now assume that $A_1, A_2 \nsubseteq Z(G)$. The idea of the following proof is to enforce the conjugacy of the bottom parts A_1 and A_2 first, and afterwards without violating it to make the finite top parts S_1/A_1 and S_2/A_2 also conjugate.

We choose two elements a_1 and a_2 such that

$$a_1 \in A_1, \qquad a_1^{n!} \notin Z(G),$$
$$a_2 \in A_2, \qquad a_2^{(n!)^2} \notin Z(G).$$

This choice is possible because A_1 and A_2 are divisible.

Consider the finite group $\mathrm{gp}\{a_1, a_2\}$. Its Sylow p-subgroups are conjugate. Hence there is an element $x_1 \in G$ such that $\mathrm{gp}\{a_1^{x_1}, a_2\}$ is a p-group, because both generators lie in one and the same Sylow p-subgroup. Now we claim that

$$[(a_1^{x_1})^{n!}, a_2^{n!}] = 1.$$

The reason is that for $p \neq \mathrm{char}\ F$ a p-group has a monomial representation of degree n and then the $n!$th powers of the elements must lie in the stabilizer, which is diagonal, hence abelian.

Let $C_1 = C_G((a_1^{x_1})^{n!})$. Then $C_1 \neq G$ by the choice of a_1 so that $\lambda(C_1) < \lambda(G)$ by the remark on levels. Therefore the Sylow p-subgroups of C_1 are conjugate by the inductive hypothesis. But clearly $A_1^{x_1} \subseteq C_1$ and we have seen above that $a_2^{n!} \in C_1$. So there is an element $x_2 \in G$ such that $\mathrm{gp}\{A_1^{x_1 x_2}, a_2^{n!}\}$ is a p-group.

Now we use once more the existence of a monomial representation for this group and the fact that A_1 is divisible to conclude that $\mathrm{gp}\{A_1^{x_1 x_2}, a_2^{(n!)^2}\}$ is abelian, being contained in the stabilizer. Let $C_2 = C_G(a_2^{(n!)^2})$. Again by the choice of a_2 the inductive hypothesis is applicable and the Sylow p-subgroups of C_2 are conjugate. Now $A_1^{x_1 x_2} \subseteq C_2$ and $A_2 \subseteq C_2$ so that there exists an element $x_3 \in G$ for which $\mathrm{gp}\{A_1^{x_1 x_2 x_3}, A_2\}$ is abelian, by the stabilizer argument and the fact that A_1 and A_2 are divisible.

Now $\mathrm{gp}\{A_1^{x_1 x_2 x_3}, A_2\}$ has a unique smallest divisible subgroup B of finite index and clearly $B = \mathrm{gp}\{A_1^{x_1 x_2 x_3}, A_2\}$. We wish to show that in fact $B = A_1^{x_1 x_2 x_3} = A_2$. Since $|S_i : A_i|$ is finite, we can write $S_i = A_i H$ with H_i finite. Let $\Gamma = \mathrm{gp}\{S_2, B\}$. Now A_2 is normal in S_2 and in B, hence in Γ. But A_2 is a p-group. Hence S_2/A_2 is a finite Sylow p-subgroup of Γ/A_2 and by our earlier remark the conjugacy theorem holds for the Sylow p-subgroups of Γ/A_2. This shows that B/A_2 is a finite p-group. But as a divisible group B has no proper subgroups of finite index and this means that $B = A_1^{x_1 x_2 x_3} = A_2$.

Finally, we look at $\mathrm{gp}\{S_1^{x_1 x_2 x_3}, S_2\} = A_2 K$, where $K = \mathrm{gp}\{H_1^{x_1 x_2 x_3}, H_2\}$ is finite and has conjugate Sylow p-subgroups. Hence there is an element $x_4 \in K$, in fact $x_4 \in N_G(A_2)$, such that $\mathrm{gp}\{H_1^{x_1 x_2 x_3 x_4}, H_2\}$ is a p-group. But then so is $\mathrm{gp}\{H_1^{x_1 x_2 x_3 x_4}, H_2\} \cdot A_2$, and this group contains both $S_1^{x_1 x_2 x_3 x_4}$ and S_2. Consequently

$$S_1^{x_1 x_2 x_3 x_4} = S_2$$

and the proof of the conjugacy theorem is complete.

A closer scrutiny of the proof shows that it is not specifically concerned with linear groups, but is applicable in a more general situation. Let G be a locally finite group and π a set of primes. Suppose that the following conditions are satisfied:

(1) The maximal π-subgroups of every finite factor H/K of G $(1 \subseteq K \lhd H \subseteq G)$ are conjugate.

(2) For each $x \notin Z(G)$ the maximal π-subgroups of the centralizer $C_G(x)$ of x in G are conjugate.

(3) For each $p \in \pi$ the p-subgroups of G satisfy the minimal condition for subgroups.

(4) There exists a positive integer $k(G)$ such that every π-subgroup of G has an abelian normal subgroup of finite index dividing $k(G)$.

The conclusion then is the

Theorem. *The maximal π-subgroups of G are conjugate.*

References

[1] RICHARD BRAUER and WALTER FEIT, An analogue of Jordan's Theorem in characteristic p, *Ann. of Math.* **84** (1966), 119–131.

[2] A. BOREL, Groupes linéaires algébriques, *Ann. of Math.* (2) **64** (1956), 20–82.

[3] K. W. GRUENBERG, The Engel structure of linear groups, *J. Algebra* **3** (1966), 291–303.

[3a] I. KAPLANSKY, *Notes on ring theory*, mimeographed lecture notes written by K. HOFFMAN; re-issued by the University of Chicago, 1965.

[4] A. I. MAL'CEV, On certain classes of infinite soluble groups, *Mat. Sb. N.S.* **28** (70) (1951), 567–588 [Russian]; *Amer. Math. Soc. Transl.* (2) **2** (1956), 1–21.

[5] V. P. PLATONOV, The structure of periodic linear groups and algebraic groups, *Dokl. Akad. Nauk SSSR* **160** (1965), 541–544 [Russian]; *Soviet Math. Doklady* **6** (1965), 144–148.

[6] I. SCHUR, Über Gruppen periodischer Substitutionen, *Sitzungsb. Preuss. Akad. Wiss.* **1911**, 619–627.

[7] D. SUPRUNENKO and V. P. PLATONOV, On a theorem of Schur, *Dokl. Akad. Nauk BSSR* **7** (1963), 510–512.

[8] H. ZASSENHAUS, Beweis eines Satzes über diskrete Gruppen, *Abh. Math. Sem. Univ. Hamburg* **12** (1938), 289–312.

Proc. Internat. Conf. Theory of Groups, Austral. Nat. Univ. Canberra, August 1965, pp. 185–190. © Gordon and Breach Science Publishers, Inc. 1967

Certain groups and homomorphisms associated with a semigroup

R. P. HUNTER* and L. W. ANDERSON*

The present discussion is concerned for the most part with certain groups which are connected with the theory of $\mathcal{H}$-class. Applications of these groups to problems of fiber spaces and homomorphisms will then be considered.

The following definitions will be needed throughout. For any semigroup S the Green equivalences are defined by

$$a \equiv b(\mathcal{R}) \rightleftharpoons a \cup aS = b \cup bS,$$

$$a \equiv b(\mathcal{L}) \rightleftharpoons Sa \cup a = Sb \cup b,$$

$$\mathcal{H} = \mathcal{R} \cap \mathcal{L}, \qquad \mathcal{D} = \mathcal{L} \circ \mathcal{R} = \mathcal{R} \circ \mathcal{L},$$

$$a \equiv b(\mathcal{F}) \rightleftharpoons a \cup Sa \cup aS \cup SaS = b \cup Sb \cup bS \cup SbS.$$

It is to be noted that if S does not have an identity one may adjoin one in the natural way. This adjunction of an identity will not disturb the Green equivalences but gives them a simpler form.

Letting A and B be subsets of S the left, right, and bilateral quotient sets are defined by

$$A \,.\, ^{\cdot} B = \{x \mid Bx \subseteq A\},$$

$$A \,^{\cdot} .\, B = \{y \mid yB \subseteq A\},$$

$$A \,.\,.\, B = \{(x, y) \mid xBy \subseteq A\}.$$

Now let h be any point of S and H_h its $\mathcal{H}$-class. Forming the sub-semigroup $H_h \,.\, ^{\cdot} H_h$ one defines the congruence $\mathcal{S}$ upon this sub-semigroup by

$$x \equiv y(\mathcal{S}) \rightleftharpoons hx = hy.$$

* With support of the National Science Foundation Grant GP 4066.

It is important to note that if H is an $\mathscr{H}$-class and a_0 is any fixed element of H then $H \cdot {}^{\cdot} H = H \cdot {}^{\cdot} a_0$. Moreover the congruence $\mathscr{S}$ may be defined by specifying the action of an element of $H \cdot {}^{\cdot} H$ at a_0 or at all points. The Schützenberger group (or more appropriately the right Schützenberger group) is now defined by

$$H \cdot {}^{\cdot} H / \mathscr{S} \,=\, \Gamma(H).$$

It can be shown that $\Gamma(H)$ is a simply transitive permutation group upon the set H. (Depending upon one's point of view, the Schützenberger group may be thought of as being associated with a single element h, its $\mathscr{H}$-class H_h, its $\mathscr{L}$-class L_h, or indeed its $\mathscr{D}$-class D_h. This is of somewhat more import if one is dealing with a representation theory such as that discussed in "Groups, homomorphisms, and the Green relations".)

It follows now that $\Gamma(H)$ acts as permutation group on the principal left ideal generated by H. In particular $\Gamma(H)$ acts simply upon the $\mathscr{H}$-class of H, each orbit being another $\mathscr{H}$-class. (See, for example [2], [3].)

In a number of problems involving certain fibrations and certain congruences one is led to consider the product of S^*—the dual of S— and S, $S^* \times S$, as a transformation semigroup of S. The action then being $s \cdot (a, b) = asb$. In this context the semigroup $(H \cdot {}^{\cdot} . H)^* \times (H \cdot {}^{\cdot} H)$ acts upon H. Upon this semigroup we now define the congruence Σ by

$$(a, b) \,\equiv\, (c, d)(\Sigma) \,\rightleftharpoons\, ahb \,=\, chd$$

for all $h \in H$.

In order to describe the above quotient semigroup it is convenient to first consider the following simple group-theoretic construction: Let G be any group and $G^* \times G$ the product of its dual with itself. Form the quotient group $G^* \times G / \Delta(Z)$ where Z is the center of G and $\Delta : G \to G^* \times G$ is the homomorphism defined by $\Delta(g) = (g^{-1}, g)$.

Theorem. *With the notation as above, there is an isomorphism between $\sigma(\Gamma(H))$ and the quotient semigroup $(H \cdot {}^{\cdot} . H)^* \times (H \cdot {}^{\cdot} H) / \Sigma$.*

(For the details of this and related results see [4].) The group $\sigma(H)$ may thus be reasonably called the bilateral Schützenberger group.

Before proceeding to consider other groups associated with an $\mathscr{H}$-class we recall the following definition due to Koch and Wallace [11]. The semigroup S is called stable if for $a, b \in S$

(i) $Sa \subseteq Sab$ implies $Sa = Sab$ and

(ii) $aS \subseteq baS$ implies $aS = baS$.

Taking S to be a semigroup with identity, the relation between stability and the above quotient sets is given by the following result [5]:

The semigroup S is stable if and only if for each $a \in S$, $H_a \cdot H_a \times H_a \cdot H_a = H_a \cdot \{a\}$. In particular, in a stable semigroup

$$H \,.\, . \,H = H \,\cdot\, . \,H \times H \,.\,\cdot\, H.$$

Thus, for a stable semigroup, the group $\sigma(H)$ can be given simply as $H \,.\, . \,H / \Sigma$.

One particularly important class of stable semigroups is the class of compact (Hausdorff) semigroups.

The classical example of an unstable semigroup is the so-called bicyclic semigroup $\mathscr{C}(p, q)$, the semigroup generated by the elements p and q subject to the relation that pq be the identity. This semigroup is not even embeddable in a stable semigroup.

We recall that the usual ordering for E—the set of idempotents—is $e \leq f$ if and only if $ef = fe = e$. In [1] it is shown that if S is a compact semigroup then if x is an arbitrary element of S and e is an idempotent, minimal with respect to satisfying $xe = x$, then $H_x \subseteq xH_e$. In other words H_x is contained in the translate of a subgroup of S. Instead of taking the semigroup to be compact it is enough to assume stability and that the semigroup be orbit-compact. The latter notion simply means that the semigroup admits a topology in which each cyclic sub-semigroup has compact closure. This includes, of course, torsion semigroups. However it is to be noted that the compactness will insure the existence of minimal idempotents as considered above. Stability and orbit-compactness will not give such assurance. Consider, for example, the following constructions: Let $[a, b]$ be an arc and c any cutpoint of $[a, b]$. Let $[a, c]$ have the multiplication of the real interval $[\frac{1}{2}, 1]$ where the multiplication is defined by the maximum of $\frac{1}{2}$ and the usual product. Let $[c, b]$ have the multiplication defined by the minimum of two elements, the order being from c to b. If $x < c < y$ then $xy = yx = x$. Form the cartesian product of $[a, b]$ with, say, some finite group G. Let S be the semigroup formed by deleting from $[a, b] \times G$ the maximal subgroup determined by c. Now S has nondegenerate $\mathscr{H}$-classes which have no minimal idempotents. Moreover S is stable and is a torsion semigroup. The above construction may be modified by taking $[a, c]$ and $[c, b]$ as usual unit intervals and deleting from the cartesian product all three maximal subgroups. The resulting semigroup has nondegenerate $\mathscr{H}$-classes which are without any idempotent right identity. In this example stability can be preserved while of course torsion cannot.

In the case of a semigroup having a compact topology the $\mathscr{H}$-class H will be compact, which means that the quotient semigroup $H \,.\, ^{\cdot} H$ will be compact. The congruence $\mathscr{S}$ defined before will now be closed, and so $\Gamma(H)$ is in a natural way a topological compact group. The action upon H is continuous and in particular H is the underlying space of a topological group. Now since the semigroup $H \,.\, ^{\cdot} H$ is compact, it has a completely simple minimal ideal $K(H \,.\, ^{\cdot} H)$. It follows that $K(H \,.\, ^{\cdot} H)$ is the disjoint union of maximal subgroups G_e. The groups G_e are characterized by e being a minimal idempotent right identity for H. (See [1].) Moreover it is known, [1], that each of the groups G_e has the form $e(H \,.\, ^{\cdot} H)e$. It thus follows that the natural homomorphism which maps $H \,.\, ^{\cdot} H$ onto $\Gamma(H)$ maps G_e onto $\Gamma(H)$. The kernel of the homomorphism $\rho\colon G_e \to \Gamma(H)$ is the subgroup $(h \,.\, ^{\cdot} h) \cap G_e$ where $h \in H$.

Thus in the above way one may associate with an element h, or perhaps preferably with its $\mathscr{H}$-class H, the following rather natural groups: The (right) Schützenberger group $\Gamma(H)$ and the group $\sigma(H)$ or bilateral group. At least in the case of a compact group one has furthermore the groups H_e where e is a primitive idempotent in $H \,.\, ^{\cdot} H$, all such groups having the property that $H \subseteq hH_e$. Next one has the subgroup G_e of H_e which has the property that $H = hG_e$. All such groups G_e are isomorphic with one another. Finally there is Ker ρ which is just the stability subgroup of G_e. Naturally, one has the left-handed versions of all of these groups which may be different, and then the group $H^{\cdot} \,.\, H \cap H \,.\, ^{\cdot} H / \mathscr{S}\mathscr{S}$ where $x \equiv y(\mathscr{S}\mathscr{S})$ if and only if $hx = hy$ and $xh = yh$ for all $h \in H$.

Among the various analogues which may be formed by using algebraic conditions we mention the following: If $H \,.\, ^{\cdot} H$ has a primitive idempotent and is regular, then, if the semigroup contains no copy of the bicyclic semigroup, the minimal ideal of $H \,.\, ^{\cdot} H$ exists and is completely simple. In particular in a torsion regular semigroup the discussion involving the groups G_e can be appropriately carried out.

For the various notions involving fiber spaces we follow Cartan [6]. As mentioned before, for a compact semigroup, $\Gamma(H)$ like H is compact. Moreover the $\mathscr{L}$-class of H is compact and the action of $\Gamma = \Gamma(H)$ defines a principal fibration upon L, the base being $L/\mathscr{H}$. This is a starting point from which a number of fibrations can be considered [3]. If L_1 and L_2 are $\mathscr{L}$-classes of the same $\mathscr{D}$-class, the resulting fibrations $L_1 \to L_1/\mathscr{H}$ and $L_2 \to L_2/\mathscr{H}$ are canonically isomorphic. A $\mathscr{D}$-class is given a fibration by both $\mathscr{L}$ and $\mathscr{R}$, and a regular $\mathscr{D}$-class is an associated fiber space of the form (L, R). In the case of a compact inverse semigroup the idempotents in a $\mathscr{D}$-class D form a full cross-section for both $D \to D/\mathscr{L}$ and $D \to D/\mathscr{R}$. Moreover for an $\mathscr{L}$-class L and an $\mathscr{R}$-class R the mapping $T: L \to R$ defined by $T(a) = a^{-1}$ is a homomorphism of L onto R such that $T(ab) = b^{-1}T(a)$.

It is well known that any homomorphism on a compact group defines a fibration. One can show in fact that if K is any compact simple semigroup and α a homomorphism onto a compact group G then K is a fiber space with base G and projection α. The restriction that G be a group is necessary. The condition that the homomorphism be onto is not restrictive since a compact simple subsemigroup of a group is a subgroup (see [4]).

Another approach to fibering a semigroup is given by the core endomorphism. One may define, [10], a homogroup as a semigroup having an ideal as one of its $\mathscr{H}$-classes. This means simply that the minimal ideal exists and is a group. Letting K be this subgroup-ideal and letting e be its identity it follows that $x \to xe$ defines a retracting endomorphism of all of S onto K. (See [10].) Let S be a compact homogroup with identity and ζ the core endomorphism. The set $H_1 J$ receives a fibering, using here $J = \{x \mid xe = e\}$ as fiber. If the core endomorphism maps H_1 onto K, then $S = H_1 J$, so that S itself is fibered. If K is a group and is the orbit of some $\mathscr{H}$-class H, then SH is fibered in rather the same way, with the Schützenberger group now playing a role like that of H_1. Another way in which the core fibration becomes important is the following: If S is a compact semigroup and K is abelian and eH_1 is not totally disconnected, then there exists a congruence on S which is degenerate outside of K and such that the quotient semigroup has the circle group as minimal ideal and the latter is an orbit of H_1. (See [10].) Thus a number of different versions of "principal" fiber space arise quite naturally. The usual version occurs via the action of the Schützenberger groups. Another arises in which the fiber is the underlying space of a semigroup with identity. Still another through a homomorphism of a completely simple semigroup onto a group.

References

[1] L. W. ANDERSON and R. P. HUNTER, The $\mathcal{H}$-equivalence in compact semigroups, *Bull. Soc. Math. Belg.* **14** (1962), 274–296.

[2] L. W. ANDERSON and R. P. HUNTER, The $\mathcal{H}$-equivalence in a compact semigroup II, *J. Austral. Math. Soc.* **3** (1963), 288–293.

[3] L. W. ANDERSON et R. P. HUNTER, Sur les espaces fibrés associés à une $\mathcal{D}$-classe d'un demi-groupe compact, *Bull. Acad. Polon. Sci. Sér. Sci. Math. Astronom. Phys.* **12** (1964), 249–251.

[4] L. W. ANDERSON et R. P. HUNTER, Une version bilatère du groupe de Schützenberger, *Bull. Acad. Polon. Sci. Sér. Math. Astronom. Phys.* **13** (1965), 527–531.

[5] L. W. ANDERSON, R. P. HUNTER, and R. J. KOCH, Some results on stability in semigroups, *Trans. Amer. Math. Soc.* **117** (1965), 521–529.

[6] H. CARTAN, Généralités sur les espaces fibrés, *Séminaire Henri Cartan de l'Ecole Normale Supérieure*, 1949/1950. *Espaces fibrés et homotopie*. 2ème éd., Paris, 1956.

[7] A. H. CLIFFORD and D. D. MILLER, Semigroups having zeroid elements, *Amer. J. Math.* **70** (1948), 117–125.

[8] A. H. CLIFFORD and G. B. PRESTON, *The algebraic theory of semigroups*, Vol. I, Math. Surveys No. 7, Amer. Math. Soc., Providence, R.I., 1961.

[9] J. A. GREEN, On the structure of semigroups, *Ann. of Math.* **54** (1951), 163–172.

[10] R. P. HUNTER, On the structure of homogroups with applications to the theory of compact connected semigroups, *Fund. Math.* **52** (1963), 69–102.

[11] R. J. KOCH and A. D. WALLACE, Stability in semigroups, *Duke Math. J.* **24** (1957), 193–195.

[12] M. P. SCHÜTZENBERGER, $\overline{\mathcal{D}}$ représentations des demi-groupes, *C.R. Acad. Sci. Paris* **244** (1957), 1994–1996.

[13] GABRIEL THIERRIN, Contributions à la théorie des équivalences dans les demi-groupes, *Bull. Soc. Math. France* **83** (1955), 103–159.

Proc. Internat. Conf. Theory of Groups, Austral. Nat. Univ. Canberra, August 1965, pp. 191–202. © Gordon and Breach Science Publishers, Inc. 1967

On transitive permutation groups of Fermat prime degree

NOBORU ITO

The purpose of this paper is to prove the following theorem.

Theorem. *Let Ω be the set of symbols $1, 2, \ldots, p$, where p is a Fermat prime: $p = 2^m + 1$. Let $\mathfrak{G}$ be a nonsolvable transitive permutation group over Ω. Then if $\mathfrak{G}$ contains an odd permutation, $\mathfrak{G}$ coincides with the symmetric group $\mathbf{S}$ over Ω.*

Notation. Let $\mathfrak{X}$ be a subgroup of $\mathfrak{G}$. Then

$Ns\mathfrak{X}$ = the normalizer of $\mathfrak{X}$ in $\mathfrak{G}$;

$Cs\mathfrak{X}$ = the centralizer of $\mathfrak{X}$ in $\mathfrak{G}$;

$Z(\mathfrak{X})$ = the center of $\mathfrak{X}$;

$\mathfrak{X}'$ = the commutator subgroup of $\mathfrak{X}$;

$\mathfrak{X}(l)$ = a Sylow l-subgroup of $\mathfrak{X}$, where l is a prime;

$A(\mathfrak{X})$ = the group of automorphisms of $\mathfrak{X}$;

$\mathfrak{X}_1$ = the maximal subgroup of $\mathfrak{X}$ leaving the symbol 1 of Ω fixed;

$\mathfrak{X}_{1,2}$ = the maximal subgroup of $\mathfrak{X}$ leaving the symbols 1 and 2 of Ω fixed individually.

§1. Roughly speaking, the proof of the theorem is divided into two stages. The purpose of the first stage is to prove the following fact: under the assumptions of the theorem, $\mathfrak{G}$ contains an involution with appropriate conditions (see the condition (∗) in (iii)).

(i) If $m = 2$, then it is seen at once that $\mathfrak{G}$ coincides with $\mathbf{S}$. Therefore it will be assumed hereafter that $m \geq 4$.

(ii) Let $\mathfrak{N}$ be the subgroup of $\mathfrak{G}$ consisting of all the even permutations in $\mathfrak{G}$. Then $\mathfrak{N}$ is also nonsolvable and hence by a well-known theorem of

14—к.   191

Burnside ([2], p. 341) $\mathfrak{N}$ is doubly transitive on Ω. Set $\mathfrak{P} = \mathfrak{G}(p)$. Then using a theorem of Sylow, it is obtained that $\mathfrak{G} = (Ns\mathfrak{P})\mathfrak{N}$. Hence $Ns\mathfrak{P}$ must contain an odd permutation. Set $\mathfrak{Q} = (Ns\mathfrak{P})_1$. Then because of $\mathfrak{P} = Cs\mathfrak{P}$, which is obvious, it is easily seen that $\mathfrak{Q}$ is a cyclic group of order 2^m generated by a 2^m-cycle Q. Since $\mathfrak{P}\mathfrak{Q}$ is a sharply doubly transitive subgroup of $\mathfrak{G}$, by a theorem of Wielandt ([8], Theorem 27.1) $\mathfrak{G}$ is triply transitive on Ω. Set $\mathfrak{H} = \mathfrak{G}_1$ and $\mathfrak{K} = \mathfrak{G}_{1,2}$. Then the following factorization of $\mathfrak{H}$ is obtained: $\mathfrak{H} = \mathfrak{Q}\mathfrak{K}$, $\mathfrak{Q} \cap \mathfrak{K} = 1$. There exists an $\mathfrak{S} = \mathfrak{G}(2)$ such that it contains $\mathfrak{Q}$ and it is contained in $\mathfrak{H}$. From the above factorization it follows that $\mathfrak{S} = \mathfrak{Q}\mathfrak{S}_{1,2}$ and $\mathfrak{Q} \cap \mathfrak{S}_{1,2} = 1$. Since $\mathfrak{Q}$ is transitive on $\Omega - \{1\}$, $Z(\mathfrak{S})$ must be contained in $\mathfrak{Q}$. It is easily seen that $Cs\mathfrak{Q} = \mathfrak{Q}$. On the other hand, it is well known that $A(\mathfrak{Q})$ is abelian of type $(2^{m-2}, 2)$ ([2], p. 115). Therefore it may be assumed that $\mathfrak{S} \supseteq Ns\mathfrak{Q} = \mathfrak{Q}(\mathfrak{S}_{1,2} \cap Ns\mathfrak{Q})$. Put $\mathfrak{R} = \mathfrak{S}_{1,2} \cap Ns\mathfrak{Q}$. Then $\mathfrak{R}$ is isomorphic to a subgroup of $A(\mathfrak{Q})$. If $\mathfrak{R} = 1$, then $Ns\mathfrak{Q} = \mathfrak{Q}$ and this implies that $\mathfrak{Q} = \mathfrak{S}$. But it contradicts the double transitivity of $\mathfrak{N}$. Thus it must hold that $\mathfrak{R} \neq 1$.

(iii) Now we claim that

(∗) $\mathfrak{R}$ contains an involution J such that $JQJ = Q^{1+2^{m-1}}$.

In order to get a contradiction, it will be assumed that the opposite holds. Then $\mathfrak{R}$ must be cyclic and it must hold that either $JQJ = Q^{-1}$ or $JQJ = Q^{-1+2^{m-1}}$, where J is the involution of $\mathfrak{R}$ ([2], p. 135). J leaves the symbols 1 and 2 of Ω fixed. Then, without loss of generality, it may be set $Q = (2, 3, \ldots, 2^m, p)$. If J leaves the symbol j ($\neq 1, 2$) of Ω fixed, then $Q^{j-2} = (2, j, \ldots) \cdots$ and $JQ^{j-2}J = (2, j, \ldots) \cdots$. Since Q is regular on $\Omega - \{1\}$, this implies that $Q^{j-2} = JQ^{j-2}J$. Then it is easily seen that $Q^{j-2} = Q^{2^{m-1}}$, which implies that $j = 2 + 2^{m-1}$. Conversely, since J is commutative with $Q^{2^{m-1}}$, J leaves the symbol $2 + 2^{m-1}$ of Ω fixed. Thus the involution J leaves just three symbols of Ω fixed, and therefore it is an odd permutation. But if the order of $\mathfrak{R}$ is not smaller than four, J must be an even permutation. Therefore the order of $\mathfrak{R}$ must be equal to two.

It will be shown that $\mathfrak{S} \cap Cs\langle Q^2 \rangle = Q$. Indeed, if $\mathfrak{S} \cap Cs\langle Q^2 \rangle \not\supseteq \mathfrak{Q}$, then $\mathfrak{S} \cap Cs\langle Q^2 \rangle \cap Ns\mathfrak{Q} \not\supseteq \mathfrak{Q}$. Choose a permutation S of $\mathfrak{S} \cap Cs\langle Q^2 \rangle \cap Ns\mathfrak{Q}$ such that $S \notin \mathfrak{Q}$ and $S^2 \in \mathfrak{Q}$. Then it follows that $S \in Ns\mathfrak{Q}$ and $S^{-1}Q^2S = Q^2$, which implies that $JQ^2J = Q^2$. Because of $m \geq 4$, this is a contradiction. Thus, in particular, $\mathfrak{Q}$ is normal in $\mathfrak{S} \cap Ns\langle Q^2 \rangle$ and $\mathfrak{S} \cap Ns\langle Q^2 \rangle = \mathfrak{S} \cap Ns\mathfrak{Q} = Ns\mathfrak{Q}$.

Now if it is assumed that $\mathfrak{Q}\mathfrak{R} \nsubseteq \mathfrak{S}$, then it follows that $\mathfrak{S} \cap Ns(\mathfrak{Q}\mathfrak{R}) \nsupseteq \mathfrak{Q}\mathfrak{R}$. But since $(\mathfrak{Q}\mathfrak{R})' = \langle Q^2 \rangle$, it follows that $\mathfrak{S} \cap Ns(\mathfrak{Q}\mathfrak{R}) = \mathfrak{S} \cap Ns\langle Q^2 \rangle = \mathfrak{S} \cap Ns\mathfrak{Q} = \mathfrak{Q}\mathfrak{R}$. This is a contradiction. Thus it must hold that $\mathfrak{S} = \mathfrak{Q}\mathfrak{R}$.

(iv) Let $\mathfrak{M}$ be a minimal normal subgroup of $\mathfrak{G}$. Then since $\mathfrak{G}$ is doubly transitive on Ω, $\mathfrak{M}$ is transitive on Ω and hence $\mathfrak{M}$ contains $\mathfrak{P}$. Since $\mathfrak{M}$, as a minimal normal subgroup, is a direct product of isomorphic simple groups, this implies that $\mathfrak{M}$ is simple. Now using a theorem of Sylow, it is obtained that $\mathfrak{G} = (Ns\mathfrak{P})\mathfrak{M} = \mathfrak{Q}\mathfrak{M}$. Thus $\mathfrak{M}$ is nonsolvable and therefore $\mathfrak{M}$ is doubly transitive on Ω by the theorem of Burnside cited above. Consider $\mathfrak{M}(2) \subseteq \mathfrak{S}$. Then $\mathfrak{M}(2)$ is of order divisible by 2^m and it consists of even permutations only. Hence it holds that $\mathfrak{M}(2) = \langle Q^2, JQ \rangle$. It holds that $(JQ)^{-1}Q^2(JQ) = Q^{-2}$. If JQ is not an involution, then $\mathfrak{M}(2)$ is a generalized quaternion group. But since $\mathfrak{M}$ is simple, by a theorem of Brauer and Suzuki [1] this is impossible. Therefore JQ must be an involution and $\mathfrak{M}(2)$ is a dihedral group.

By using the transfer theorem of Grün ([9], p. 134) it is easily seen that $\mathfrak{M} \cap Cs\langle Q^{2^{m-1}} \rangle$ is 2-nilpotent. Furthermore, since $\mathfrak{M}(2)$ $(\subseteq Cs\langle Q^{2^{m-1}} \rangle)$ is transitive on $\Omega - \{1\}$, it follows that $\mathfrak{M} \cap Cs\langle Q^{2^{m-1}} \rangle = \mathfrak{M}(2)$.

Now by a result of Gorenstein and Walter ([4], Theorem 1) $\mathfrak{M}$ is isomorphic to a linear fractional group $LF(2, q)$ with q odd or to the alternating group of degree seven. But since $\mathfrak{M}$ is a transitive permutation group of Fermat prime degree, by a theorem of Galois ([3], Section 262) it is easily seen that $\mathfrak{M}$ is isomorphic to $LF(2, 5)$ against the assumption $m \geq 4$. Thus it has been proved that

(*) $\mathfrak{R}$ contains an involution J such that $JQJ = Q^{1+2^{m-1}}$.

§2. Now we are in the second stage of the proof. Let **A** be the alternating group over Ω. Then, first of all, we notice the following result of W. A. Manning ([6] I, Theorem; II, Theorems 1 and 2): if the minimal degree of $\mathfrak{G}$ is less than $(p+1)/3$, then $\mathfrak{G}$ contains **A**. Hence in order to prove the theorem, it suffices to show that the subgroup $\langle P, J \rangle$ contains a permutation $(\neq 1)$ whose degree is less than $(p+1)/3$, where P is a generator of $\mathfrak{P}$.

(i) In order to facilitate the calculation, taking $\mathbf{Z}(\bmod p)$, where $\mathbf{Z}$ denotes the domain of rational integers, as Ω, P, Q, and J will be represented analytically. $P(x) = x + 1$ is taken as analytic form of P. Let $Q(x)$ be the analytic form of Q. Then since $\mathfrak{P}$ is transitive on $\mathbf{Z}(\bmod p)$, it can be assumed that $Q(0) = 0$. It holds that $PQ = QP^g$, where g is a primitive root modulo p. From this it follows that $Q(x+1) = Q(x) + g$. Then it is easily seen that $Q(x) = gx$. Now let $J(x)$ be the analytic form of J. Then the relation $JQ^2 = Q^2 J$ implies that $J(0) = 0$. Since Ω is transitive on $\mathbf{Z}(\bmod p) - \{0\}$, it may be assumed that $J(1) = 1$. Now the relation $JQ = Q^{1 + 2^{m-1}} J$ implies that $J(gx) = -gJ(x)$. Hence it holds that $J(x) = x$ if x is a quadratic residue modulo p or $-x$ if x is a quadratic nonresidue modulo p. Here it is noticed that -1 is a quadratic residue modulo p.

(ii) Let s be a function on $\mathbf{Z}(\bmod p)$ such that $s(a) = +$ if a is a quadratic residue modulo p or $= -$ if a is a quadratic nonresidue modulo p. 0 will be regarded as a residue. But we must be always rather careful about 0. Let $(a_1, a_2, \ldots, a_r)$ be an ordered sequence in $\mathbf{Z}(\bmod p)$. Then it will be defined that

$$s(a_1, a_2, \ldots, a_r) = (s(a_1), s(a_2), \ldots, s(a_r)).$$

Now it might be convenient to have the following figure in mind.

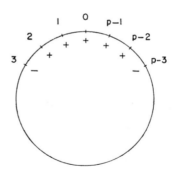

(iii) The cycle structure of the permutation PJ will be examined. Here it is noticed that because of $m \geq 4$, 2 is a quadratic residue modulo p. Only the cases which will be needed explicitly will be listed. They reveal, however, that the situation is rather mechanical.

If $s(x, x+1) = (-, -)$, then the cycle structure of PJ contains the transposition $(x, -(x+1))$. Indeed, since $\frac{1}{2}(p-1)$ is a quadratic residue modulo p, it follows that $x \neq -(x+1)$.

If $s(x, x+1, x+2) = (-, +, -)$, then the cycle structure of PJ contains the 4-cycle $(x, x+1, -(x+2), -(x+1))$. Indeed, since $p-1$ is a residue, it follows that $x \neq -(x+2)$.

If $s(x, x+1, x+2, x+3) = (-, +, +, -)$, then the cycle structure of PJ contains the 6-cycle $(x, x+1, x+2, -(x+3), -(x+2), -(x+1))$. The exception occurs if $x = \frac{1}{2}(p-3)$, since 3 is a nonresidue modulo p. In the exceptional case the cycle structure of PJ contains the 3-cycle $(x, x+1, x+2)$.

If $s(x, x+1, x+2, x+3, x+4) = (-, +, +, +, -)$, then the cycle structure of PJ contains the 8-cycle

$$(x, x+1, x+2, x+3, -(x+4), -(x+3), -(x+2), -(x+1)).$$

Indeed, since $p-2$ is a residue, it follows that $x \neq -(x+4)$.

If $s(x, x+1, \ldots, x+5, x+6) = (-, +, \ldots, +, -)$, then the cycle structure of PJ contains the 12-cycle

$$(x, x+1, \ldots, x+5, -(x+6), -(x+5), \ldots, -(x+1)).$$

The exception occurs if $x = p-3$, in which case the cycle structure of PJ contains the 6-cycle $(x, x+1, \ldots, x+5)$.

If $s(x, x+1, \ldots, x+11, x+12) = (-, +, \ldots, +, -)$, then the cycle structure of PJ contains the 24-cycle

$$(x, x+1, \ldots, x+11, -(x+12), -(x+11), \ldots, -(x+1)).$$

Indeed, the exception might occur when $x = p-6$. But since $(p-6)+9 = 3$ is a nonresidue, it does not occur.

(iv) Let A_j be the number of different x's such that

$$s(x, x+1, \ldots, x+j, x+j+1) = (-, \underbrace{+, \ldots, +}_{j}, -) \qquad (j = 0, 1, 2, \ldots).$$

Similarly let B_j be the number of different x's such that

$$s(x, x+1, \ldots, x+j, x+j+1) = (+, \underbrace{-, \ldots, -}_{j}, +) \qquad (j = 0, 1, 2, \ldots).$$

Then the following two equalities hold:

(1.1) $$A_0 + A_1 + A_2 + \cdots = \tfrac{1}{2}(p-1)$$

and

(1.2) $$B_0 + B_1 + B_2 + \cdots = \tfrac{1}{2}(p+1).$$

Consider an x such that

$$s(x, x+1, \ldots, x+j, x+j+1) = (-, \underbrace{+, \ldots, +}_{j}, -),$$

and let $(x, x+1, \ldots, x+j)$ correspond to x $(j=0, 1, 2, \ldots)$. Then the following equality is obtained:

(2.1) $$A_0 + 2A_1 + 3A_2 + \cdots = p.$$

Similarly it is obtained that

(2.2) $$B_0 + 2B_1 + 3B_2 + \cdots = p.$$

(v) **Lemma 1.** *It holds that*

(3.1) $$A_0 = \tfrac{1}{4}(p-1)$$

and

(3.2) $$B_0 = \tfrac{1}{4}(p+3).$$

Proof ([3], Section 64). In fact, A_0 is equal to a quarter of the number of solutions in $\mathbf{Z}(\bmod p)$ of the equation $nX^2 + 1 = nY^2$, where n is a nonresidue modulo p. The equality (3.2) can be similarly proved.

(vi) Because of $m \geq 4$, 2 is a residue modulo p. Hence by the similar argument as in (v) ([3], Section 64) it will be seen that the number of different x's such that $s(x, x+1, x+2) = (-, *, -)$ is equal to $\tfrac{1}{4}(p-1)$, and that of different x's such that $s(x, x+1, x+2) = (+, *, +)$ is equal to $\tfrac{1}{4}(p+3)$. Then the following two equalities can be easily obtained:

(4.1) $$A_1 + \sum_{j \geq 3} (j-2)B_j = \tfrac{1}{4}(p-1)$$

and

(4.2) $$B_1 + \sum_{j \geq 3} (j-2)A_j = \tfrac{1}{4}(p+3).$$

Substituting (3.1) into (1.1),

(5) $$A_1 + A_2 + \cdots = \tfrac{1}{4}(p-1).$$

Subtracting (1.1) from (2.1),

(6) $$A_1 + 2A_2 + \cdots = \tfrac{1}{2}(p+1).$$

Subtracting (5) from (6),

(7) $$A_2 + 2A_3 + \cdots = \tfrac{1}{4}(p+3).$$

Substituting (7) into the right-hand side of (5),

(8.1) $$1 + A_1 = A_3 + 2A_4 + 3A_5 + \cdots.$$

Similarly it is obtained that

(8.2) $$B_1 = B_3 + 2B_4 + 3B_5 + \cdots.$$

From (4.1) and (8.2) it follows that

(9) $$A_1 + B_1 = \tfrac{1}{4}(p-1).$$

(vii) **Lemma 2.** *It holds that*

(10) $$A_1 = B_1 = (p-1)/8.$$

Proof. On account of (9) it suffices to prove that $A_1 = B_1$. Now as in (v) it is easily seen that A_1 is equal to one eighth of the number of solutions in $\mathbf{Z}(\bmod p)$ of the simultaneous equations

(11.1) $$\begin{cases} Y^2 - nX^2 = 1 \\ nZ^2 - Y^2 = 1 \end{cases} \quad \text{where } n \text{ is a nonresidue.}$$

Similarly the following simultaneous equations have to be considered for B_1:

(11.2) $$\begin{cases} nV^2 - U^2 = 1 \\ W^2 - nV^2 = 1 \end{cases} \quad \text{where } n \text{ is a nonresidue.}$$

But (11.2) has solutions of the form $(U, V, W) = (*, 0, *)$, which do not correspond to B_1. Thus in this case $8B_1 + 4$ is equal to the number of solutions of (11.2) in $\mathbf{Z}(\bmod p)$.

Now let K denote the number of solutions of (11.1) in $\mathbf{Z}(\bmod p)$. Then it will be shown that $K \equiv -1 (\bmod p)$. For this purpose, the following theorem of Segre ([7], Teorema 1) is available: set

$$g(X, Y, Z) = \{(Y^2 - nX^2 - 1)^{p-1} - 1\}\{(nZ^2 - Y^2 - 1)^{p-1} - 1\}.$$

A term $X^j Y^k Z^l$ of $g(X, Y, Z)$ is called of type ω if j, k, and l are positive multiples of $p-1$. Then it holds that the sum of coefficients of terms of type ω is congruent to $-K$ modulo p.

Now it is easily seen that

$$g(X, Y, Z) = \sum_{a=1}^{p-1} \sum_{b=1}^{p-1} \binom{p-1}{a} \binom{p-1}{b} \sum_{c=0}^{a} \sum_{d=0}^{b} (-1)^{a+b-c-d}$$
$$\times \binom{a}{c}\binom{b}{d} (nX^2)^c (Y^2)^{a-c+d} (nZ^2)^{b-d}.$$

Therefore a term of $g(X, Y, Z)$ is of type ω if and only if

(12)
$$\begin{cases} c = \tfrac{1}{2}(p-1) & \text{or} \quad p-1, \\ a-c+d = \tfrac{1}{2}(p-1) & \text{or} \quad p-1, \\ b-d = \tfrac{1}{2}(p-1) & \text{or} \quad p-1. \end{cases}$$

(12) has three "easy" solutions:

   (I) $a = b = p-1$, $c = \tfrac{1}{2}(p-1)$, $d = 0$,
   (II) $a = b = p-1$, $c = d = \tfrac{1}{2}(p-1)$,
   (III) $a = b = c = p-1$, $d = \tfrac{1}{2}(p-1)$.

The coefficients of terms of $g(X, Y, Z)$ corresponding to (I), (II), and (III) are congruent to -1, 1, and -1 modulo p, respectively. Further there also exists another solution of (12) of the form $a+d=p-1$, $b-d=c=\tfrac{1}{2}(p-1)$. The sum of coefficients of terms of $g(X, Y, Z)$ corresponding to these solutions is easily seen to be equal to

$$S = \sum_{a=\frac{1}{2}(p-1)}^{p-1} (-1)^{a-\frac{1}{2}(p-1)} \binom{p-1}{a} \binom{a}{\frac{1}{2}(p-1)} \binom{p-1}{3(p-1)/2-a}$$
$$\times \binom{3(p-1)/2-a}{p-1-a}$$
$$= \sum_{a=\frac{1}{2}(p-1)}^{p-1} (-1)^{a-\frac{1}{2}(p-1)} \binom{\frac{1}{2}(p-1)}{a-\frac{1}{2}(p-1)}^2$$
$$= \sum_{b=0}^{\frac{1}{2}(p-1)} (-1)^b \binom{\frac{1}{2}(p-1)}{b}^2.$$

In the identity $(X-1)^m (1/X+1)^m = (X^2-1)^m / X^m$ comparing the coefficients of constant terms of both sides, it is obtained that

$$\sum_{c=0}^{m} (-1)^c \binom{m}{c}^2 = (-1)^{\frac{1}{2}m} \binom{m}{\frac{1}{2}m}.$$

Thus it follows that

$$S = (-1)^{\frac{1}{4}(p-1)}\binom{\frac{1}{2}(p-1)}{\frac{1}{4}(p-1)}.$$

Now about this number the following theorem of Eisenstein ([5], Section 5, Zusatz) is available: let $\mathbf{Q}$ be the field of rational numbers and let $\mathbf{i}$ be the imaginary unit. p can be decomposed into the product of two primary prime numbers in the field $\mathbf{Q}(\mathbf{i})$: $p = p_1 p_2$, where $p_1 = 1 + 2^{\frac{1}{2}m}\mathbf{i}$ and $p_2 = 1 - 2^{\frac{1}{2}m}\mathbf{i}$. Then it holds that

(13) $$p_2 \equiv S \pmod{p_1}.$$

Multiplying (13) with p_2,

$$p_2^2 \equiv p_2 S \pmod{p}.$$

Because of $p = 2^m + 1$, it follows that

$$(S-2)p_2 \equiv 0 \pmod{p}.$$

This implies that

$$S - 2 \equiv 0 \pmod{8_1}.$$

Since S is rational,

$$S - 2 \equiv 0 \pmod{p_2}.$$

Hence finally it is obtained that

$$S - 2 \equiv 0 \pmod{p}.$$

Thus it has been proved that

$$-K \equiv -1 + 1 - 1 + 2 = 1 \pmod{p}.$$

Set $8A_1 = K = K^*p - 1$. Then K^* is a positive integer. On account of (9) it is obtained that $8B_1 = (2 - K^*)p - 1$. This implies that $K^* = 1$. The proof of Lemma 2 is now completed.

(viii) Substituting (10) into (5) and (8.1), it is obtained that

(14) $$A_2 + A_3 + \cdots = (p-1)/8$$

and

(15) $$A_3 + 2A_4 + \cdots = (p+7)p/.$$

Eliminating A_3 from (14) and (15),

(16) $$1 + A_2 = A_4 + 2A_5 + \cdots.$$

Finally from (14) and (16) follows the inequality:

(17) $1 + 2A_2 + A_3 \geq (p-1)/8$.

(ix) Now assume that $(PJ)^{24} \neq 1$. Then by (iii) $(PJ)^{24}$ leaves at least $A_0 + 2A_1 + 3 + 3(A_2 - 1) + 4A_3$ symbols of Ω fixed. By (3.1), (10) and (17) it follows that

$$A_0 + 2A_1 + 3A_2 + 4A_3 \geq \tfrac{1}{4}(p-1) + \tfrac{1}{4}(p-1) + 3(p-9)/16$$
$$= (11p - 35)/16.$$

Thus the minimal degree of $\mathfrak{G}$ is less than $p - (11p - 35)/16 = (5p + 35)/16$. If $(p+1)/3 > (5p+35)/16$, then, as it is remarked at the beginning of this section, by a result of Manning cited there, $\mathfrak{G}$ coincides with $\mathbf{S}$. Thus it can be assumed that $(p+1)/3 \leq (5p+35)/16$. This implies that $p \leq 89$. Since p is a Fermat prime bigger than 5, $p = 17$ is only possible. Then it is not so difficult to check that $\mathfrak{G}$ coincides with $\mathbf{S}$. Therefore it can be assumed that $(PJ)^{24} = 1$. From the observation in (iii) it is easily seen that every A_j must vanish except for $j = 0, 1, 2, 3, 5$ and 11. Therefore using (3.1) and (10), it is obtained from (1.1) and (2.1) that

(18) $A_2 + A_3 + A_5 + A_{11} = (p-1)/8$

and

(19) $3A_2 + 4A_3 + 6A_5 + 12A_{11} = \tfrac{1}{2}(p+1)$.

Subtracting three times of (18) from (19),

(20) $A_3 + 3A_5 + 9A_{11} = (p+7)/8$.

(x) $A_{11} = 0$ can be assumed. In fact, assume that A_{11} does not vanish. Then it follows that $(PJ)^{12} \neq 1$. $(PJ)^{12}$ leaves at least $A_0 + 2A_1 + 3A_2 + 6A_5$ symbols of Ω fixed. Thus the minimal degree of $\mathfrak{G}$ is not greater than

$$p - (A_0 + 2A_1 + 3A_2 + 6A_5) = \tfrac{1}{2}(p-1) - 3A_2 - 6A_5.$$

If this number is less than $(p+1)/3$, again by a result of Manning cited above $\mathfrak{G}$ coincides with $\mathbf{S}$. Therefore it can be assumed that

$$(p+1)/3 \leq \tfrac{1}{2}(p-1) - 3A_2 - 6A_5,$$

namely,

(21) $3A_2 + 6A_5 \leq (p-5)/6$.

Substituting (21) into (19), it is obtained that

(22) $$4A_3 + 12A_{11} \geq (p+4)/3.$$

Similarly because of $A_{11} \neq 0$ it follows that $(PJ)^8 \neq 1$. $(PJ)^8$ leaves at least $A_0 + 2A_1 + 4A_3$ symbols of Ω fixed. Thus the minimal degree of $\mathfrak{G}$ is not greater than

$$p - (A_0 + 2A_1 + 4A_3) = \tfrac{1}{2}(p+1) - 4A_3.$$

If this number is less than $(p+1)/3$, then by a result of Manning cited above $\mathfrak{G}$ coincides with S. Therefore, it can be assumed that

$$(p+1)/3 \leq \tfrac{1}{2}(p+1) - 4A_3,$$

or

(23) $$4A_3 \leq (p+1)/6.$$

Substituting (23) into (22), it is obtained that

(24) $$9A_{11} \geq (p+7)/8.$$

Then from (20) it follows that $A_5 = 0$. On the other hand, since $s(-3, -2, -1, 0, 1, 2, 3) = (-, +, +, +, +, +, -)$, it must hold that $A_5 = 1$. This is a contradiction. Therefore $A_{11} = 0$ can be assumed.

(xi) $A_3 = 0$ can be assumed. First it is noticed that since $A_5 = 1$ it follows that $(PJ)^8 \neq 1$. Therefore, as in (x) it can be assumed that

(23) $$4A_3 \leq (p+1)/6.$$

Now assume that $A_3 \geq 1$. Then by (iii) it follows that $(PJ)^{12} \neq 1$. Therefore, since $A_{11} = 0$, from (22) it can be assumed that

$$4A_3 \geq (p+4)/3,$$

which contradicts (23). Thus $A_3 = 0$ can be assumed.

(xii) If $p = 17$, then it is not so difficult to check that $\mathfrak{G}$ coincides with S. Hence it may be assumed that $p > 17$. If $p = 257$, then

$$s(14, 15, 16, 17, 18, 19) = (-, +, +, +, +, -),$$

which contradicts the assumption $(PJ)^{24} = 1$. Thus it may be assumed that $p > 257$.

Set $F_1 = 5$ and $F_j = (F_{j-1} - 1)^2 + 1$ $(j = 2, 3, \ldots)$. Then the set of F_j's contains all Fermat primes (≥ 5). Now it is not so troublesome to check the following facts:

(a) Every F_j $(j \geq 4)$ is a quadratic residue modulo 13, 17, and 97, respectively. (For instance, $F_4 \equiv 62$ (mod 97), $F_5 \equiv 36$ (mod 97), $F_6 \equiv 62$ (mod 97), ...)

(b) Every F_j $(j \geq 4)$ is a quadratic nonresidue modulo 3, 5, 7, and 41, respectively.

(c) F_j $(j \geq 4)$ is a quadratic residue modulo 19 if and only if it is a nonresidue modulo 29.

(d) The residue–nonresidue property of F_j $(j \geq 4)$ is just the same for 11, 31, and 61.

Since $s(14, 15, 16, 17, 18) = (-, +, +, +, +)$, by assumption it must hold that $s(19) = +$. By (c) this implies that $s(29) = -$. If $s(61) = +$, then by (d) $s(58, 59, 60, 61, 62, 63) = (-, *, +, +, +, -)$, which contradicts either $(PJ)^{24} = 1$ or $A_3 = 0$. Thus $s(11) = s(31) = s(61) = -$. Now since $s(286, 287, 288, 289, 290) = s(13.11.2, 41.7, 3^2.2^5, 17^2, 29.5.2) = (-, +, +, +, +)$, it must hold that $s(291) = s(97.3) = +$. This implies that $s(97) = -$. This contradicts (a).

Remark. The second stage of the proof may be modified to hold for a wider class of prime numbers.

References

[1] RICHARD BRAUER and MICHIO SUZUKI, On finite groups of even order whose 2-Sylow group is a quaternion group, *Proc. Nat. Acad. Sci. U.S.A.* **45** (1959), 1757–1759.

[2] W. BURNSIDE, *Theory of groups of finite order*, 2nd ed., Cambridge Univ. Press, Cambridge, 1911.

[3] L. E. DICKSON, *Linear groups with an exposition of the Galois field theory*, Teubner, Leipzig, 1901.

[4] DANIEL GORENSTEIN and JOHN H. WALTER, On finite groups with dihedral Sylow 2-subgroups, *Illinois J. Math.* **6** (1962), 553–593.

[5] G. EISENSTEIN, Einfacher Beweis und Verallgemeinerung des Fundamentaltheorems für die biquadratischen Reste, *J. Reine Angew. Math.* **28** (1844), 223–245.

[6] W. A. MANNING, The degree and class of multiply transitive groups, I, *Trans. Amer. Math. Soc.* **18** (1917), 463–479; II, *Trans. Amer. Math. Soc.* **31** (1929), 643–653.

[7] BENIAMINO SEGRE, Sulla teoria delle equazioni e delle congruenze algebriche, II, *Atti Accad. Naz. Lincei. Rend. Cl. Sci. Fis. Mat. Nat.* (8) **27** (1959), 303–311.

[8] HELMUT WIELANDT, *Finite permutation groups*, Academic Press, New York, 1964.

[9] H. ZASSENHAUS, *Lehrbuch der Gruppentheorie, I*, Teubner, Leipzig–Berlin, 1937.

Proc. Internat. Conf. Theory of Groups, Austral. Nat. Univ. Canberra,
August 1965, pp. 203–204. © Gordon and Breach Science Publishers, Inc. 1967

On the orthogonal groups of ramified lattices over local fields

D. G. JAMES

Let K be a local field, i.e., a field complete with respect to a discrete, nonarchimedean, nontrivial valuation. Let R be the ring of integers in K, π a fixed prime of R, and F the finite residue class field. Let V be a finite dimensional, nonsingular lattice over R, i.e., a free R-module with nondegenerate scalar product $\alpha \cdot \beta \in R$ ($\alpha, \beta \in V$). An isometry φ of V is a one–one, linear transformation of V such that $\varphi(\alpha) \cdot \varphi(\beta) = \alpha \cdot \beta$ for all $\alpha, \beta \in V$. The set of all such isometries forms the orthogonal group $O(V, R)$ of V over R.

Let $\nu(x)$ denote the order of $x \in R$. Assume $\nu(\pi) = 1$ and $\nu(2) = 0$. Define, for $\alpha \in V$, $\nu(\alpha) = \min \nu(\alpha \cdot \beta)$, the minimum being taken over all $\beta \in V$. Then $V = \langle \alpha \rangle \oplus W$, W a sublattice, if and only if $\nu(\alpha) = \nu(\alpha^2)$. A vector α with this property is called a *splitting vector*. V has a Jordan splitting $V = W_1 \oplus \cdots \oplus W_s$ where $W_i = \langle \xi_{i1} \rangle \oplus \cdots \oplus \langle \xi_{in_i} \rangle$ with $\nu(\xi_{ij}) = t_i$, $1 \leq j \leq n_i$, and $0 \leq t_1 < t_2 < \cdots < t_s$. With each component W_i we associate a space $U_i = \langle \eta_{i1} \rangle \oplus \cdots \oplus \langle \eta_{in_i} \rangle$ over F with scalar product $\eta_{ij}^2 \equiv \pi^{-t_i} \xi_{ij}^2$ (mod π). An isometry φ of V induces isometries on each of the spaces U_i. Let $\Omega(V, R)$ be the commutator subgroup of $O(V, R)$. Then

$$O(V, R)/\Omega(V, R) \cong \prod_{i=1}^{s} O(U_i, F)/\Omega(U_i, F).$$

Furthermore, $\Omega(V, R)$ contains a subgroup $N(V, R)$ such that

$$\Omega(V, R)/N(V, R) \cong \prod_{i=1}^{s} \Omega(U_i, F).$$

The center $Z(V, R)$ of $O(V, R)$ contains only the two isometries $\pm I$, I the identity. The center is contained in the commutator subgroup if and only if $n_i \equiv 0$ (mod 2) and det $W_i \in K^2$, $1 \leq i \leq s$.

References

[1] EMIL ARTIN, *Geometric algebra*, Interscience, New York, 1957.
[2] O. T. O'MEARA, *Introduction to quadratic forms*, Springer, Berlin–Göttingen–Heidelberg, 1963.

Proc. Internat. Conf. Theory of Groups, Austral. Nat. Univ. Canberra,
August 1965, pp. 205–208. © Gordon and Breach Science Publishers, Inc. 1967

A characterization of a new simple group

ZVONIMIR JANKO

A proof of the following result will be discussed.

Theorem. *Let G be a finite nonabelian simple group with abelian
2-Sylow subgroups. If G does not have a doubly transitive permutation
representation, then G has order $11(11^3 - 1)(11 + 1)$. There is one and
only one isomorphism class of simple groups of that order.*

Proof. By a celebrated theorem of Feit and Thompson the order of
G must be even. Hence G contains elements of order 2 called involutions.
If the centralizer of every involution in G is soluble, then by a result of
D. Gorenstein the group G is isomorphic to $PSL(2, q)$ for $q > 3$ and
$q \equiv \pm 3 \pmod 8$ or $q = 2^n$, $n > 1$. But the groups $PSL(2, q)$ have a doubly
transitive permutation representation. It follows that G must contain
at least one involution t with the nonsoluble centralizer $C(t)$. Using
here a result of J. Walter we see that $C(t) = \langle t \rangle \times PSL(2, q)$ for some
prime power $q \geq 5$.

If $q > 5$, then I have shown in a joint paper with J. G. Thompson
that $q = 3^{2n+1}$, $n \geq 1$ and in addition if u is any nontrivial element of
$C(t)$ having an order prime to 6, then $C(u) \leq C(t)$. This result was
obtained by eliminating the case q is even and then taking a closer look
at the structure of the normalizer of a 2-signalizer in G. A 2-signalizer of
G is any odd order subgroup of G normalized by a 2-Sylow subgroup of
G. Supposing that q is not a power of 3 we show that any 2-signalizer
has an order prime to 3 and then the normalizer of every nontrivial
2-signalizer is soluble and must contain an element of order 3 which
acts fixed-point-free on the 2-signalizer. This gives at once

$$q \equiv -3 \pmod 8.$$

Now using the exceptional character theory of $C(t)$ combined with
the 2-block theory of G we show that the order of G must be $q^3(q - 1) \cdot f$,

where f is the degree of an irreducible character in the principal 2-block. This order formula simplifies the group-theoretic arguments tremendously. If R is a subgroup of order $q = p^n$ (p prime) contained in $C(t)$, then $N(R)$ has a normal 2-complement M of order $\frac{1}{4}q^3(q-1)$ which is a Frobenius group. Also a p-Sylow subgroup of G is elementary abelian and the normalizer of any nontrivial subgroup of R has an order prime to 3. This fact combined with some previous results produces a contradiction.

Knowing that q must be an odd power of 3, it is easy to proceed. For example it is possible to show that if H is any soluble subgroup of G and $|H| > 2$, then $N(H)$ is soluble. Using also the Theorem B of Hall and Higman at one point we conclude quickly that if u is any nontrivial element in $C(t)$ having an order prime to 6, then $C(u) \leq C(t)$.

After obtaining all these informations some results of N. Ward can be applied to show that G then has a doubly transitive permutation representation.

But in the case $q = 5$ the situation is completely different. Here considering arbitrary odd order subgroups normalized by four-groups one realizes at once that the elements u of odd order in $C(t)$ do not satisfy the condition $C(u) \leq C(t)$. In fact one finds out by some simple group-theoretic and character-theoretic arguments that if u is an element of order 3 or 5 in $C(t)$, then $C(u)$ has order 30. Note that $PSL(2, 5) \simeq A_5$ does not have subgroups of order 15 and so $C(u) \nleq C(t)$.

The group $N\langle u \rangle$ of order 60 is obviously very convenient for a successful character theoretic attack. In fact combining this with the exceptional character theory of $C(t)$ I was able to compute the order of G which turns out to be 175,560. But there is still a long way to go.

First step is to determine the subgroup structure. This is a trivial matter except that it seems to be impossible to decide at this stage whether $PSL(2, 11)$ is or is not a subgroup of G.

Next I have determined the ordinary character table of G. It turns out that all 15 irreducible characters are in fact real and that 56 is the smallest degree of such a character. Here it is easy to show that G does not have a doubly transitive permutation representation but it remains to show the existence and the uniqueness of the group G. This is the hardest part of the proof.

Taking a closer look at the character table of G one finds out that the principal 11-block of G contains precisely 11 ordinary and 10 absolutely irreducible modular characters. Because all ordinary irreducible characters are real it follows that all modular irreducible characters of G are real. This together with the fact that 11 divides the order of G to the first power only shows that "the Brauer's graph" which corresponds to the principal block B of G is an open polygon. It is possible to partition the ordinary characters in B in two subsets S_1 and S_2 in such a way that no two ordinary characters of a fixed set S_i have a modular irreducible constituent in common and every modular character of B appears as a modular constituent in precisely two ordinary irreducible characters of B. Also every ordinary irreducible character of B has at most two modular constituents. It follows that two ordinary characters in B are modular-irreducible and any other ordinary character in B has precisely two modular constituents.

Fourteen possibilities arise. Eleven possibilities can be ruled out and in the remaining three possibilities the minimal degree of a modular character is 7 and moreover the values of this character φ can be written down uniquely. It follows also at once that the modular 7-dimensional representation ρ of G corresponding to the character φ can be realized in the field F of 11 elements and that this is in fact the smallest possible degree of any faithful representation of G in any field.

Because it is still impossible to decide here whether $PSL(2, 11)$ is or is not a subgroup of G I drew my attention to a 2-Sylow normalizer H of G which is a soluble group of order 168 and which is also a maximal subgroup of G. Because the restriction of ρ to H is an ordinary irreducible representation of H with the known character $\varphi|_H$ it is possible to write down the matrices which represent H. From the known subgroup structure of G follows that there exists an involution $\tau \in G \setminus H$ such that $\tau\mu = \mu\tau$, $\tau\nu\tau = \nu^{-1}$, and $(t_1\tau)^5 = 1$, where μ, ν, and t_1 are certain suitable elements in H of orders 3, 7, and 2, respectively. The matrices which represent μ, ν, and t_1 are known and the above conditions determine the matrix which represents the involution τ uniquely if we use also the fact: trace $\rho(\tau) = -1$. Hence the group G is constructed and moreover the uniqueness of G is shown.

It remains to show the existence of G. A simple character-theoretic argument shows that G has in fact 11 double cosets modulo H. Denote by G_1 the matrix group generated by the matrices which represent μ, ν, τ, t_1. An ingenious computation due to M. A. Ward assisted by T. M. Gagen shows that G_1 has precisely 11 double cosets modulo H_1,

15—K.

where H_1 is the matrix group generated by the matrices which represent μ, ν, t_1. The matrices of orders 5, 11, and 19 in G_1 can be found explicitly and this together with the previous result shows that $|G_1| = 175{,}560$. It is then an easy matter to show that $C_{G_1}(\rho t_1) = \langle \rho t_1 \rangle \times F$, where $F \cong A_5$ and that all involutions are conjugate in G_1. It follows that G_1 is simple and so the existence of G is established. Working in the matrix group G_1 one finds the matrices a, b, and c satisfying $a^5 = b^5 = c^5 = (ab)^2 = (bc)^2 = (ca)^2 = (abc)^2 = 1$ and so by a result of Coxeter the group $\langle a, b, c \rangle \cong PSL(2, 11)$. Hence $PSL(2, 11)$ is in fact a subgroup of G. W. A. Coppel has shown that G is a subgroup of the Chevalley simple group $G_2(11)$ but according to R. Ree there seems to be no way of constructing a larger class of simple groups having G as the first member.

Proc. Internat. Conf. Theory of Groups, Austral. Nat. Univ. Canberra, August 1965, pp. 209–215. © Gordon and Breach Science Publishers, Inc. 1967

On Huppert's characterization of finite supersoluble groups

OTTO H. KEGEL

In [4] Huppert showed that the finite group G is supersoluble if and only if every maximal subgroup M of G has index $|G:M|$ a prime number. So, in particular, a supersoluble group G has the following property:

(∗) Every maximal subgroup of G admits a supplement which is cyclic and of prime-power order.

The purpose of this note is to show that this property characterizes a class of finite groups only slightly wider than the class of supersoluble groups: *A finite group G is supersoluble if and only if it has property* (∗) *and does not map homomorphically onto the symmetric group on four letters.*

In course of the proof rather precise information on the structure of the non-Frattini chief factors of a group with property (∗) is obtained: they are either cyclic or of order four. That a chief factor of a group with property (∗) need not be cyclic or of order four is shown by an example; in fact, one can give examples of groups with property (∗) and chief factors of arbitrarily large rank. This example may be used to show that the class of groups with property (∗) is not, in general, a formation in the sense of Gaschütz [3]. On the other hand, the class of all finite groups with chief factors either cyclic or of order four which have property (∗) is a formation.

1. The principal step of the proof, which reduces it so that standard methods become applicable, is the simple

Proposition 1. *If the finite group G has property* (∗), *then G is soluble.*

209

Remark. Although the author could prove only the above very special case, it seems reasonable to conjecture that a finite group G is soluble if every maximal subgroup of G admits an abelian supplement.

Proof. Assume the statement is false, and let G be a counter-example of minimal order. Then G possesses a unique minimal normal subgroup $N \neq 1$, and G/N is soluble. Since N is not nilpotent, there is a maximal subgroup M of G not containing N; choose M of minimal index. By property (∗) there is a cyclic subgroup S of prime-power order p^n in G with $MS = G$. Now, one has $M \cap S = 1$; for otherwise the normal closure of $S \cap M$ in G, which is contained in M, would be a nontrivial normal subgroup not containing N, and this would contradict the uniqueness of N. Hence G is a permutation group of degree p^n (e.g., acting on the right cosets Mx of G) and thus a subgroup of the symmetric group $\mathfrak{S}_{p^n}$ on p^n letters.

Let P be a Sylow p-subgroup of N; then $G = N \cdot \mathcal{N}_G P$, by the Frattini argument. Let M_1 be a maximal subgroup of G containing $\mathcal{N}_G P$, then— by property (∗)—there is a cyclic supplement T of prime-power order of M_1 in G.

By the choice of M and M_1 one has

$$|T| > |S|.$$

But, since the order of a cyclic subgroup of prime-power order in $\mathfrak{S}_{p^n}$ is equal to the length of its longest orbit, one has

$$|T| \leq |S|;$$

a contradiction. Hence G is soluble.

2. Call the chief factor M/N of the finite group G a *Frattini chief factor* if the minimal normal subgroup M/N of G/N is contained in the Frattini subgroup $\Phi(G/N)$ of G/N; call it a *non-Frattini chief factor* otherwise.

For any group G and any homomorphism λ of G onto $\mathfrak{S}_4$, the symmetric group on four letters, let $D_\lambda(G)$ denote the largest subgroup of G mapped by λ onto the normal subgroup of order four of $\mathfrak{S}_4$. Define $D(G)$ to be $\bigcap D_\lambda(G)$ with λ ranging over all homomorphisms of G onto $\mathfrak{S}_4$, if there are any, and $D(G) = G$ otherwise.

Theorem. *For the finite soluble group G the following statements are equivalent:*

(a) *G has property* (∗).

(b) *Every maximal subgroup of G has a cyclic supplement in G.*

(c) *Every non-Frattini chief factor of G is either cyclic or of order four, and $D(G/X)$ is supersoluble for every normal subgroup X of G.*

Proof. Since maximal subgroups of finite soluble groups have prime-power index, the equivalence of (a) and (b) is trivial. That (a) implies (c) will be the contents of the next proposition. (c) implies (b): Suppose this were not so, and let G be a counter-example of minimal order with a maximal subgroup M which does not admit a cyclic supplement in G. Since, by the minimality of G, in every proper homomorphic image of G every maximal subgroup will admit a cyclic supplement, the largest normal subgroup M_G of G contained in M equals 1. So, for any minimal normal subgroup N of G, one has

$$MN = G \quad \text{and} \quad M \cap N = 1.$$

Since G is a counter-example to (b), N is not cyclic. So by (c), $|N| = 4$. If now $|M| = 3$, then $G \simeq \mathfrak{A}_4$, and $G = D(G)$; but $D(G)$ was assumed to be supersoluble, a contradiction. Thus $|M| = 6$, and $G \simeq \mathfrak{S}_4$. But $\mathfrak{S}_4$ has property $(*)$. Hence, there cannot be such a counter-example.

Proposition 2. *The finite group G with property $(*)$ has the following properties:*

(a) *Every non-Frattini chief factor of G is cyclic or of order four.*

(b) *If L is the smallest normal subgroup of G with supersoluble factor group G/L, then L is a 2-subgroup of $G' \cap D(G)$.*

(c) *The subgroup $G' \cap D(G)$ is nilpotent.*

(d) *The subgroup $D(G)$ is supersoluble.*

Proof. (a) Suppose the statement (a) is false, and let G be a counter-example of minimal order. Let N be a minimal normal subgroup of G with $N \nleq \Phi(G)$. By Proposition 1, N is abelian, and since G is a counter-example to (a), one may assume that N is noncyclic of order larger than four. There is a maximal subgroup M in G complementary to N. By the minimality of G, one has $M_G = 1$, where M_G is the largest normal subgroup of G contained in M. By property $(*)$, there is a cyclic supplement S for M in G; and $M \cap S = 1$ since $M_G = 1$. Now, a theorem of Ritt [5] yields either $S \simeq N$ or $|N| = |S| = 4$, a contradiction. Hence there cannot be any counter-example to (a).

(b) $G' \cap D(G) \supseteq L$ is evident, since G/G' and $G/D(G)$ and hence also $G/(G' \cap D(G))$ are supersoluble. So let G be a counter-example of minimal order to the statement that L is a 2-group, and let M be maximal among the normal subgroups of G properly contained in L. By Huppert ([4], Satz 10) $L/M \nsubseteq \Phi(G/M)$, and by (a) one has $|L/M| = 4$. Let N be a minimal normal subgroup of G contained in L, then—by the minimality of G—N is the unique minimal normal subgroup of G (in L); and N is of odd order; L/N is a 2-group. Let T be a Sylow 2-subgroup of L. T cannot be normal in L, since otherwise there would be a normal subgroup of G properly contained in L and of odd index in L, which is impossible. Thus one has $L = NT$, and, by the Frattini argument,

$$G = L \cdot \mathscr{N}_G T = N \cdot \mathscr{N}_G T, \qquad N \nsubseteq \mathscr{N}_G T.$$

But since N is a minimal normal subgroup of G, one has $N \cap \mathscr{N}_G T = 1$, and $\mathscr{N}_G T$ is a maximal subgroup of G. Thus $N \nsubseteq \Phi(G)$, and—by (a)—N is cyclic of odd order. But then N is in the center of L. Hence T is normal in L. But this cannot be the case. This contradiction establishes (b).

(c) Assume (c) proved for groups of order smaller than $|G|$. Let N be a minimal normal subgroup of G contained in $G' \cap D(G)$. If $N \nsubseteq \Phi(G)$ then, by (a), N is either cyclic or of order four. In either case N is contained in the center of $G' \cap D(G)$. But by assumption $(G/N)' \cap D(G/N)$ is nilpotent, and since $D(G/N) \supseteq D(G)/N$, one has that $(G' \cap D(G))/N$ is nilpotent. But then $G' \cap D(G)$ is also nilpotent. If $N \subseteq \Phi(G)$, then $(G/N)' \cap D(G/N)$ is nilpotent by induction hypothesis, and a theorem of Gaschütz ([2], Satz 10) yields that the inverse image I of $(G/N)' \cap D(G/N)$ in G is nilpotent. But $G' \cap D(G)$ is a subgroup of I, and thus also nilpotent.

(d) Assume (d) is false, and let G be a counter-example of minimal order. Let N be the smallest normal subgroup of $D(G)$ so that $D(G)/N$ is supersoluble. As a characteristic subgroup of $D(G)$, N is normal (even characteristic) in G. By the minimality of G, and because of $D(G/X) \supseteq X \cdot D(G)/X$ for every normal subgroup X of G, one has that N is the unique minimal normal subgroup of G in $D(G)$ and, in fact, even in G. By (b), N is a 2-group. Let F be the largest nilpotent normal subgroup of $D(G)$, then F is—by the unicity of N—a 2-group. F contains its own centralizer in $D(G)$ (cf. Fitting [1], p. 106, Hilfssatz 12).

Now, F is elementary abelian; for if $\Phi(F) \neq 1$, then $D(G/\Phi(F))$ is supersoluble, and so is $D(G)/\Phi(F)$, but $\Phi(F) \subseteq \Phi(D(G))$ (Gaschütz [2], Satz 10), and then, by Huppert ([4], Satz 10), $D(G)$ is supersoluble—a contradiction.

Since F is a 2-group and since, by (c), $F \supseteq D(G) \cap G'$, it is obvious that F is a Sylow 2-subgroup of $D(G)$. Let K be a 2-complement of $D(G)$. By Maschke's theorem there is a subgroup M of F with

$$M^K = M, \qquad F = NM, \quad \text{and} \quad N \cap M = 1.$$

Since $D(G/N) \supseteq D(G)/N$ is supersoluble, and hence has a Sylow tower, one may choose M to commute elementwise with K. In fact, $M = \mathcal{N}_F K$; for if $L = N \cap \mathcal{N}_F K$ were different from 1, then $L = N$ by the minimality of N, and $D(G)$ would be nilpotent, a contradiction. But now, since

$$G = D(G)\mathcal{N}_G K = F \cdot \mathcal{N}_G K = N \cdot \mathcal{N}_G K,$$

the subgroup M of F is normal in G; by the unicity of N, this is only possible if $M = 1$.

But now, since the minimal normal subgroup N of G is complemented by $\mathcal{N}_G K$, one has $N \nsubseteq \Phi(G)$, and, by (a), $|N| = 4$. But then $|\mathcal{N}_G K|$ is either 3 or 6. If $|\mathcal{N}_G K| = 3$, then $G \simeq \mathfrak{A}_4$, the alternating group on four letters. This group does not have property (∗); so $|\mathcal{N}_G K| = 6$, and $G \simeq \mathfrak{S}_4$. But $D(\mathfrak{S}_4)$ is a 2-group and hence supersoluble, a contradiction. Hence such a counter-example cannot exist, and (d) is established.

3. In view of Proposition 2(a) one may wonder whether every chief factor of a finite group with property (∗) is cyclic or of order four. This is not the case, as is shown by the following example of a group with property (∗) and a Frattini chief factor of order 16. By the same method one may show that, in fact, there are groups with property (∗) and a minimal normal 2-subgroup of arbitrarily high order.

Example 1. Consider the group

$$X = \langle x_1, x_2, x_3, x_4 \mid x_i^2 = (x_i \circ x_j)^2 = x_1 \circ x_2 = x_3 \circ x_4$$
$$= x_i \circ (x_j \circ x_k) = 1 \quad \text{for } i, j, k = 1, 2, 3, 4 \rangle,$$

where $g \circ h = g^{-1}h^{-1}gh$. Obviously, the group X is a 2-group of exponent four and class two, $|X| = 2^{10}$. One has $X^2 = \Phi(X) = X'$. Let $\bar{X} = X/X'$, and let $\bar{x}_i$ be the image of x_i under the canonical homomorphism of X

onto $\bar{X}$. The automorphism groups A_1 of $\langle \bar{x}_1, \bar{x}_2 \rangle$ resp. A_2 of $\langle \bar{x}_3, \bar{x}_4 \rangle$ are isomorphic to $\mathfrak{S}_3$. So let A be the automorphism group of $\bar{X}$ which induces A_1 in $\langle \bar{x}_1, \bar{x}_2 \rangle$ and A_2 in $\langle \bar{x}_3, \bar{x}_4 \rangle$ and is isomorphic to $A_1 \times A_2$. One checks readily that the group A of automorphisms of $\bar{X}$ may be lifted to a group of automorphisms of X, and that the vector space X' is irreducible under the action of A. Now let $G = XA$ be the split extension of X by A. Then $X' = \Phi(X)$ is a minimal normal subgroup of G. By Gaschütz ([2], Satz 5), $X' \subseteq \Phi(G)$. Since $G/X' = \mathfrak{S}_4 \times \mathfrak{S}_4$, one has that G/X' has property $(*)$, but then also G has property $(*)$.

It may be of interest that this example allows one to show that the class of all finite groups with property $(*)$ is not a formation in the sense of Gaschütz [3].

Example 2. Let Y be the direct product of the group X of Example 1 by a group M isomorphic to X'. Then A is in a natural way a group of automorphisms of Y such that M and X' are isomorphic even as A-modules. Let N be a diagonal of $M \times X'$, as A-modules; then N too is an A-module. Now let H be the split extension of Y by A, then the group $G = X \cdot A$ of Example 1 is a maximal subgroup of H, and $H/M \simeq G \simeq H/N$ have property $(*)$, whereas H does not have property $(*)$ since the minimal normal subgroups M and N of H are complemented in H (i.e., $M, N \not\subseteq \Phi(H)$), and $|M| = |N| = 16$.

On the other hand, one has

Proposition 3. *The class of all those finite groups which have all chief factors cyclic or of order four and enjoy property $(*)$ is a formation.*

That this formation is not saturated (cf. [3]) is clear from Example 1.

Proof. That this class is a formation simply means that if for a finite group G with normal subgroups M and N the factor groups G/M and G/N are in this class, then so is the factor group $G/(M \cap N)$. It is evident that the condition on the chief factors of G/M and G/N carries over to the chief factors of $G/(M \cap N)$. So all that remains to be shown is that $G/(M \cap N)$ has property $(*)$. Suppose this were not so, and let G be a counter-example of least order; then $N \cap M = 1$, and one may assume that M and N are minimal normal subgroups of G. By assumption there is a maximal subgroup K of G which does not have a cyclic supplement. This implies

$$G = MK = NK, \qquad M \cap K = N \cap K = 1.$$

But then $|M| = |N| = 4$. Let C denote the centralizer of M in K; then C is normal in G, and $|K/C| = 3$ or 6. If $K/C \simeq \mathfrak{S}_3$ then $G/C \simeq \mathfrak{S}_4$, and an element g of G which maps onto an element of order four under the natural homomorphism of G onto G/C generates a subgroup $\langle g \rangle$ with $K\langle g \rangle = G$, a contradiction. Thus $|K/C| = 3$. But then there is a pair of normal subgroups $R \supset S$ of G contained in $D(G)$ such that $G/C \simeq R/S$ qua operator groups with G a sa group of operators; in fact, R is contained in the kernel of every homomorphism of G onto $\mathfrak{S}_4$. Now let V/N (resp. W/M) be the intersection of all the kernels of homomorphisms of G/N (resp. G/M) onto $\mathfrak{S}_4$. Then the intersection of all the kernels of homomorphisms of G onto $\mathfrak{S}_4$ will be contained in $V \cap W$. But since— by Proposition 2—$(V \cap W)/(V \cap W \cap M)$ and $(V \cap W)/(V \cap W \cap N)$ are supersoluble, $(V \cap W)/(V \cap W \cap M \cap N) = V \cap W$ is supersoluble. Thus R/S and hence G/C are supersoluble. Since this is not the case, we have arrived at a contradiction showing that no such counter-example can exist.

References

[1] HANS FITTING, Beiträge zur Theorie der Gruppen endlicher Ordnung, *Jber. Deutsch. Math.-Verein.* **48** (1938), 77–141.
[2] WOLFGANG GASCHÜTZ, Über die Φ-Untergruppe endlicher Gruppen, *Math. Z.* **58** (1953), 160–170.
[3] WOLFGANG GASCHÜTZ, Zur Theorie der endlichen auflösbaren Gruppen, *Math. Z.* **80** (1963), 300–305.
[4] BERTRAM HUPPERT, Normalteiler und maximale Untergruppen endlicher Gruppen, *Math. Z.* **60** (1954), 409–434.
[5] J. F. RITT, On algebraic functions which can be expressed in terms of radicals, *Trans. Amer. Math. Soc.* **24** (1923) 21–30.

Proc. Internat. Conf. Theory of Groups, Austral. Nat. Univ. Canberra, August 1965, pp. 217–219. © Gordon and Breach Science Publishers, Inc. 1967

Varieties and the Hall–Higman paper

L. G. KOVÁCS

A class $\mathfrak{X}$ of groups is said to have the Burnside property if $\mathfrak{X}$ consists of finite groups and, for each positive integer k, $\mathfrak{X}$ contains only finitely many (isomorphism classes of) k-generator groups. For each positive integer e, let $\mathfrak{R}_e$ denote the class of finite groups of exponent dividing e, and $\mathfrak{S}_e$ the class of soluble groups in $\mathfrak{R}_e$. The restricted Burnside conjecture (R_e) for exponent e is that $\mathfrak{R}_e$ has the Burnside property, and the corresponding soluble Burnside conjecture (S_e) asserts the same for $\mathfrak{S}_e$. A group will be called simply monolithic if the intersection of its nontrivial normal subgroups is a nonabelian simple group. The factors of a group are the factor groups of its subgroups.

P. Hall and G. Higman [1] gave some reduction theorems for the Burnside conjectures and remarked (p. 39 in [1]) that "theorems of absolute validity" could also be deduced by their method. The purpose of this note is to put on record some facts about varieties which follow from, or are closely related to, the paper of Hall and Higman. The following statement appears to express the full force of their method:

(A) *If $\mathfrak{X}$ is a class with the Burnside property and $\mathfrak{Y}$ is a class of finite groups whose Sylow subgroups and simply monolithic factors all lie in $\mathfrak{X}$, then $\mathfrak{Y}$ also has the Burnside property.*

One further observation will be relevant; this relies on more recent developments. It can be checked that each class $\mathfrak{R}_e$ contains only a finite number of the simple groups which are known at present (in fact, each large number which is known to occur as the order of a finite simple group is divisible by some large prime). Let an A^*-group be a locally finite group whose Sylow subgroups are all abelian. According to Z. Janko, there exist now (mostly unpublished) results (due to D. Gorenstein, Z. Janko, J. G. Thompson, J. H. Walter, and N. Ward) which show that all finite simple A^*-groups are known. These facts and (A) together imply:

(B)   *For each positive integer e, the class of finite A^*-groups of exponent dividing e has the Burnside property.*

For a class $\mathfrak{X}$ of groups, let $\mathfrak{X}_0$ denote the class of the finitely generated subgroups of the groups in $\mathfrak{X}$. (The variety var $\mathfrak{X}$ generated by $\mathfrak{X}$ is obviously the same as var $\mathfrak{X}_0$.) In a conversation with M. F. Newman it was observed that var $\mathfrak{X}$ is locally finite if and only if $\mathfrak{X}_0$ has the Burnside property. Thus (R_e) is equivalent to the statement that the class of all locally finite groups of exponent dividing e is a variety, and (S_e) is equivalent to: "the class of all locally finite-and-soluble groups of exponent dividing e is a variety." In particular, A. I. Kostrikin's result [2] that (R_p) is true for every prime p means that the class of all locally finite groups of exponent dividing p is a variety: call it the Kostrikin variety $\mathfrak{K}_p$ of exponent p.

The first of the four facts to be mentioned here concerns a special case of the finite basis problem. From Schreier's Theorem it follows easily that the laws of a locally finite variety $\mathfrak{V}$ are finitely based if and only if $\mathfrak{V}$ can be defined by its k-variable laws for some positive integer k. Thus

(1)   *The laws of $\mathfrak{K}_p$ are not finitely based if and only if there exist, to each positive integer k, infinite $(k+1)$-generator groups of exponent p in which all k-generator subgroups are finite.*

(2)   *If $\mathfrak{V}$ is a locally finite variety, then the locally soluble groups in $\mathfrak{V}$ form a subvariety $\mathfrak{V}_{LS}$, and the locally nilpotent groups form a subvariety $\mathfrak{V}_{LN}$. If $\mathfrak{V}_{LN}$ is soluble, so is $\mathfrak{V}_{LS}$.*

The proof of the first statement is straight-forward; the second follows from Theorem 3.6.2 of [1].

(3)   *If $\mathfrak{U}$ is a locally finite variety and $\mathfrak{V}$ is the class of those groups whose nilpotent factors and finitely generated simply monolithic factors all belong to $\mathfrak{U}$, then $\mathfrak{V}$ is also a locally finite variety.*

This is proved from (A). Similarly, one obtains from (B) the following:

(4)   *For each positive integer e, the class $\mathfrak{A}_e^*$ of A^*-groups of exponent dividing e is a (locally finite) variety.*

References

[1] P. HALL and G. HIGMAN, On the p-length of p-soluble groups and reduction theorems for Burnside's problem, *Proc. London Math. Soc.* (3) **6** (1956), 1–42.
[2] A. I. KOSTRIKIN, On Burnside's problem, *Izv. Akad. Nauk SSSR Ser. Mat.* **23** (1959), 3–34 [Russian].

Proc. Internat. Conf. Theory of Groups, Austral. Nat. Univ. Canberra, August 1965, pp. 221–223. © Gordon and Breach Science Publishers, Inc. 1967

Just-non-Cross varieties

L. G. KOVÁCS and M. F. NEWMAN

It is convenient here to define a *Cross variety* as a variety generated by a single finite group: we shall not use the rather deep result (Sheila Oates and M. B. Powell [9]) that this definition is equivalent to the usual one. It is easy to see (e.g., from Lemma 4.3 of Graham Higman [3]) that the nilpotent groups of a Cross variety are of bounded class and the chief factors of the finite groups of a Cross variety are of bounded order. Let $\mathfrak{W}$ be a variety whose laws are finitely based; then the laws of every subvariety of $\mathfrak{W}$ are finitely based if and only if every set of subvarieties of $\mathfrak{W}$ has a minimal element (with respect to partial order by inclusion). The question to be considered here is a very special case of the suggested equivalent of the finite basis problem: For what varieties $\mathfrak{V}$ is it true that among the non-Cross subvarieties of $\mathfrak{V}$ there are minimal ones? Call a non-Cross variety *just-non-Cross* if every proper subvariety of it is Cross, and rephrase the question: What non-Cross varieties have just-non-Cross subvarieties? Also, find as many just-non-Cross varieties as possible.

Notation. For a nonnegative integer e, $\mathfrak{A}_e$ denotes the variety of abelian groups of exponent dividing e. If q is an odd prime, $\mathfrak{Q}_q$ denotes the variety of nilpotent groups of class (at most) 2 and exponent dividing q; $\mathfrak{Q}_2$ is the variety of groups of exponent dividing 4 with central derived groups of exponent dividing 2. The letters p, q, r will denote arbitrary but distinct primes.

The just-non-Cross varieties we know are $\mathfrak{A}_0$, $\mathfrak{A}_p\mathfrak{A}_p$, $\mathfrak{A}_p\mathfrak{Q}_q$, $\mathfrak{A}_p\mathfrak{A}_q\mathfrak{A}_r$. The claim that $\mathfrak{A}_0$ is one is trivial. To see that $\mathfrak{A}_p\mathfrak{A}_p$ is non-Cross, note that it contains nilpotent wreath products of arbitrary large class. Using results of N. D. Gupta and the second author [2], we prove in [6] that every proper subvariety of $\mathfrak{A}_p\mathfrak{A}_p$ is Cross. Higman's paper [3] shows that $\mathfrak{A}_p\mathfrak{Q}_q$ is just-non-Cross and $\mathfrak{A}_p\mathfrak{A}_q\mathfrak{A}_r$ is not Cross. The fact that all proper subvarieties of $\mathfrak{A}_p\mathfrak{A}_q\mathfrak{A}_r$ are Cross has been proved by P. J. Cossey [1].

It may be conjectured that each soluble non-Cross variety must contain one of these just-non-Cross varieties. It has been shown [6] that if the nilpotent groups of a soluble variety are not of bounded class, then it must contain an $\mathfrak{A}_p\mathfrak{A}_p$. Cossey has proved [1] that if the nilpotent groups of a soluble variety are all abelian, then it contains either $\mathfrak{A}_0$ or an $\mathfrak{A}_p\mathfrak{A}_q\mathfrak{A}_r$. However, the gap is still wide.

In both of the results just quoted, the assumption of solubility is essentially used. It is hard to see what, if any, extension of the first may hold in the locally soluble case: the variety $\mathfrak{K}_p$ of locally soluble groups of exponent dividing p (where p is a prime; cf. [5]) presents a case we dare not hope to handle. On the other hand, the prospects of generalizing Cossey's result to cover all insoluble cases seem good; he is giving attention to this.

Our original reason for suspecting that the product varieties mentioned above are just-non-Cross was Theorem 6.3 of A. L. Šmel'kin [10]: The product of two nontrivial varieties $\mathfrak{U}$, $\mathfrak{V}$ is Cross if and only if (i) the exponents of $\mathfrak{U}$ and $\mathfrak{V}$ are coprime, (ii) $\mathfrak{U}$ is nilpotent, and (iii) $\mathfrak{V}$ is abelian. The three types of just-non-Cross product varieties can be obtained by minimally violating one of these conditions and at the same time maximally satisfying the others. The "if" part of the theorem is proved with an argument of D. C. Cross (cf. Higman [4]). Šmel'kin proved the "only if" part with the help of verbal wreath products. Peter M. Neumann has a proof (only partly published in [8]) which he obtained from estimates of orders of finite relatively free groups. We indicate yet another proof for the "only if" part and at the same time show that there are no further just-non-Cross product varieties.

Let $\mathfrak{U}$, $\mathfrak{V}$ be as above. If (i) is false, $\mathfrak{U}\mathfrak{V}$ obviously contains an $\mathfrak{A}_p\mathfrak{A}_p$. If $\mathfrak{U}$ contains at least one non-nilpotent finite group, then a subvariety generated by a finite non-nilpotent group of least order in $\mathfrak{U}$ is an $\mathfrak{A}_p\mathfrak{A}_q$; so if at the same time (i) holds, then $\mathfrak{U}\mathfrak{V}$ contains an $(\mathfrak{A}_p\mathfrak{A}_q)\mathfrak{A}_r$. If $\mathfrak{V}$ contains at least one non-abelian finite group, then a subvariety generated by a finite non-abelian group of least order in $\mathfrak{V}$ is either an $\mathfrak{A}_q\mathfrak{A}_r$ or a $\mathfrak{Q}_q$; so if at the same time (i) holds, then $\mathfrak{U}\mathfrak{V}$ contains an $\mathfrak{A}_p(\mathfrak{A}_q\mathfrak{A}_r)$ or an $\mathfrak{A}_p\mathfrak{Q}_q$. Suppose that $\mathfrak{U}\mathfrak{V}$ is Cross. Then $\mathfrak{U}\mathfrak{V}$ cannot contain any $\mathfrak{A}_p\mathfrak{A}_p$ or $\mathfrak{A}_p\mathfrak{Q}_q$ or $\mathfrak{A}_p\mathfrak{A}_q\mathfrak{A}_r$, and so (i) must hold; moreover, as every

Cross variety is locally finite (cf. B. H. Neumann [7]), the above argument implies that (ii) and (iii) must also hold. Suppose that $\mathfrak{U}\mathfrak{V}$ is just-non-Cross. Then $\mathfrak{U}$ and $\mathfrak{V}$ are Cross and by the "if" part of the theorem at least one of (i)–(iii) must fail: so the above argument gives that either $\mathfrak{U} = \mathfrak{A}_p$ and $\mathfrak{V} = \mathfrak{A}_q \mathfrak{A}_r$ or $\mathfrak{O}_q$, or $\mathfrak{U} = \mathfrak{A}_p \mathfrak{A}_q$ and $\mathfrak{V} = \mathfrak{A}_r$.

We started from the finite basis problem. Although the results reported on are very incomplete, in one direction some hope is already showing that these investigations may lead back to the positive solution of some special cases of that problem.

References

[1] P. J. COSSEY, On varieties of A-groups, *these Proc.*, p. 71.

[2] N. D. GUPTA and M. F. NEWMAN, On metabelian groups, *these Proc.*, pp. 111–113.

[3] GRAHAM HIGMAN, Some remarks on varieties of groups, *Quart. J. Math. Oxford Ser.* (2) **10** (1959), 165–178.

[4] G. HIGMAN, Identical relations in finite groups, *Conv. Internaz. di Teoria dei Gruppi Finiti* (*Firenze*, 1960), pp. 93–100. *Edizioni Cremonese*, Rome, 1960.

[5] L. G. KOVÁCS, Varieties and the Hall–Higman paper, *these Proc.*, 217–219.

[6] L. G. KOVÁCS and M. F. NEWMAN, On non-Cross varieties of groups [in preparation].

[7] B. H. NEUMANN, Identical relations in groups, I., *Math. Ann.* **114** (1937), 506–525.

[8] PETER M. NEUMANN, Some indecomposable varieties of groups, *Quart. J. Math. Oxford Ser.* (2) **14** (1963), 46–50.

[9] SHEILA OATES and M. B. POWELL, Identical relations in finite groups, *J. Algebra* **1** (1964), 11–39.

[10] A. L. ŠMEL'KIN, Wreath products and varieties of groups, *Izv. Akad. Nauk SSSR Ser. Mat.* **29** (1965), 149–170 [Russian].

Proc. Internat. Conf. Theory of Groups, Austral. Nat. Univ. Canberra, August 1965, pp. 225–231. © Gordon and Breach Science Publishers, Inc. 1967

Ordering of extensions

F. LOONSTRA

§1. Introduction

In the following only abelian groups are considered. The extensions of a fixed group A can be ordered, defining for two extensions $G(A)$, $G'(A)$

(1) $$G(A) < G'(A)$$

if there is a homomorphism

(2) $$\eta \colon G \to G'$$

leaving invariant the elements of A. Two extensions $G(A)$ and $G'(A)$ are called related with respect to A,

(3) $$G(A) \sim G'(A),$$

if $G(A) < G'(A)$ and $G'(A) < G(A)$. An extension $G(A)$ therefore defines a class $\{G(A)\}$ of related extensions. For the classes $\{G(A)\}$ and $\{G'(A)\}$ an order relation can be defined by means of a representative of each class. The system $V(A)$ of all classes $\{G(A)\}$ of extensions of A has a minimal element, the class $\{A\}$ of all extensions having A as a direct summand. The maximal element is the class $\{D(A)\}$ represented by a divisible group $D(A)$ containing A. $V(A)$ has lattice properties: if $G_i(A)$, $i \in I$, is a set of extensions, then

(4) $$\bigcap_{i \in I} \{G_i(A)\} = \Big\{ \sum_{i \in I}{}^{*} G_i(A) \Big\},$$

$\sum_{i \in I}^{*} G_i(A)$ being the cartesian product of the $G_i(A)$;

(5) $$\bigcup_{i \in I} \{G_i(A)\} = \Big\{ \sum_{i \in I} G_i(A)/\mathfrak{A} \Big\},$$

$\sum_{i \in I} G_i(A)$ being the (restricted) direct product of the $G_i(A)$, while $\mathfrak{A}$

225

is the subgroup of $\sum_i G_i(A)$ consisting of the elements $(\ldots, a_1, \ldots, a_k, \ldots)$ with $\cdots + a_1 + \cdots + a_k + \cdots = 0$.

In the following we shall study first the order structure of the system $V_b(A)$ of classes of extensions $G(A)$ under the condition that A is torsionfree and $G(A)/A$ bounded. In this case it turns out that the ordering of the system $V_b(A)$ is the same as the ordering of a class of subgroups (by inclusion) in a minimal divisible extension of A. A similar result holds for the structure $V_f(A)$ of classes under the restrictions that A is torsionfree and the factor groups $G(A)/A$ are finite.

I wish to express my thanks to Professor R. Baer and Professor L. Fuchs for their contributions!

§2. The system V_b

Among the divisible groups $D(A)$ containing A there are minimal ones; in the following we suppose that $D(A)$ is a minimal divisible extension of A. Now we make two restrictions:

R1. *A is torsionfree; then $D(A)$ is a direct sum of groups of the rational numbers.*

R2. *We only consider those classes $\{G(A)\}$ of extensions with the property that $G(A)/A$ is bounded.*

The system of these classes will be denoted by $V_b(A)$. It follows that $mG(A) \subseteq A$ for a certain integer m and we have

$$G(A) = T \oplus G_1(A),$$

where T is a torsion group and $G_1(A)$ is a torsionfree extension of A. Therefore we have $G(A) \sim G_1(A)$; that means that the class $\{G(A)\}$ contains a torsionfree extension $G_1(A)$. From the fact that $D(A)$ represents the maximal class follows the existence of an A-homomorphism

$$\eta \colon G_1(A) \to D(A).$$

If

$$g_1 \to g_1\eta = g_1', \qquad mg_1 = a,$$

we have

$$mg_1 = mg_1',$$

therefore

$$g_1 = g_1'.$$

That means that $D(A)$ contains an isomorphic copy $G_1^*(A)$ of a representative $G_1(A)$ of the class $\{G(A)\}$. One proves easily that this copy $G_1^*(A) \subseteq D(A)$ is uniquely defined by the class $\{G(A)\}$. In fact: if $\{G_1(A)\} < \{G_2(A)\}$, where $G_1(A)$, $G_2(A)$ are now supposed to be torsion-free, we have for the copies $G_1^*(A)$, $G_2^*(A)$ in $D(A)$ also

$$G_1^*(A) \; < \; G_2^*(A);$$

from this it follows directly that

$$G_1^*(A) \; \subseteq \; G_2^*(A).$$

If therefore $G_1(A) \sim G_2(A)$ then we have in $D(A)$

$$G_1^*(A) \; = \; G_2^*(A).$$

Every class $\{G(A)\}$ in $V_b(A)$ therefore defines in a unique way a subgroup $G^*(A)$ with

$$A \subseteq G^*(A) \subseteq D(A), \qquad G^*(A)/A \quad \text{bounded}$$

and conversely any such group $G^*(A)$ defines uniquely a class $\{G^*(A)\}$ of V_b. Therefore we have

Theorem 1. *Suppose that $D(A)$ is a minimal divisible extension of the torsionfree abelian group A and $V_b(A)$ the system of the classes of related extensions of A with the condition that $G(A)/A$ is bounded; then $V_b(A)$ is order isomorphic with the system of those subgroups $X(A) \subseteq D(A)$ with*

$$A \subseteq X(A) \subseteq D(A), \qquad X(A)/A \quad \text{bounded},$$

and ordered by inclusion.

The last result includes the possibility that $G(A)/A$ is finite. In that case the system of corresponding classes will be denoted by V_f. The structure of V_f can also be obtained using the notion of essential extension of A and it is for this reason that we give another method.

§3. Essential extensions

An extension $G(A)$ of A is called an essential extension of A if the following condition is satisfied:

(E) *If X is a subgroup of G and $X \cap A = 0$, then $X = 0$.*

Equivalent with (E) is:

(E′) *$G(A)$ is an essential extension of A if a homomorphism α of $G(A)$, inducing on A an isomorphism, is itself an isomorphism.*

(E) $\Rightarrow$ (E′): If $K(\alpha)$ is the kernel of α, then $K(\alpha) \cap A = 0$, therefore $K(\alpha) = 0$ and α is an isomorphism.

(E′) $\Rightarrow$ (E): Suppose X is subgroup of $G(A)$ and $A \cap X = 0$. If $D(A)$ is a divisible extension of A we define a homomorphism

$$\varphi \colon X \oplus A \to D(A)$$

such that $(x + a)\varphi = a$; therefore $X\varphi = 0$ and $a\varphi = a$ ($a \in A$). This homomorphism can be extended to a homomorphism $\bar{\varphi} \colon G(A) \to D(A)$, such that $a\bar{\varphi} = a\varphi = a$. Then $\bar{\varphi}$ must be an isomorphism of $G(A)$ in $D(A)$, whence $X = 0$.

Not every extension is essential. Example: Let $A = \{a\} \cong C(p)$, $G_0 = \{A; g_1, g_2, \ldots, g_k, \ldots\}$, such that $p^k g_k = a$ ($k = 1, 2, \ldots$). Then G_0 is not essential, for G_0 contains $X = \{b\}$, with

$$b = (p - 1)g_1 + pg_2;$$

$pb = 0$. We have $X \cap A = 0$, $X \neq 0$.

If $W(A)$ is an essential extension, then $G_0(A)$ cannot be related with $W(A)$. This is shown by the following indirect argument. If

$$\eta \colon W(A) \to G_0(A), \qquad \xi \colon G_0(A) \to W(A)$$

are two homomorphisms leaving invariant the elements of A, then $K(\eta) \cap A = 0$, and η is an isomorphism, $W\eta \cong W$ is a subgroup of $G_0(A)$. Furthermore we have $K(\xi) \cap A = 0$ and also $K(\xi) \cap W\eta = 0$, for $W\eta$ is essential. It follows that

$$G_0(A) = W\eta \oplus K(\xi),$$

for if $g \in G_0(A)$, then $g\xi = (w\eta)\xi$ and therefore

$$(g - w\eta)\xi = 0$$

and we have

$$g - w\eta = k, \qquad k \in K(\xi)$$

and

$$g = w\eta + k.$$

The elements $a, g_1, g_2, \ldots, g_k, \ldots$ are in $W\eta$, for $a\xi \neq 0$ and $g_k \xi \neq 0$, for $(p^k g_k)\xi = a$. That means that $G_0(A) \subseteq W\eta$, therefore $K(\xi) = 0$ and $G_0(A) \cong W(A)$; this gives a contradiction, for $G_0(A)$ is not essential and $W(A)$ is.

§4. The system V_f

Among the divisible groups containing A there are minimal ones. Kulikov[1] proved: A divisible group $D(A)$ is a minimal divisible extension of A if and only if $D(A)$ is an essential (divisible) extension of A. We shall suppose that $D(A)$ is a minimal divisible extension of A. Thus we have:

1. Any subgroup $X(A)$ with $A \subseteq X(A) \subseteq D(A)$ is an essential extension of A.

2. Any essential extension $H(A)$ of A is isomorphic with a subgroup $X(A)$ of $D(A)$ with $A \subseteq X(A) \subseteq D(A)$.

As we suppose that condition R1 of Section 2 holds, we know that $X(A)$ is uniquely determined by $H(A)$; moreover, if $H_1(A)$ and $H_2(A)$ are two related essential extensions of A, then the corresponding subgroups $X_1(A)$ and $X_2(A)$ in $D(A)$ are equal, while if $H_1(A) < H_2(A)$ we have $X_1(A) \subseteq X_2(A)$.

We now suppose that among the extensions related with $G(A)$ there are extensions $X(A)$ with finite factor group $X(A)/A$. The system $V_f(A)$ of the corresponding classes is a subset of $V(A)$. Then $\{G(A)\}$ contains an extension $X(A)$ with minimal factor group. This extension is essential. Indeed: any homomorphism α of $X(A)$, leaving invariant each element of A must be an isomorphism. In the contrary case $X(A)\alpha$ would be an extension of A with a smaller factor group. Therefore we have the following:

Theorem 2. *The system $V_f(A)$ is order-isomorphic with the system of the subgroups $X(A)$ of $D(A)$, $A \subseteq X(A) \subseteq D(A)$, with finite factor group $X(A)/A$, the ordering of the subgroups $X(A)$ being by inclusion.*

§5. Examples of V_f

If A is an infinite cyclic group, then $D(A)$ is the group R_0 of the rational numbers. A subgroup $X(A)$ with $A \subseteq X(A) \subseteq R_0$ and $X(A)/A$ finite must be an infinite cyclic group $\{1/m\}$, where m is any natural number. Conversely any cyclic group $X(A) = \{p/q\}$ containing A is generated by a rational $1/q$ and we have for $X_m(A) = \{1/m\}$, $X_n(A) = \{1/n\}$

$$X_m(A) \subseteq X_n(A)$$

---

[1] See L. Fuchs, *Abelian groups*, Akadémiai Kiadó, Budapest, 1958; p. 66.

if and only if $m|n$. $V_f(A)$ is therefore order isomorphic with the set of positive integers ordered by divisibility.

If A is a direct sum of two infinite cyclic groups, then $D(A) = R_0 \oplus R_0$. Suppose we have $A \subseteq X(A) \subseteq D(A)$, $X(A)/A$ being finite, and denote the components of $X(A)$ in $D(A)$ by X_1, X_2. Then we have for $X(A)/A$ a subdirect sum

$$X(A)/A \cong X_1/C \pm X_2/C \qquad (C = \text{group of integers});$$

therefore, X_1/C and X_2/C must be finite and it follows that X_1 and X_2 are two infinite cyclic groups G_m, G_n:

$$X_1 = G_m = \left\{\frac{1}{m}\right\}, \qquad X_2 = G_n = \left\{\frac{1}{n}\right\} \quad (m, n \text{ integers}).$$

Any subdirect sum $X(A)$ of X_1 and X_2 containing $C \oplus C$ must be a direct sum of two infinite cyclic groups and can be found by means of a third group F and two epimorphisms

$$\alpha \colon X_1 \to F, \qquad \beta \colon X_2 \to F;$$

X is the subgroup of $X_1 \oplus X_2$ of all pairs (x_1, x_2) with

$$x_1\alpha = x_2\beta.$$

For F we have only the possibility $F = C(k)$ (finite cyclic group, $k \geq 1$); moreover we have the conditions

$$k|m, \qquad k|n.$$

If $\alpha \colon X_1 \to C(k)$ is the mapping

$$\frac{kv+s}{m} \to s \qquad (0 \leq s \leq k-1)$$

and $\beta \colon X_2 \to C(k)$ is defined by

$$\frac{1}{n}\beta = r, \qquad (r, k) = 1,$$

then X is a direct sum of two infinite cyclic groups

$$X = \left\{\left(\frac{1}{m}, \frac{r'}{n}\right)\right\} \oplus \left\{\left(0, \frac{k}{n}\right)\right\}, \qquad k|m, \ k|n, \ (r', k) = 1.$$

Choosing for m, n, r', k integers satisfying the mentioned conditions we find all extensions $X(A)$.

If

$$X(A) = \left\{ \left(\frac{1}{m}, \frac{r}{n} \right) \right\} \oplus \left\{ \left(0, \frac{k}{n} \right) \right\}, \qquad k|m, \; k|n, \; (r, k) = 1$$

and

$$X'(A) = \left\{ \left(\frac{1}{m'}, \frac{r'}{n'} \right) \right\} \oplus \left\{ \left(0, \frac{k'}{n'} \right) \right\}, \qquad k'|m', \; k'|n', \; (r', k') = 1,$$

it follows from the condition $X(A) \subseteq X'(A)$ that

$$(6) \qquad m|m', \quad n|n', \quad \frac{m}{k} \Big| \frac{m'}{k'}, \quad \frac{n}{k} \Big| \frac{n'}{k'}, \quad k' \Big| \left(\frac{m'}{m} r' - \frac{n'}{n} r \right).$$

These results directly follow from the condition that the generating elements $(1/m, r/n)$ and $(0, k/n)$ of $X(A)$ are in $X'(A)$. If conversely $X(A)$ and $X'(A)$ satisfy the conditions (6), then we see that $X(A) \subseteq X'(A)$.

Proc. Internat. Conf. Theory of Groups, Austral. Nat. Univ. Canberra,
August 1965, pp. 233–239. © Gordon and Breach Science Publishers, Inc. 1967

Subplanes of projective planes

P. J. LORIMER

If a finite projective plane Π is a Veblen–Wedderburn plane, then it is well known that the order of Π is a power of a prime number p. A natural definition of the characteristic of Π can be given so that this characteristic is p and so that Π has a projective subplane (the fundamental subplane) which is isomorphic to the linear desarguesian plane over the Galois field of order p.

Using an algebraic device, this idea of characteristic has been extended to arbitrary finite projective planes [3]. However, this generalization has two disadvantages. Firstly, the characteristic is not necessarily a prime number, and secondly, little can be said about the existence of subplanes having the characteristic as order.

The object of this paper is to present another generalization of the characteristic of a plane. If Π is a projective plane of order n and characteristic p, then the definition will ensure that the following are true:

(1) $p \geq 2$ is a prime number;

(2) Π has an affine subplane which is isomorphic to the linear affine plane over the Galois field of order p;

(3) $p-1$ divides $n-1$ and $p(p-1)$ divides $n(n-1)$;

(4) the definition reduces to the usual one for Veblen–Wedderburn planes.

1. Co-ordinates in projective planes

Definition 1.1.   Let R be a set of permutations on a set Σ which contains n symbols. Then R will be said to be sharply doubly transitive on Σ if R contains $n(n-1)$ permutations, the identity is a member of R, and whenever (a, b) and (c, d) are two pairs of symbols of Σ, then there is exactly one $r \in R$ with the properties $r(a)=c$ and $r(b)=d$.

Definition 1.2.   A set R of permutations on a set Σ is said to be a ternary ring of Hall if R is sharply doubly transitive on Σ and the relation $\sim$ on R defined by $r_1 \sim r_2$ *if and only if* $r_1 = r_2$ *or* $r_1^{-1} r_2$ *fixes no symbol of* Σ is an equivalence relation.

Because Σ is a finite set the following theorem holds (see [1]).

Theorem 1.3.   *Every sharply doubly transitive set of permutations is a ternary ring.*

The following theorems are easily proved.

Theorem 1.4.   *If* $a \in \Sigma$, *then* R *contains exactly* $n - 1$ *elements* r *with the property* $r(a) = a$.

Theorem 1.5.   *Exactly* $n - 1$ *elements of* R *fix no symbols of* Σ *and these, together with the identity, form an equivalence class under* $\sim$.

Theorem 1.6.   *There are* $n - 1$ *equivalence classes and each contains* n *permutations.*

We now turn to the basic relationship between ternary rings and projective planes. Although the following theorem is standard, see for example [1], or [2], p. 353, we prefer to give a proof here for two reasons. Because it is easier to consider perspectivities between lines than it is to consider perspectivities between pencils of lines we prefer to work in dual planes, so that the proof given here is the dual of the usual one. Also, this proof makes the relationship between ternary rings and projective planes seem more natural.

Theorem 1.7.   *Every projective plane can be given coordinates from a ternary ring, and every ternary ring can be used as a coordinate system for a projective plane.* (The sense in which we use the term "coordinate system" will be made clear from the proof.)

Proof.   Let R be a ternary ring on a set Σ of n symbols. We build a projective plane in the following way.

Firstly there are to be two (as yet disjoint) lines L_1 and L_2 each containing n points. The points of each of the lines will be labeled by the n symbols of Σ. For each pair (a_1, a_2) of points on L_1 and L_2 respectively there is to be a line passing through a_1 and a_2. $n(n-1)$ further points are added and each of these is labeled by a permutation of R. The point labeled by $r \in R$ is to lie on the line passing through the points a_1 and a_2 on L_1 and L_2 if and only if $r(a_1) = a_2$.

There is one further point to be added. It is the intersection of L_1 and L_2 and will be labeled by the symbol ∞. We assume ∞ is not a member of Σ.

Let R^* be an equivalence class of R. If $r_1, r_2 \in R^*$, then r_1 and r_2 are not yet joined by a line, for otherwise $r_1(a) = r_2(a)$ for some $a \in \Sigma$. We add another line and it is to contain the point ∞ and all the points labeled by the permutations of R^*.

When a line has been added for every equivalence class of R, we obtain a projective plane of order n.

Conversely, let Π be a projective plane of order n, and let L_1 and L_2 meeting in a point ∞ be any two lines of Π. Let Σ be the set of points on L_1 excluding ∞ and let I be any point not on L_1 or L_2.

Each of the $n(n-1)$ points P of Π which do not lie on L_1 or L_2 determine a permutation of Σ in the following way. Project L_1 onto L_2 through the point I and then project L_2 back onto L_1 through P. The resulting mapping of Σ onto itself is a permutation and it follows easily that the $n(n-1)$ permutations thus defined form a ternary ring. Each of the points on L_1 and L_2 is then labeled by the permutation corresponding to it.

If P_1 and P_2 are labeled by the permutations r_1 and r_2, then it is clear that $r_1 \sim r_2$ if and only if $P_1 = P_2$ or P_1, P_2 and ∞ are collinear.

Definition 1.8.   The lines L_1 and L_2 referred to in the previous theorem will be called the coordinate axes, and the point I will be called the unit point.

2. Ternary rings

If R is a ternary ring on a set Σ containing n elements, then the elements of R are permutations on Σ. As such they generate a subgroup $G(R)$ say, of the symmetric group of degree n. The properties of $G(R)$ are investigated in this section. Where theorems follow easily, their proofs are omitted.

Theorem 2.1. $G(R)$ *contains a sharply doubly transitive subset, viz. R.*

Corollary 2.2. $G(R)$ *is doubly transitive on Σ.*

Definition 2.3.   If $a, b \in \Sigma$, denote the stabilizer of a by H_a and denote the stabilizer of a and b by H_{ab}.

Theorem 2.4. $H_{ab} = H_a \cap H_b$.

Theorem 2.5. *If* $a, b \in \Sigma$, *then* $|G(R)| = n|H_a| = n(n-1)|H_{ab}|$ *where* $|K|$ *is the order of the group* K.

Definition 2.6. If $a, b \in \Sigma$, denote by $\lambda = \lambda(a, b)$ the subset of Σ which is fixed pointwise by every permutation of H_{ab}.

Suppose $\lambda(a, b)$ contains q symbols.

Theorem 2.7. $q \geq 2$.

Theorem 2.8. *The permutations of* R *form a complete set of representatives of the cosets of* H_{ab} *in* $G(R)$.

Theorem 2.9. *If* $g \in G(R)$ *and* $g(a) = c$, $g(b) = d$, *then* $g H_{ab} g^{-1} = H_{cd}$.

Theorem 2.10. $H_{ab} = H_{cd}$ *if and only if* $c, d \in \lambda(a, b)$.

Theorem 2.11. *If* $r \in R$, *then* $r \in N(H_{ab})$ *if and only if* $r(a)$, $r(b) \in \lambda(a, b)$. $N(H_{ab})$ *is the normalizer of* H_{ab} *in* $G(R)$.

Theorem 2.12. $N(H_{ab})$ *contains* $q(q-1)$ *elements of* R.

Corollary 2.13. $|N(H_{ab})| = q|H_a \cap N(H_{ab})| = q(q-1)|H_{ab}|$.

Theorem 2.14. *The factor group* $N(H_{ab})/H_{ab}$, *regarded as a permutation group on the* q *symbols of* $\lambda(a, b)$ *is doubly transitive and only the identity fixes two symbols of* $\lambda(a, b)$.

Theorem 2.15. *The* $q(q-1)$ *symbols of* $R \cap N(H_{ab})$, *regarded as permutations on the* q *symbols of* $\lambda(a, b)$ *form a sharply doubly transitive set and only the identity fixes two symbols of* $\lambda(a, b)$.

Proof. If $r \in R \cap N(H_{ab})$, then the mapping $r \to r H_{ab}$ is an isomorphism of $R \cap N(H_{ab})$ onto $N(H_{ab})/H_{ab}$ when both sets are regarded as permutations on $\lambda(a, b)$.

Definition 2.16. Denote the set $R \cap N(H_{ab})$ regarded as permutations on $\lambda(a, b)$ by R'.

Corollary 2.17. q *is a prime power, say* $q = p^m$, *and* R' *is isomorphic to a group of linear substitutions* $x \to xm + b$ *in a near field.*

Proof. See [4], or [2], p. 382.

Theorem 2.18. $q - 1$ *divides* $n - 1$ *and* $q(q-1)$ *divides* $n(n-1)$.

Proof. By 2.5 and 2.13.

3. Subplanes of projective planes

Because we are working in dual planes, affine subplanes of projective planes are defined as follows:

Definition 3.1.   Let Π be a projective plane, and Π' a subset of the points and lines of Π. Then Π' is an affine subplane of Π if and only if

(1) the intersection of every two lines of Π' is a point of Π';

(2) if P is a point of Π' and L is a line of Π' which does not pass through P, then there is exactly one point of Π' which lies on L and is not joined to P by a line of Π';

(3) there exist four lines in Π' no three of which pass through any point.

Definition 3.2.   Let R be a ternary ring of permutations on a set Σ. If R' is a subset of R, λ a subset of Σ; and

(1) $r(\lambda) = \lambda$ for every $r \in R'$;

(2) R' is a ternary ring of permutations on λ when the elements of R are regarded as permutations on λ;

then R' will be called a ternary subring of R.

Ternary subrings and affine subplanes are intimately related as the following theorem shows.

Theorem 3.3.   *Let R be a ternary ring on a set Σ of n symbols and suppose that R is a system of coordinates for a projective plane as in 1.7. Suppose that $R' \subseteq R$, $\lambda \subseteq \Sigma$ and that the permutations of R' acting on λ form a ternary subring of R. If Π' consists of the points on the coordinate axes which are labeled by the points of λ, the lines of Π joining any pair of these points, excluding the coordinate axes, and the points of Π labeled by the permutations of R', then Π' is an affine subplane of Π.*

Proof.   By 1.7 and 3.2.

It is well known that every affine plane can be extended to a projective plane. The next theorem states the conditions under which an affine subplane Π' which arises in a projective plane Π as in 3.3 can be extended to a subplane of Π.

Theorem 3.4.   *Let R, R', Π, Π', Σ, and λ be as in 3.3. Suppose λ contains m symbols. As in 1.2 equivalence relations are defined on R and R'. Denote these by $r_1 \sim r_2(R)$ and $r_1 \sim r_2(R')$ respectively. Then Π' is* *an affine subplane of a projective subplane of order m of Π if and only if $r_1 \sim r_2(R')$ implies $r_1 \sim r_2(R)$.*

4. The fundamental subplane

Theorem 4.1. *Let Π be a projective plane of order n. Then there is a prime power q with the properties:*

(1) *$q - 1$ divides $n - 1$, and $q(q - 1)$ divides $n(n - 1)$;*

(2) *Π contains an affine subplane of order q which is isomorphic to a linear affine plane over a near field.*

Proof. By 2.17, 3.2, and 3.3.

Every linear affine plane over a near field of order p^m, p prime, contains an affine subplane which is isomorphic to the linear affine plane over the Galois field of order p. Hence

Theorem 4.2. *Let Π be a projective plane of order n. Then there is a prime number p with the properties:*

(1) *$p - 1$ divides $n - 1$ and $p(p - 1)$ divides $n(n - 1)$.*

(2) *Π contains an affine subplane Π' which is isomorphic to the linear affine plane over the Galois field of order p.*

Definition 4.3. If p and Π' are as in 4.2, p is to be called the characteristic of the plane Π, and Π' the fundamental subplane of Π.

Theorem 4.4. *Let Π be a Veblen–Wedderburn plane of order p^m, p prime. Then the characteristic of Π is p, and the fundamental subplane of Π can be extended to a projective plane in Π.*

Proof. In this case, $G(R)$ is the semidirect product of the elementary abelian group A of order p^m and its group G of automorphisms. $G \times A$ is regarded as a permutation group on the elements of A. Study of $G \times A$ reveals that q (see 2.6) must be a power of p so that Π has characteristic p. The rest of the theorem also follows from the structure of $G \times A$.

5. Conclusions

1. The number q (see 2.6 and 4.1) is not necessarily a prime number. For example, if Π is a linear plane over a field or near field of order p^n, then $q = p^n$. However, in the Veblen–Wedderburn planes that the author has considered, q is already a prime. In a sense, the planes over fields

have the smallest group $G(R)$ so that these considerations would suggest that if Π is not a linear plane over a field or near field, then q itself is a prime so that Theorem 4.1 reduces immediately to Theorem 4.2.

2. Theorem 4.1 does not preclude any integers from being orders of projective planes, for if n is any integer we always have 1 divides $n-1$ and 2 divides $n(n-1)$.

3. The characteristic of a plane is not necessarily independent of the ternary ring used for the coordinates of the plane. If different ternary rings can be used as coordinates for the same plane, then it may happen that the different ternary rings define different characteristics. However, this may sometimes be precluded by the order of the plane itself. For example, if there is a projective plane of order 10, then it follows from 4.2 that its characteristic must be 2.

References

[1] MARSHALL HALL, JR., Projective planes, *Trans. Amer. Math. Soc.* **54** (1943), 229–277.

[2] MARSHALL HALL, JR., *The theory of groups*, Macmillan, New York, 1959.

[3] BENJAMINO SEGRE, *Lectures on modern geometry*, with an appendix by LUCIO LOMBARDO-RADICE, Edizioni Cremonese, Rome, 1961.

[4] HANS ZASSENHAUS, Kennzeichnung endlicher linearer Gruppen als Permutationsgruppen, *Abh. Math. Sem. Univ. Hamburg* **11** (1935), 17–40.

Proc. Internat. Conf. Theory of Groups, Austral. Nat. Univ. Canberra,
August 1965, pp. 241–249. © Gordon and Breach Science Publishers, Inc. 1967

A theorem about critical p-groups

I. D. MACDONALD

This note is a report of attempts to find explicit relations among critical groups each of which generates the same variety. A well-known instance of this situation is provided by the variety generated by either of the non-abelian groups of order 8.

Our main result is:

Theorem 2. *If* $\mathrm{Var}(G) = \mathrm{Var}(H)$ *where G and H are critical p-groups, then* $\mathrm{Var}(\mathsf{QS-I})G = \mathrm{Var}(\mathsf{QS-I})H$.

Here $\mathrm{Var}(\mathsf{QS-I})X$ means the variety generated by the set of proper factors of the group X.

Two metabelian varieties of some interest are presented as examples. The first is generated by a finite p-group but *not* by any critical p-group, while the second is generated by any one of n critical p-groups, n being an arbitrary positive integer given in advance.

The first theorem and its proof are due to Professor Hanna Neumann (unpublished).

Theorem 1. *If G is a critical group with minimally n generators and H is a group with m generators such that* $\mathrm{Var}(G) = \mathrm{Var}(H)$, *then $n \leq m$.*

Proof. It is well known (see for instance [1], Lemma 4.1) that, since H lies in $\mathrm{Var}(G)$, H can be presented as an epimorphic image of a subgroup of the cartesian product of certain subgroups of G; and since H has m generators it is possible to select these subgroups in such a way that each has m generators. Now suppose $m < n$. Each of the subgroups in question then lies in $\mathrm{Var}(\mathsf{QS-I})G$. Therefore H lies in $\mathrm{Var}(\mathsf{QS-I})G$. But since G is critical we cannot now have G in $\mathrm{Var}(H)$. This contradiction shows that $n \leq m$.

Corollary. *The minimal number of generators of a critical group is an invariant of the variety it generates.*

Proof of Theorem 2.   Put $\mathfrak{V} = \mathrm{Var}(G)$ and $\mathfrak{W} = \mathrm{Var}(\mathsf{QS}-\mathsf{I})G$. Suppose that G has class c, has exponent p^e, and has minimally n generators.

Having taken a law $w(x_1, \ldots, x_m) = 1$, we show that whether it holds in $\mathfrak{W}$ or does not hold in $\mathfrak{W}$ is determined solely by the laws in $\mathfrak{V}$ and the fact that a critical group generates $\mathfrak{V}$. Note that $w(x_1, \ldots, x_m) = 1$ holds in $\mathfrak{W}$ if and only if this law holds in the maximal factor group of G and in every maximal subgroup of G.

The monolith N of the critical p-group G is central and has order p. Therefore $w(x_1, \ldots, x_m) = 1$ holds in the maximal factor group G/N of G if and only if both the following laws hold in G:

$$(w(x_1, \ldots, x_m), x_{m+1}) = 1, \qquad w(x_1, \ldots, x_m)^p = 1.$$

So the laws of $\mathfrak{V}$ determine whether or not $w(x_1, \ldots, x_m) = 1$ holds in the maximal factor group of G.

Examination of maximal subgroups is similar in principle but more complicated in detail. Suppose firstly that $w(x_1, \ldots, x_m) = 1$ is a law in every maximal subgroup of G. We shall replace each x_i by some variable of the form

$$(*) \qquad y_1^{\alpha_{i,1}} \cdots y_{n-1}^{\alpha_{i,n-1}} y_n^{p\alpha_{i,n}} B_i(y_1, \ldots, y_n)$$

where $0 \le \alpha_{i,j} < p^e$ for $1 \le j < n$, $0 \le \alpha_{i,n} < p^{e-1}$, and $B_i(y_1, \ldots, y_n)$ is the product of basic commutators in $y_1, \ldots, y_n$; in this product the basic commutators are to occur in their proper order, and we allow a given commutator to occur at most $p^e - 1$ times, and not at all if its weight exceeds $c + 1$. Thus we obtain one of finitely many possible laws in $y_1, \ldots, y_n$, and which we obtain depends of course on our choices of the $\alpha_{i,j}$ and the $B_i(y_1, \ldots, y_n)$. We assert that in any case the resulting law holds in G, and so in $\mathfrak{V}$. For when we substitute fixed elements of G for the y_i, say $y_i = g_i$, the corresponding values of x_i all lie in the subgroup $\mathrm{gp}\{g_1, \ldots, g_{n-1}, \Phi(G)\}$, which lies in some maximal subgroup M of G, and we know that $w(x_1, \ldots, x_m) = 1$ is a law in M.

Conversely, suppose that we obtain a law $w = 1$ in G whenever we take the word $w(x_1, \ldots, x_m)$ and replace each x_i by some variable of the form $(*)$. Let M be an arbitrary (but fixed) maximal subgroup of G. We may then choose elements $g_1, \ldots, g_n$ in G such that $G = \mathrm{gp}\{g_1, \ldots, g_n\}$ and $M = \mathrm{gp}\{g_1, \ldots, g_{n-1}, \Phi(G)\}$. We assert that $w(x_1, \ldots, x_m) = 1$ is a law in M. For an arbitrary element of M has the form $(*)$ with y_i replaced by g_i for $1 \le i \le n$, and so substitution of elements of M for the

x_i in $w(x_1, \ldots, x_m)$ yields 1 by hypothesis. So if replacement of the x_i in $w(x_1, \ldots, x_m) = 1$ by variables of the form $(*)$ yields a law in G, then $w(x_1, \ldots, x_m) = 1$ is itself a law in every maximal subgroup of G.

We have now shown that $w(x_1, \ldots, x_m) = 1$ holds in every maximal subgroup of G if and only if replacement of each x_i by variables of the form $(*)$ yields a law in G. So the laws of $\mathfrak{V}$ determine whether or not $w(x_1, \ldots, x_m) = 1$ holds in every maximal subgroup of G.

Therefore $\mathfrak{W}$ is completely determined by $\mathfrak{V}$ and by the fact that a critical group generates $\mathfrak{V}$. Theorem 2 as stated has now been proved.

The obvious converse of Theorem 2 is false, for Theorem 5.4 of [2] makes it clear that critical p-groups of different classes may have the same variety of proper factors. There is nevertheless a rather unsatisfactory converse of Theorem 2:

Theorem 3. *If G and H are critical p-groups with $\mathrm{Var}(\mathrm{QS}-\mathrm{I})G = \mathrm{Var}(\mathrm{QS}-\mathrm{I})H$ and if G and H are isologic with respect to the latter variety, then $\mathrm{Var}(G) = \mathrm{Var}(H)$.*

Proof. If $\mathfrak{W}$ is any variety and G is any group, let $W(G)$ be the verbal subgroup of G corresponding to $\mathfrak{W}$ and let $M_W(G)$ be the associated marginal subgroup of G. We say that G and H are isologic with respect to $\mathfrak{W}$ if there is an isomorphism between $G/M_W(G)$ and $H/M_W(H)$ which induces an isomorphism between $W(G)$ and $W(H)$. For further details of the isologic concept see [2] and the references given there. In particular we need the fact, shown on p. 98 of [2], that if G and H are isologic with respect to $\mathfrak{W}$ then H is isomorphic to a factor group of a subgroup of the direct product of G and some group in $\mathfrak{W}$.

Now let $\mathfrak{W} = \mathrm{Var}(\mathrm{QS}-\mathrm{I})G = \mathrm{Var}(\mathrm{QS}-\mathrm{I})H$. The theorem follows at once.

It is not true that, if G is isologic to H with respect to some variety, G and H being critical p-groups, then $\mathrm{Var}(G) = \mathrm{Var}(H)$; for instance, any two abelian groups are isoclinic. Nor is it true, as we shall see, that if G and H are critical p-groups with $\mathrm{Var}(G) = \mathrm{Var}(H)$ then G and H are isologic with respect to $\mathrm{Var}(\mathrm{QS}-\mathrm{I})G$.

In view of Theorem 2 a most pleasing situation would arise if it were true that every variety generated by a finite p-group is generated by a critical p-group. But this is by no means so, even in the metabelian case.

Example 1. *There is a variety generated by a metabelian 2-group but not by any one critical group.*

Proof.   Let $\mathfrak{B}$ be the metabelian variety, nilpotent of class 6, defined by the laws (1)–(6):

(1) $$(x, y)^8 = 1,$$

(2) $$(x, y, z)^4 = 1,$$

(3) $$(x, y, z, u, v)^2 = 1,$$

(4) $$(x, y, y, y)^2 = 1,$$

(5) $$(x, y, y, y, y, y) = 1,$$

(6) $$(x, y)^4(x, y, x, y)^2(x, y, x, x, y, y) = 1.$$

Our variety will be a certain subvariety $\mathrm{Var}(G)$ of $\mathfrak{B}$.

We prove later that

(i) *in any group of $\mathfrak{B}$ the values of $(x, y)^4$, $(x, y, x, y)^2$, (x, y, x, x, y, y) are all central; and*

(ii) *in any two-generator group of $\mathfrak{B}$ the verbal subgroups associated with $(x, y)^4$, $(x, y, x, y)^2$, (x, y, x, x, y, y) respectively have order at most 2.*

We shall also construct a group $G = \mathrm{gp}\{a, b\}$ such that

(iii) $G \in \mathfrak{B}$; *and*

(iv) $(a, b)^4$, $(a, b, a, b)^2$, (a, b, a, a, b, b) *each has order 2 while the subgroup they generate has order 4.*

Then we argue as follows. Suppose that $\mathrm{Var}(G) = \mathrm{Var}(C)$ where C is critical. Because G has two generators, Theorem 1 implies that C has two generators, and so C may be represented as a factor group F/N of the free group F on two generators of $\mathrm{Var}(G)$. Let $V_1(F)$, $V_2(F)$, $V_3(F)$ be the verbal subgroups of F associated with $(x, y)^4$, $(x, y, x, y)^2$, (x, y, x, x, y, y) respectively. Each $V_i(F)$ has order 2 by (ii), (iii), and (iv); together they generate a central subgroup Z_0 of F having order 4 by (6), (i), and (iv), and clearly Z_0 is noncyclic. Since C is critical and so monolithic its subgroup Z_0N/N must have order 2 or 1, which implies that N contains one of the verbal subgroups $V_i(F)$ of F. Therefore $F \notin \mathrm{Var}(C) = \mathrm{Var}(G)$. This contradiction shows that C cannot exist.

Since (i)–(iv) are proved by calculation, we shall omit all but the most important details. Statement (i) follows from (2) and (3), the metabelian property being used. A key fact in (ii) and later is that there is a linearity property when any variable in $(x, y)^4$, $(x, y, x, y)^2$ or

(x, y, x, x, y, y) is replaced by a product of variables and an expansion is carried out; this results from (2), (3), and the class 6 property. Take $G = \mathrm{gp}\{a, b\}$ in $\mathfrak{V}$ and let $x \equiv a^\xi b^\eta$, $y \equiv a^\zeta b^\omega$ modulo $\Phi(G)$; it will be found that

$$(x, y)^4 = (a, b)^{4(\xi\omega - \eta\zeta)},$$
$$(x, y, x, y)^2 = (a, b, a, b)^{2(\xi\omega + \eta\zeta)(\xi\omega - \eta\zeta)},$$
$$(x, y, x, x, y, y) = (a, b, a, a, b, b)^{(\xi^2\omega^2 + \eta^2\zeta^2)(\xi\omega - \eta\zeta)}.$$

Here (4) and (5) have been used, as well as the law

$$(x, y, u, v) = (x, y, v, u)$$

which holds in every metabelian group. Now (ii) follows from (1), (2), and (3).

We outline a construction for the group G which we have in mind. The abelian group A generated by

$$c, d_1, d_2, e_1, e_2, e_3, f_1, f_2, f_3, f_4, g_1, g_2$$

with the relations

$$c^8 = 1, \quad d_i^4 = 1, \quad e_1^2 = e_2^4 = e_3^2 = 1, \quad f_j^2 = 1, \quad g_k^2 = 1,$$

for $1 \leq i \leq 2$, $1 \leq j \leq 4$, $1 \leq k \leq 2$ has order 2^{17}. It may be verified that an automorphism α of A, with order 4, is obtained from the mapping

$$
\begin{aligned}
c\alpha &= cd_1, \\
d_1\alpha &= d_1 e_1, & d_2\alpha &= d_2 e_2, \\
e_1\alpha &= e_1 f_1, & e_2\alpha &= e_2 f_2, & e_3\alpha &= e_3 f_3, \\
f_1\alpha &= f_1, & f_2\alpha &= f_2 g_1, & f_3\alpha &= f_3 c^4 e_2^2, & f_4\alpha &= f_4 g_2, \\
g_1\alpha &= g_1, & g_2\alpha &= g_2.
\end{aligned}
$$

Define B to be the splitting extension of A by a group generated by an element a, with order 8, which induces α in A. It may be verified that an automorphism β of B, with order 8, is obtained from the mapping

$$
\begin{aligned}
a\beta &= ac, \\
c\beta &= cd_2, \\
d_1\beta &= d_1 e_2, & d_2\beta &= d_2 e_3, \\
e_1\beta &= e_1 f_2, & e_2\beta &= e_2 f_3, & e_3\beta &= e_3 f_4, \\
f_1\beta &= f_1 g_1, & f_2\beta &= f_2 c^4 e_2^2, & f_3\beta &= f_3 g_2, & f_4\beta &= f_4, \\
g_1\beta &= g_1, & g_2\beta &= g_2.
\end{aligned}
$$

Define G to be the splitting extension of B by a group generated by an element b, with order 8, which induces β in B. Thus G has order 2^{23}, and may be described by generators and defining relations. It is easy to see that $G = \mathrm{gp}\{a, b\}$.

We proceed to (iii). It may be verified that $G' = A$, so that G is metabelian; that G is nilpotent of class 6; and that the exponents of $\gamma_2(G), \gamma_3(G), \gamma_5(G)$ are 8, 4, 2 respectively. This leaves (4), (5), (6) to be established. As for (4) we have

$$(a, b, a, a)^2 = e_1^2 = 1,$$
$$(a, b, b, b)^2 = e_3^2 = 1$$

in G, which enables us to deduce that, if $x \equiv a^\xi b^\eta$, $y \equiv a^\zeta b^\omega$ modulo $\Phi(G)$, then

$$(x, y, y, y)^2 = (a, b, a, b)^{4\zeta\omega\theta} = e_2^{4\zeta\omega\theta} = 1,$$

θ being $\xi\omega - \eta\zeta$. To prove (5) we note that

$$(a, b, a, a, a, a) = (a, b, b, b, b, b) = 1;$$

the same substitutions for x and y as above give the value of

$$(x, y, y, y, y, y)$$

as

$$g_1^{4\zeta^3\omega\theta}(c^4 e_2^2)^{6\zeta^2\omega^2\theta} g_2^{4\zeta\omega^3\theta},$$

which is clearly 1. Similarly the value of

$$(x, y)^4 (x, y, x, y)^2 (x, y, x, x, y, y)$$

is found to be

$$c^{4\theta}(e_1^{\xi\zeta} e_2^{\xi\omega + \eta\zeta} e_3^{\eta\omega})^{2\theta}(c^4 e_2^2)^{(\xi^2\omega^2 + \eta^2\zeta^2)\theta}.$$

Since the values of ξ, η, ζ, ω are essentially integers modulo 2 we have $\xi^2\omega^2 + \eta^2\zeta^2 \equiv \xi\omega + \eta\zeta \equiv \theta$, and $\theta^2 \equiv \theta$, so it follows that (6) holds. This proves (iii): $G \in \mathfrak{B}$.

Turning to (iv) we find that

$$(a, b)^4 = c^4,$$
$$(a, b, a, b)^2 = e_2^2,$$
$$(a, b, a, a, b, b) = c^4 e_2^2;$$

each of these elements in G has order 2, and together they generate a subgroup of order 4. Therefore (iv) holds.

We have thus completed our justification of the claims made for Example 1. It seems likely that the same method would serve to construct varieties generated by a finite p-group (p being any prime) but not by any critical group. Such a construction has hardly enough interest or merit to make it worth the trouble, especially as the class of the variety would no doubt far exceed p. We note that very many varieties are not generated by any critical group, for example the abelian variety of exponent 6.

The purpose of the second example is to demonstrate that the content of our theorems is by no means trivial.

Example 2. *For each integer n greater than 1 there is a variety generated by any one of n nonisomorphic critical metabelian p-groups, p being any prime greater than n.*

Proof. After constructing in outline a certain group G we exhibit n nonisomorphic critical generators G_m of $\mathfrak{B} = \mathrm{Var}(G)$, where $1 \leq m \leq n$.

The abelian group A generated by

$$c_{ij} \qquad (0 \leq i \leq n, 0 \leq j \leq n-i)$$

with the relations

$$c_{ij}^p = 1 \qquad (0 \leq i \leq n, 0 \leq j \leq n-i)$$

has order $p^{(n+1)(n+2)/2}$. It may be verified that, as $p > n$, an automorphism α of A, with order p, is obtained from the mapping

$$c_{ij}\alpha = c_{ij}c_{i+1,j} \qquad (0 \leq i+j < n),$$
$$c_{k,n-k}\alpha = c_{k,n-k} \qquad (0 \leq k \leq n).$$

Define B to be the splitting extension of A by a group generated by an element a, with order p, which induces α in A. It may be verified that an automorphism β of B, with order p, is obtained from the mapping

$$a\beta = ac_{00},$$
$$c_{ij}\beta = c_{ij}c_{i,j+1} \qquad (0 \leq i+j < n),$$
$$c_{k,n-k}\beta = c_{k,n-k} \qquad (0 \leq k \leq n).$$

Define G to be the splitting extension of B by a group generated by an element b, with order p, which induces β in B. It is easy to see that $G = \mathrm{gp}\{a, b\}$, that $G' = A$ which implies that G is metabelian, and that G has class $n+2$. If $p > n+2$ then G, being a regular p-group generated by elements of order p, even has exponent p.

Next we define N_m for $1 \leq m \leq n$ to be the subgroup generated by the following elements of G:

$$c_{n-m+i-j+2,j-1}^{-1} c_{n-m+i-j+1,j} \qquad (1 \leq i \leq m-1, 1 \leq j \leq i),$$
$$c_{i,m+j} \qquad (0 \leq i \leq n-m, 0 \leq j \leq n-m-i).$$

It is readily verified that N_m is a normal subgroup of G, and we define G_m to be G/N_m. Thus $G_m \in \mathrm{Var}(G)$ for $1 \leq m \leq n$.

However, it is also true that $G \in \mathrm{Var}(G_m)$. It is easy to see that G is isomorphic to a subgroup of the direct product of several copies of G_m. Thus each G_m is a generator of $\mathfrak{B} = \mathrm{Var}(G)$.

The order of G_m will be found to be $p^{2+(n-m+2)m}$ for $1 \leq m \leq n$, so we have n nonisomorphic groups provided there is no isomorphism between G_m and G_{n-m+2} with $m < n-m+2$. Suppose there is an isomorphism between these groups according to which x, b in G_m map onto a, $a^\xi b^\eta g$ in G_{n-m+2}, where $g \in \gamma_2(G_{n-m+2})$. Since $c_{i,m+j} = 1$ in G_m we have $(a, (m+1)a^\xi b^\eta g) = 1$ in G_{n-m+2}. Therefore we have

$$c_{0m}^{\eta^{m+1}} c_{1,m-1}^{m\xi\eta^m} \cdots c_{m-1,1}^{m\xi^{m-1}\eta^2} c_{m0}^{\xi^m\eta} \equiv 1$$

modulo $\gamma_{m+3}(G_{n-m+2})$ in G_{n-m+2}, and this implies that $\eta \equiv 0$ modulo p. After similarly examining the action of the supposed automorphism on a, b in G_m, we come to a contradiction. Therefore no two of the groups G_m are isomorphic.

It remains to prove that each G_m is critical. Since G_m has two generators it is sufficient to prove that its center Z_m is cyclic. We shall denote $aN_m, bN_m, c_{ij}N_m$ $(0 \leq i \leq n-m, 0 \leq j \leq m-1)$ by a, b, c_{ij} respectively, and $c_{n-m+i-j+1,j}N_m$ $(1 \leq i \leq m-1, 1 \leq j \leq i)$ by $c_{n-m+i+1}$.

We show that $Z_m \leq \gamma_2(G_m)$. If $a^\eta b^\zeta g \in Z_m$ where $g \in \gamma_2(G_m)$ then $(a^\eta b^\zeta g, c_{00}) = 1$, and examination of this relation modulo $\gamma_3(G_m)$ gives $\eta \equiv \zeta \equiv 0$ modulo p. So we may take an arbitrary central element to be

$$x = \prod c_{ij}^{\xi_{ij}} \cdot \prod c_{n-m+k+1}^{\xi_k}$$

where the products are taken over $0 \leq i \leq n-m$, $0 \leq j \leq m-1$, $0 \leq k \leq m-1$. Form (x, b), and find that

$$\prod c_{i,j+1}^{\xi_{ij}} \cdot \prod c_{n-m+k+2}^{\xi_k} = 1,$$

where $0 \leq i \leq n-m$, $0 \leq j \leq m-2$, $0 \leq k \leq m-2$. It follows that the only nonzero ξ's are essentially $\xi_{i,m-1}$ and ξ_{m-1}, so we shall take

$$x = \prod c_{i,m-1}^{\xi_{i,m-1}}$$

s c_n is central anyway. Form (x, a), and find that

$$\prod c_{i+1,m-1}^{\xi_{i,m-1}} = 1$$

here $0 \leq i \leq n-m-1$. It follows that each $\xi_{i,m-1} \equiv 0$ modulo p. Hence $G_m = \text{gp}\{c_n\}$, which is certainly cyclic; each G_m is critical.

This completes our consideration of Example 2.

It should be mentioned in this context that Professor B. H. Neumann as produced (unpublished) a set of 2-groups akin to Example 2, amely the critical groups G_m generating the same variety for $1 \leq m \leq n$, here

$$G_1 = \text{gp}\{a, b \mid a^{2^{17}} = b^2 = 1, b^{-1}ab = a^{-1}\},$$
$$G_m = \text{gp}\{a, b \mid a^{2^n} = 1, b^{2^{m-1}} = a^{2^{n-1}}, b^{-1}ab = a^{-1}\},$$

or $2 \leq m \leq n$. Note that G_1 has order 2^{n+1} while G_m has order 2^{n+m-1} or $2 \leq m \leq n$, so that the groups are nonisomorphic.

Finally, we mention that if we take $n=2$ in Example 2 then G_1 is ot isologic to G_2 with respect to the common proper factor variety. 'or otherwise G_2 would be isomorphic to a factor group of a subgroup f one copy of G_1 and some group in $\text{Var}(\text{QS}-\text{I})G_1$, and an easy calulation leads to a contradiction. Indeed, if $n=2$ and $p>3$ then every roper subvariety of $\text{Var}(G)$ has class 3; and the same method of proof hows that G_1 is not isologic to G_2 with respect to any proper subvariety f $\text{Var}(G)$.

I thank the Professors Neumann for showing me two of their npublished results which I have included. I must record special thanks o Professor Hanna Neumann who has been responsible for persistent timulation in the preparation of these results.

References

1] GRAHAM HIGMAN, Some remarks on varieties of groups, *Quart. J. Math. Oxford Ser.* (2) **10** (1959), 165–178.

2] PAUL M. WEICHSEL, On critical p-groups, *Proc. London Math. Soc.* (3) **14** (1964), 83–100.

Proc. Internat. Conf. Theory of Groups, Austral. Nat. Univ. Canberra,
August 1965, pp. 251–259. © Gordon and Breach Science Publishers, Inc. 1967

Varieties of groups

HANNA NEUMANN

arieties may be considered as special categories, with algebras of a
ertain type as objects and operation-preserving mappings as morph-
ims. The role of varieties in universal algebra or category theory is not
ur concern here; we are interested in varieties of groups only, for the
iformation on groups that their study affords. Accordingly, for us a
ariety is a suitable class of groups.

The subject was initiated in the 1930's by G. Birkhoff's paper on
quationally defined classes of algebras [5] and B. H. Neumann's paper
20] on identical relations in groups. The name "variety" for such classes
f algebras is due to Philip Hall whose courses on universal algebra in
ambridge were for several years the only further contribution to the
ibject in general; the more specific study of groups in the setting of
eneral algebra seems to have been entirely dormant until the recent
urst of activity which started no more than fifteen years ago. What
llows is a survey of problems and results leading up to current
vestigations as far as they are known to me.

Let w be a *word*, that is an element of the free group X freely generated
y the alphabet $\mathfrak{x} = \{x_1, x_2, \ldots\}$. If A is a group and $\alpha \colon X \to A$ a
omomorphism, then $w\alpha$ is a *value* of w in A. To compute it, let $x_i\alpha = a_i$
r $x_i \in \mathfrak{x}$, $w = w(x_1, \ldots, x_n)$, then $w\alpha = w(a_1, \ldots, a_n)$ where the right-
and side is evaluated in the group A.

If $\mathfrak{v}$ is a set of words, then the corresponding *verbal subgroup* of A
defined by

$$\mathfrak{v}(A) = \mathrm{gp}(w\alpha : w \in \mathfrak{v}, \alpha \in \mathrm{Hom}(X, A)).$$

or example, for $\mathfrak{v} = \{[x, y]\}$—we write $[x, y] = x^{-1}y^{-1}xy$—we have
$(A) = A'$, the derived group of A. If θ is a homomorphism defined on
., then clearly $\mathfrak{v}(A\theta) = \mathfrak{v}(A)\theta$.

251

The *variety* defined by $\mathfrak{v}$ is the class of all groups A with $\mathfrak{v}(A) = \{1$
If then we say that a word is a *law* in A if all its values in the group A
are equal to one, then every word of $\mathfrak{v}$ is a law in every group of th
variety defined by $\mathfrak{v}$. So, for example, $[x, y]$ defines the variety of a
abelian groups.

Put $\mathfrak{v}(X) = V$, the set of words that are *consequences* of $\mathfrak{v}$. Then
is a fully invariant subgroup of X. Using that every countable group
has a presentation as an epimorphic image of X, $A = X\alpha$ for suitabl
$\alpha \in \mathrm{Hom}(X, A)$, one confirms

$$\mathfrak{v}(A) = \mathfrak{v}(X\alpha) = \mathfrak{v}(X)\alpha = V\alpha = V(A).$$

Hence the sets of words $\mathfrak{v}$ and $V = \mathfrak{v}(X)$ determine the same verb
subgroup of A for all A. Consequently one has the following chain
equivalent statements:

$$\mathfrak{v}(A) = \{1\} \Leftrightarrow V(A) = \{1\} \Leftrightarrow V(X)\alpha = \{1\} \Leftrightarrow V(X) \le \ker \alpha,$$

showing that the variety $\mathfrak{B}$ defined by $\mathfrak{v}$ consists of all epimorphi
images of $F/V(F) \cong X/V$ (where F is a free group of countably infinit
rank) as far as its countable groups are concerned. As these determin
the variety we neglect the others for the purposes of this lecture.

If F_n is free of rank n, then $F_n/V(F_n) = F_n(\mathfrak{B})$ is called the $\mathfrak{B}$-fre
group of rank n. Every group of $\mathfrak{B}$ is an epimorphic image of som
$\mathfrak{B}$-free group, just as every group is an epimorphic image of som
absolutely free group.

$\mathfrak{B}$-free groups can be defined intrinsically: A group is called *relativel*
free (P. Hall [9]) if it possesses a set of generators such that ever
mapping of these generators into the group can be continued to a
endomorphism of the group. One finds that the relatively free group
are precisely the groups that are $\mathfrak{B}$-free in some variety $\mathfrak{B}$.

These relatively free groups, as a class, are worth investigating. Th
group G of $\mathfrak{B}$ is called *projective in* $\mathfrak{B}$ if to every pair of groups A and
in $\mathfrak{B}$ and homomorphisms $\alpha: A \twoheadrightarrow B$, $\beta: G \to B$, there exists a hom
morphism $\gamma: G \to A$ such that $\gamma\alpha = \beta$. In 1950, Saunders MacLane [1
pointed out that every $\mathfrak{B}$-free group is projective in $\mathfrak{B}$ and that, if
has the *Schreier property*, conversely all projective groups in $\mathfrak{B}$ are fre
Here a variety is said to have the Schreier property if all subgroups
its free groups are free in the same variety; the varieties $\mathfrak{O}$ (all groups
$\mathfrak{A}$ (all abelian groups), $\mathfrak{A}_p$ (all abelian groups of prime exponent p) a
examples of such varieties.

In 1954, P. Hall [9] set out to determine all projective groups in a variety hoping—I think this is fair to say—that almost always these would be free in the variety, although "very few varieties have the Schreier property" (P. Hall, *loc. cit.*). He found few varieties which are such that all their projective groups are free in the variety and those that have the property seem to have it by accident, as it were. There is still a chance that the question, asked in P. Hall's paper, whether in all varieties of exponent zero or a prime power the projective groups are free in the variety, may have a positive answer. But this remains a conjecture. The other problem suggested by these observations, namely to determine all Schreier varieties, has recently been solved: The varieties $\mathfrak{O}$, $\mathfrak{A}$, and $\mathfrak{A}_p$ are in fact the only varieties with the Schreier property (Peter M. Neumann and James Wiegold [26]). The proof raised some questions that have been answered, at least partly, since this lecture was given: Peter M. Neumann and M. F. Newman [25] succeeded in making the argument in the case of varieties of finite exponent independent of A. I. Kostrikin's deep result on finite groups of prime exponent; M. F. Newman also simplified substantially the proof in the case of varieties of exponent zero (*loc. cit.*) where, moreover, theorems of Auslander and Lyndon type are now no longer needed. We turn next to these theorems, and to problems related to them.

In 1955, M. Auslander and R. C. Lyndon [1] showed: If R and S are normal subgroups of an absolutely free group F, then $R' \subseteq S'$ implies $R \subseteq S$. B. H. Neumann [21] generalized this in 1962 by showing that $V(R) \subseteq V(S)$ implies $R \subseteq S$ for certain verbal subgroups including in particular the terms of the lower central series, and Peter M. Neumann [24], in 1965, extended still further the class of verbal subgroups for which the implication can be shown true. But the general problem remains open: Does $V(R) \subseteq V(S)$ with $\mathfrak{B} \neq \mathfrak{O}$ imply $R \subseteq S$? A special case of this conjecture has recently been proved (Peter M. Neumann [24]): If $V(R) = V(F)$ and $\mathfrak{B} \neq \mathfrak{O}$, then $R = F$. Tekla Taylor (cf. M. J. Dunwoody [6]) found that when F has infinite rank, or also under some special assumptions on $\mathfrak{B}$, the assumption that R is normal in F is superfluous. One hopes, therefore, that for every subgroup R of F and $\mathfrak{B} \neq \mathfrak{O}$, $V(R) = V(F)$ is possible only when $R = F$.

The proofs of the results just mentioned are generally complicated. More information on free subgroups of free groups would be helpful—this seems natural if one observes, for example, that $V(R) = V(F)$

implies that the $\mathfrak{V}$-free group $F/V(F)$ contains the $\mathfrak{V}$-free subgroup $R/V(R)$.

Much useful work has been done by A. W. Mostowski [19], A. L. Šmel'kin [31], Gilbert Baumslag [3], [4], and others to collect information on conditions under which a subset of a $\mathfrak{V}$-free group generates freely a $\mathfrak{V}$-free subgroup. All the results obtained so far relate to residually nilpotent free groups of various kinds. There is a complete blank in our knowledge on free subgroups of other relatively free groups. A closer look at the existing results shows that in all but the absolutely free groups the rank of a free subgroup is never greater than the rank of the whole group. Results by Peter M. Neumann [24] lead to the conjecture that if for some $n > 1$, $F_{n+1}(\mathfrak{V})$ can be embedded in $F_n(\mathfrak{V})$ then $\mathfrak{V} = \mathfrak{O}$—there, of course, this is possible: even the absolutely free group of rank two contains free subgroups of every finite and countably infinite rank. Again, a proof of this conjecture would be of considerable help in other problems, but is still outstanding.

I need now an alternative characterization of varieties by means of *closure properties* (G. Birkhoff [5]): A class $\mathfrak{V}$ of groups is a variety if and only if it is closed under the operations of taking subgroups, epimorphic images and cartesian products. This characterization shows at once that varieties form a lattice under the natural partial ordering by inclusion, and the lattice of the corresponding fully invariant subgroups V of X is anti-isomorphic to that of the varieties under the natural correspondence where $\mathfrak{V}$ is the variety defined by the set V of words. The lattice is complete, $\mathfrak{U} \wedge \mathfrak{V}$ and $\mathfrak{U} \vee \mathfrak{V}$ being the intersection and the least variety containing $\mathfrak{U}$ and $\mathfrak{V}$ respectively. The variety generated by a set of groups is now defined; we write var(A) or var$\{A_\lambda\}$. One knows (B. H. Neumann [20]) that every free group of the variety var$\{A_\lambda\}$ is a subgroup of a cartesian product of groups in $\{A_\lambda\}$. Hence the relevance of residual properties in varietal contexts: for example, if $\mathfrak{V}$ is generated by its finite groups, then every $\mathfrak{V}$-free group is residually finite, and conversely.

The characterization by closure properties shows without trouble that the *product* $\mathfrak{U}\mathfrak{V}$ of two varieties, defined as the class of all groups that are extensions of a group in $\mathfrak{U}$ by a group in $\mathfrak{V}$, is itself a variety. With respect to this multiplication, varieties form a semigroup, with identity $\mathfrak{E}$ (consisting of the trivial group only) and zero element $\mathfrak{O}$. *Indecom-*

posable varieties are then defined in the obvious way, and on account of a theorem by F. W. Levi [14] on chains of characteristic subgroups in an absolutely free group, every variety is a product of a finite number of indecomposable factors. In fact this decomposition is unique (A. L. Šmel'kin [30]; B. H., Hanna, and Peter M. Neumann [22]), so that the semigroup of varieties is free, freely generated by the indecomposable varieties. Large classes of varieties are indecomposable (Peter M. Neumann [23]); whether their number is countable or not, or equivalently, whether the set of varieties is countable or not, is an open problem to which we refer back later in a different context.

The proofs, found simultaneously but independently, of the uniqueness of decomposition both use wreath products in exactly the same way. The reason underlying the argument is that the wreath product of $F(\mathfrak{U})$ by $F(\mathfrak{V})$, both groups taken of infinite rank, generates the variety $\mathfrak{U}\mathfrak{V}$. In fact it even *discriminates* it in the following sense: A group D discriminates the variety $\mathfrak{V}$ if to every finite set $\mathfrak{w}$ of words that are not laws in $\mathfrak{V}$ there exists a single homomorphism $\delta: X \to D$ such that $1 \notin \mathfrak{w}\delta$. This concept of discrimination is a strong tool and has been used for many purposes; Gilbert Baumslag [2] used it to prove that common residual properties of $F(\mathfrak{U})$ and $F(\mathfrak{V})$ are normally inherited by $F(\mathfrak{U}\mathfrak{V})$, thus giving a new and simple proof of Karl Gruenberg's results on residual properties of free polynilpotent groups ([8]).

A. L. Šmel'kin has used the *verbal wreath product*, whose base group is a verbal product rather than the direct product, to obtain information on the free groups of product varieties, in particular also in the polynilpotent case. The fact that wreath products have proved an effective tool for the investigation of product varieties means that product varieties are on the whole more tractable than other non-nilpotent varieties. We remark, however, that a different approach to polynilpotent varieties is at present being used by M. A. Ward who generalizes the commutator collecting process to investigate a wider class of varieties than the traditional collecting process can be applied to (see his report in these Proceedings).

The earliest, and still the most exciting problem in the theory of varieties is the *finite basis problem*: Are all the laws of a variety consequences of a finite number amongst them? Note that the term *basis* merely means that all laws follow from those in the basis; no minimality or irredundance is implied. As every variety can be generated by a single group (for instance its free group of countably infinite rank) we may ask equivalently: Are the laws of every group finitely based?

The first significant result is due to R. C. Lyndon [15] who proved in 1952 that the laws of every nilpotent variety have a finite basis. In 1959, Graham Higman [10] extended the method and the result to product varieties $\mathfrak{U}\mathfrak{B}$ where $\mathfrak{U}$ is nilpotent and the laws of $\mathfrak{B}$ are finitely based. Very little later Higman's pupil, D. C. Cross, could show that the laws of certain finite soluble groups have a finite basis [11]. His methods were extended by M. B. Powell [29] to cover every finite soluble group and, in 1964, Sheila Oates and M. B. Powell [28] succeeded in proving the result for every finite group, thus providing the answer to a question asked by B. H. Neumann in 1937 [20]. Finally D. E. Cohen very recently proved that the laws of every metabelian variety are finitely based.

Before reporting in more detail on these two most recent developments, we remark that if the laws of every variety have a finite basis, then the number of varieties is countably infinite. But the construction of a variety that is not finitely based, or—equivalently—of a properly descending infinite chain of varieties, would not by itself answer the question whether the set of distinct varieties is, or is not, countable.

We return to the theorem by Oates and Powell on *the laws of a finite group*. We call *factor* of a group any factor group of a subgroup; the factor is proper if it is not isomorphic to the whole group. The crucial concept introduced by Cross is that of a *critical group*, that is a finite group which does not belong to the variety generated by its proper factors. Now call *Cross variety* a variety $\mathfrak{B}$ with the following three properties:

(i) finitely generated groups in $\mathfrak{B}$ are finite,

(ii) $\mathfrak{B}$ contains only a finite number of nonisomorphic critical groups,

(iii) its laws are finitely based.

Then one easily shows that every subvariety of a Cross variety is itself a Cross variety. The problem of showing that the laws of a finite group are finitely based now is equivalent to proving that every finite group is contained in some Cross variety.

The rest of the original proof has recently been replaced by a considerably simpler argument due to L. G. Kovács and M. F. Newman [12], who have proved:

The class $\mathfrak{C}$ of all groups such that

(a) the exponent divides a fixed positive integer e,

(b) every chief series has factors whose orders are at most m for some fixed positive integer m,

(c) every nilpotent factor has class at most c for some positive integer c,

is a Cross variety. Since every finite group clearly belongs to some such class, this completes the argument.

It is easy to establish that $\mathfrak{C}$ has the property (i) of Cross varieties. To confirm (ii) one has to show that the order of a critical group in $\mathfrak{C}$ can be bounded in terms of e, m, and c; this requires a simplified version of the relevant part of the Oates–Powell argument. It follows now that the variety generated by $\mathfrak{C}$ is in fact generated by a single finite group, say G, in $\mathfrak{C}$. It is known that all groups in the variety $\mathfrak{B}$ defined by a sufficiently large finite set of laws of G satisfy (a) and (c). The next step is to show that, for suitable n, G satisfies the law u_n defined recursively by

$$u_3 = [(x_1^{-1}x_2)^{x_1.2}, (x_1^{-1}x_3)^{x_1.3}, (x_2^{-1}x_3)^{x_2.3}]$$

and

$$u_n = [u_{n-1}, (x_1^{-1}x_n)^{x_1.n}, \ldots, (x_{n-1}x_n)^{x_{n-1}.n}] \quad \text{for } n > 3,$$

and that, if u_n is among the defining laws of $\mathfrak{B}$, then every group in $\mathfrak{B}$ satisfies (b) as well, at least with a suitable m' in place of m. Consequently, in this case $\mathfrak{B}$ satisfies also (i) and (ii), so that $\mathfrak{B}$ is a Cross variety. By definition, $\mathfrak{C} \subset \mathfrak{B}$. The final step is to show that $\mathfrak{C}$ is a variety, and this is now easy.

The investigations of the structure of critical groups by L. G. Kovács and M. F. Newman [13], Sheila Oates [27], P. M. Weichsel [32], [33] are leading on in various directions. The continued interest of these is due to the fact that every variety that is generated by its finite groups is generated by its finite critical groups. We refer for current investigations to the various reports, in this volume, by the authors just quoted, to Graham Higman's report on his investigation, along different lines, of locally finite varieties, and to the reports by W. Brisley, P. J. Cossey, and I. D. Macdonald.

D. E. Cohen's result that all metabelian varieties are finitely based is achieved by entirely different methods. He proves the maximal condition for verbal subgroups—in fact for a larger class of subgroups—in metabelian free groups by translating the problem into one on chains of operator modules, using Magnus' matrix representation of metabelian groups [17]; cf. also R. H. Fox [7]. From here the problem is further reduced to one on chain conditions for certain subsemigroups of a semigroup and is finally solved in this form. M. F. Newman has remarked that a slight modification of, and addition to, the argument shows that every subvariety of $\mathfrak{N}_c\mathfrak{A} \cap \mathfrak{A}\mathfrak{N}_c$ is finitely based, that is every variety that is both nilpotent-by-abelian and abelian-by-nilpotent. This is as far as some of us would expect matters to go: there are those who believe that even the variety of all center-extended-by-metabelian groups will contain infinite properly descending chains of subvarieties.

References

[1] MAURICE AUSLANDER and R. C. LYNDON, Commutator subgroups of free groups, *Amer. J. Math.* **77** (1955), 929–931.

[2] GILBERT BAUMSLAG, Wreath products and extensions, *Math. Z.* **81** (1963), 286–299.

[3] GILBERT BAUMSLAG, Some subgroup theorems for free ʋ-groups, *Trans. Amer. Math. Soc.* **108** (1963), 516–525.

[4] GILBERT BAUMSLAG, A subgroup theorem for some product varieties, *Arch. Math.* **6** (1965), 337–341.

[5] GARRETT BIRKHOFF, On the structure of abstract algebras, *Proc. Cambridge Philos. Soc.* **31** (1935), 433–454.

[6] M. J. DUNWOODY, On verbal subgroups of free groups, *Arch. Math.* **16** (1965), 153–157.

[7] RALPH H. FOX, Free differential calculus. I. Derivation in the free group ring, *Ann. of Math.* (2) **57** (1953), 547–560.

[8] K. W. GRUENBERG, Residual properties of infinite soluble groups, *Proc. London Math. Soc.* (3) **7** (1957), 29–62.

[9] P. HALL, The splitting properties of relatively free groups, *Proc. London Math. Soc.* (3) **4** (1954), 343–356.

[10] GRAHAM HIGMAN, Some remarks on varieties of groups, *Quart. J. Math. Oxford Ser.* (2) **10** (1959), 165–178.

[11] GRAHAM HIGMAN, Identical relations in finite groups, *Conv. Internaz. di Teoria dei Gruppi Finiti (Firenze, 1960)*, pp. 93–100, Edizioni Cremonese, Rome, 1960.

[12] L. G. KOVÁCS and M. F. NEWMAN, Cross varieties of groups, *Proc. Roy. Soc. London Ser. A* **292** (1966), 530–536.

[13] L. G. KOVÁCS and M. F. NEWMAN, On critical groups, *J. Austral. Math. Soc.* **6** (1966), 237–250.

[14] FRIEDRICH LEVI, Über die Untergruppen der freien Gruppen. II., *Math. Z.* **37** (1933), 90–97.

[15] R. C. LYNDON, Two notes on nilpotent groups, *Proc. Amer. Math. Soc.* **3** (1952), 579–583.

[16] SAUNDERS MACLANE, Duality for groups, *Bull. Amer. Math. Soc.* **56** (1950), 485–516.

[17] WILHELM MAGNUS, On a theorem of Marshall Hall, *Ann. of Math.* (2) **40** (1939), 764–768.

[18] S. MORAN, A subgroup theorem for free nilpotent groups, *Trans. Amer. Math. Soc.* **103** (1962), 495–515.

[19] A. W. MOSTOWSKI, Nilpotent free groups, *Fund. Math.* **49** (1961), 259–269.

[20] B. H. NEUMANN, Identical relations in groups, I, *Math. Ann.* **114** (1937), 506–525.

[21] B. H. NEUMANN, On a theorem of Auslander and Lyndon, *Arch. Math.* **13** (1962), 4–9.

[22] B. H. NEUMANN, HANNA NEUMANN, and PETER M. NEUMANN, Wreath products and varieties of groups, *Math. Z.* **80** (1962), 44–62.

[23] PETER M. NEUMANN, Some indecomposable varieties of groups, *Quart. J. Math. Oxford Ser.* (2) **14** (1963), 46–50.

[24] PETER M. NEUMANN, On word subgroups of free groups, *Arch. Math.* **16** (1965), 6–21.

[25] PETER M. NEUMANN and M. F. NEWMAN, Schreier varieties of groups [to appear].

[26] PETER M. NEUMANN and JAMES WIEGOLD, Schreier varieties of groups, *Math. Z.* **85** (1964), 392–400.

[27] SHEILA OATES, Identical relations in groups, *J. London Math. Soc.* **38** (1963), 71–78.

[28] SHEILA OATES and M. B. POWELL, Identical relations in finite groups, *J. Algebra* **1** (1964), 11–39.

[29] M. B. POWELL, Identical relations in finite soluble groups, *Quart. J. Math. Oxford Ser.* (2) **15** (1964), 131–148.

[30] A. L. ŠMEL'KIN, The semigroup of group manifolds, *Dokl. Akad. Nauk SSSR* **149** (1963), 543–545 [Russian] = *Soviet Math. Doklady* **4** (1963), 449–451.

[31] A. L. ŠMEL'KIN, Free polynilpotent groups, *Dokl. Akad. Nauk SSSR* **151** (1963), 73–75; *Izvestia Akad. Nauk SSSR, Ser. Math.* **28** (1964), 91–122 [Russian].

[32] P. M. WEICHSEL, A decomposition theory for finite groups with applications to *p*-groups, *Trans. Amer. Math. Soc.* **102** (1962), 218–226.

[33] P. M. WEICHSEL, On critical *p*-groups, *Proc. London Math. Soc.* (3) **14** (1964), 83–100.

Proc. Internat. Conf. Theory of Groups, Austral. Nat. Univ. Canberra,
August 1965, pp. 261–264. © Gordon and Breach Science Publishers, Inc. 1967

Identical relations in a small number of variables

SHEILA OATES

§1. *Introduction.* In his paper [3] P. M. Weichsel shows that if G is a critical group on n generators, then there is an identical relation involving precisely n variables which is satisfied by every proper factor of G but not by G itself. The converse is also true, since, if we have an identical relation $w(x_1, \ldots, x_n)$ satisfied by every proper factor (or even by every proper subgroup) of G but not by G itself, then any n elements of G which deny w must generate G.

Since a finite simple group is necessarily critical, this provides a possible means of determining the number of its generators. M. B. Powell (unpublished) has a short and elegant proof that a minimal simple group is on at most five generators, and, in the course of trying to reduce this number to two (which it is known to be, all minimal simple groups having been classified by J. G. Thompson), he has proved:

Theorem 1. *For a given exponent n, and a given prime p dividing n, there is a set of two generator laws such that a finite soluble group of exponent n has p-length l if and only if it satisfies these laws.*

In the general case, where the simple group has other simple groups as proper factors, he defines the complexity of a simple group by:

a simple group has complexity 1 if and only if it is minimal;

a simple group X has complexity $k+1$ if k is the largest complexity occurring among the simple groups which are proper factors of X;

and seeks to establish some relation between the complexity of a simple group and its number of generators.

One result we have in this direction is:

Theorem 2. *Let X be a finite simple group and $\mathfrak{U}$ the variety generated by its proper factors. If, among some irredundant set of critical generators*

261

for $\mathfrak{U}$, *we have a simple group* Y *on* d *generators, then* X *is on at most* $2d + 1$ *generators.*

(A variety $\mathfrak{U}$ is said to be irredundantly generated by a set of critical groups $C_1, \ldots, C_r$ if

$$\mathfrak{U} = \mathrm{var}(C_1, \ldots, C_r), \quad \mathfrak{U} \neq \mathrm{var}(C_i, \ldots, C_{i-1}, C_{i+1}, \ldots, C_r).)$$

The proof of this theorem is contained in the next section. However, this result is not very helpful, as it is not obvious that for any given X such a Y exists, and anyway it gives the number of generators of a group of complexity k as $O(2^k)$ which is far larger than one might hope. Section 3 contains an outline of another line of approach (suggested by Dr Powell).

§2.   It will be recalled that if $\mathfrak{V}$ is a variety, then $\mathfrak{V}^{(n)}$ denotes the variety satisfying the identical relations of $\mathfrak{V}$ involving at most n variables. By a result of B. H. Neumann [1], this is also the variety containing all those groups whose n-generator subgroups belong to $\mathfrak{V}$.

Theorem 2 is an immediate consequence of:

Theorem 3.   *Let* Y *be a simple group on* d *generators, and* $\mathfrak{A}$ *a Cross variety containing all proper factors of* Y *(but not* Y *itself). If* $\mathfrak{U} = \mathrm{var}(\mathfrak{A}, Y)$ *then there exists a relation* $w(x_1, \ldots, x_d)$ *such that* $w(Y) = Y$ *and a finite group in* $\mathfrak{U}^{(2d+1)}$ *either satisfies* $w = 1$ *or its* w-*subgroup is a direct product of groups isomorphic to* Y.

For the hypotheses of this theorem are satisfied by the groups occurring in Theorem 2, if we take $\mathfrak{A}$ as the variety generated by the critical generators of $\mathfrak{U}$ other than Y, together with all proper factors of Y. Then $\mathfrak{U} = \mathrm{var}(\mathfrak{A}, Y)$ and so, by Theorem 3, there exists a set of identical relations in $2d + 1$ variables satisfied by every proper factor of X such that any finite group satisfying them either satisfies $w = 1$, or its w-subgroup is a direct product of groups isomorphic to Y. Clearly X can fulfil neither of these conditions, so must fail to satisfy at least one of these $2d + 1$ variable identical relations, and it follows that X is on at most $2d + 1$ generators.

The proof of Theorem 3 uses a simplified version of the methods used in Section 3 of [2], taking $B = M = Y$. We first note that there is an identical relation $w(x_1, \ldots, x_d) = 1$ which holds in $\mathfrak{A}$ but not in Y, for if Y satisfied every d-variable relation of $\mathfrak{A}$, Y would belong to $\mathfrak{A}^{(d)}$, but

Y is a d-generator group, and d-generator groups in $\mathfrak{A}^{(d)}$ belong to $\mathfrak{A}$, contrary to hypothesis. We use this relation in place of the basis for the identical relations of $\mathfrak{A}$ used in [2] to obtain Theorem 3. (The only point in the proof in [2] where the fact that w was actually the basis for $\mathfrak{A}$ was used was in showing that the w-subgroup of a finitely generated group in $\mathfrak{U}^{(n)}$ was generated by a finite number of groups isomorphic to Y. Since we are only interested in the finite groups in $\mathfrak{U}^{(2d+1)}$ the fact that there can only be a finite number of such groups in the w-subgroup is trivial.)

§3. Let X be a simple group of exponent $n = \prod_{i=1}^{r} p_i^{\alpha_i}$, and having $S_1, \ldots, S_s$ as the simple groups which occur among its proper factors. Let $C_1, \ldots, C_k$ be an irredundant set of critical generators for $\mathfrak{U}$ (the variety of proper factors of X). We recall that a critical group has a monolith (unique minimal normal subgroup) which must either be elementary abelian or a direct product of isomorphic non-abelian simple groups. Let H_i be the direct product of all C_j whose monolith is of order a power of p_i, and K_i the direct product of all those having monolith a direct product of groups isomorphic to S_i, identifying the direct product of the empty set of groups with the trivial group.

If $H = \prod_{i=1}^{r} H_i$ and $K = \prod_{i=1}^{s} K_i$, then $\mathfrak{U} = \operatorname{var}(H \times K)$. Now $F(H_i)$ (the Fitting subgroup) is a group of order a power of p_i, and $F(H \times K) = \prod_{i=1}^{r} F(H_i)$.

Let W be the set of two-generator laws which hold in $H \times K$, and u a minimal two generator law not in W (minimal in the sense that if v is any other two generator law such that $v(H \times K) < u(H \times K)$, then $v(H \times K) = 1$).

Consider $u(H \times K) = u(H) \times u(K)$.

Case I. Suppose $u(H \times K) \leq F(H)$ so that $u(H_i)$ is a p_i-group. Then we can assume that $u(H_i) > 1$ for only one i. For suppose $u(H \times K) = u(H_1) \times \cdots \times u(H_i) \times \cdots \times u(H_r)$ where $u(H_i) \neq 1$. Then, if $\Pi_i = \prod_{j \neq i} p_j^{\alpha_j}$, we have that $u^{\Pi_i}(H_j) = 1$ $(i \neq j)$, $u^{\Pi_i}(H_i) \neq 1$. And so if $u(H_j) \neq 1$, $u^{\Pi_i}(H \times K) < u(H \times K)$ and $u^{\Pi_i}(H \times K) \neq 1$, contradicting the minimality of u.

Thus we have $u(H \times K)$ a normal p_i-subgroup of $H \times K$, and so, if z is any element in $H \times K$, the identical relations

$$u^{p_i^{\alpha_i}} = 1 \quad \text{and} \quad (uz^{\Pi_i})^{p_i^{\alpha_i}} = 1,$$

hold in $H \times K$. If X satisfies both these relations then either $u(X) = 1$, or it is a p_i-group, neither of which is possible. Hence X denies at least one of these relations, and so is on at most three generators.

Case II. If $u(H) \le F(H)$ and $|u(K)|$ is prime to the order of $u(H_i)$ for some i, then by replacing u by u^{π_i} we would obtain a smaller word and so this case cannot arise.

The remaining cases, where $|u(K)|$ is not prime to the order of any $u(H_i)$, or $u(H) \nleq F(H)$, are proving more difficult to handle and so far we have made little progress.

References

[1] B. H. NEUMANN, Identical relations in groups, I, *Math. Ann.* **114** (1937), 506–525.
[2] SHEILA OATES and M. B. POWELL, Identical relations in finite groups, *J. Algebra* **1** (1964), 11–39.
[3] P. M. WEICHSEL, On critical p-groups, *Proc. London Math. Soc.* (3) **14** (1964), 83–100.

Proc. Internat. Conf. Theory of Groups, Austral. Nat. Univ. Canberra, August 1965, pp. 265–277. © Gordon and Breach Science Publishers, Inc. 1967

Quelques problèmes généraux de la théorie des groupes et les groupes libres modulo n

1. *Introduction*

Soit G un groupe multiplicatif, pas nécessairement commutatif, dont 1 est l'élément neutre. Soit A un ensemble de générateurs de G et soit F une famille de relations caractéristiques qui les lie. Toute composition finie $f(a_1, \ldots, a_k)$ d'éléments de A est un produit de la forme

$$(1) \qquad a_{i_1}^{j_1} \cdots a_{i_r}^{j_r}$$

où $r \geq k$, $a_{i_s} \in A^* = \{a_1, \ldots, a_k\} \subset A$, $s = 1, \ldots, r$, les exposants $j_1, \ldots, j_r$ sont des entiers quelconques et où plusieurs facteurs peuvent être des puissances entières d'un même élément de A^*.

Toute relation de la famille F est de la forme $f(a_1, \ldots, a_k) = 1$ où f est une composition finie d'éléments de A et où, au second membre, 1 est l'élément neutre du groupe G.

Toute relation qui relie les éléments de A est une conséquence des relations de la famille F ainsi que de celles qui découlent des axiomes de groupe multiplicatif.

Tout groupe multiplicatif peut, comme on sait, être défini par un ensemble A de générateurs et une famille F de relations caractéristiques qui les lie.

Nous nous sommes posés les problèmes suivants.

1°. Trouver des caractères qui peuvent être communs à toutes les relations entre les éléments de certains ensembles de générateurs de groupes multiplicatifs.

2°. Etant donné un ensemble de propriétés qui peuvent être communes

a toutes les relations entre les éléments de certains ensembles de générateurs d'un groupe multiplicatif, étudier la classe des groupes multiplicatifs dont chacun possède au moins un ensemble de générateurs liés par des relations qui jouissent toutes des propriétés données.
Cette méthode d'investigation s'est trouvée très puissante. Elle a permis de refaire la théorie des groupes libres et elle a conduit à la découverte des groupes quasi libres, des groupes pseudo-libres et des groupes libres, quasi libres et pseudo-libres modulo n, quel que soit l'entier $n \geq 2$. Les trois dernières de ces classes de groupes, limitées à des groupes dont la puissance ne dépasse pas un nombre cardinal transfini donné quelconque $\mathfrak{m}$, présentent une structure de treillis et il existe des relations intéressantes entre ces diverses classes de groupes.

Dans un tout autre ordre d'idées, nous avons classifié tous les groupes en groupes fondamentaux et non fondamentaux.

Pour faire cette classification, on commence par introduire la notion de réductibilité d'un ensemble quelconque M d'éléments d'un groupe multiplicatif G. On dit, notamment, lorsque M est de puissance > 1, que M est réductible s'il existe un sous-ensemble M^* de M, de puissance finie $k \geq 2$ ainsi qu'un sous-ensemble B^* de G, de puissance comprise entre 1 et $k-1$, tels que l'ensemble $(M - M^*) \cup B^*$ engendre, par composition finie, tous les éléments de M (et éventuellement encore d'autres éléments de G). Ainsi, par exemple, dans le groupe symétrique $\mathfrak{S}_6$, de degré 6, dont les éléments permutent les nombres 1, 2, 3, 4, 5, 6, l'ensemble M formé des trois transpositions (1, 2), (3, 4), (5, 6) est réductible. Ces trois transpositions sont engendrées par le couple de substitutions (1, 2, 3, 4, 5, 6), (1, 2), qui constituent, comme on sait, une base de $\mathfrak{S}_6$. Par contre, on dit que l'ensemble M est irréductible si quels que soient les entiers k et l ($1 \leq k < l$) et quels que soient les sous-ensembles $M^* = \{a_1, \ldots, a_l\}$ de M et $B^* = \{b_1, \ldots, b_k\}$ de G, l'ensemble $(M - M^*) \cup B^*$ n'engendre pas, par composition finie, tous les éléments de M.

Un groupe G est appelé fondamental s'il possède au moins un ensemble irréductible A de générateurs.

2. *Réduction d'une composition finie d'éléments de G, compte tenu uniquement des axiomes de groupe multiplicatif*

Soit A un ensemble de générateurs d'un groupe multiplicatif G, dont 1 est l'élément neutre, et soit $f(a_1, \ldots, a_k)$ une composition finie quelconque des éléments d'un sous-ensemble finie quelconque $A^* = \{a_1, \ldots, a_k\}$ de A. f est un produit de la forme (1). On réduit f, compte tenu des seuls axiomes de groupe multiplicatif, en effectuant alternativement et autant de fois que c'est possible les deux opérations élémentaires suivantes. 1°. On remplace tout produit de facteurs consécutifs qui sont des puissances entières d'un même élément de A par un facteur unique dont l'exposant est la somme des exposants de tous les facteurs remplacés. 2°. On supprime tout facteur qui est une puissance nulle d'un élément de A, à moins que tout le produit envisagé ne se réduise à un tel facteur, auquel cas on pose ce produit égal a 1.

Après un nombre fini d'opérations élémentaires, le résultat de la réduction est soit 1, auquel cas on dit que f est totalement réductible, soit un produit de la forme

(2)
$$a_{u_1}^{v_1} \cdots a_{u_t}^{v_t},$$

où $a_{u_s} \in A^*$, v_s est un entier $\neq 0$, quel que soit $s = 1, \ldots, t$, et où $a_{u_s} \neq a_{u_{s+}}$, $s = 1, \ldots, t-1$. (2) est appelé la forme réduite de f, compte tenu des seuls axiomes de groupe multiplicatif.

3. *Réduction d'une composition finie de générateurs d'un groupe multiplicatif selon un module n donné*

Soit, à présent, n un entier fixe ≥ 2 et soit $f(a_1, \ldots, a_k)$ une composition finie donnée d'éléments d'un ensemble A de générateurs d'un groupe multiplicatif G. Supposons que f a déjà été réduit sur la base des axiomes de groupe multiplicatif et qu'elle a été mise sous la forme (2). La réduction de f modulo n consiste en l'application répétée des opérations élémentaires suivantes: 1°. Réduction des exposants modulo n. 2°. Suppression de tout facteur dont l'exposant, réduit modulo n, est nul. 3°. Réduction d'un produit fini de puissances d'éléments de A, compte tenu uniquement des axiomes de groupe multiplicatif. Deux cas peuvent se présenter: ou bien, après un nombre fini d'opérations élémentaires de réduction, il ne reste plus aucun facteur. On dit alors que f est totalement réductible modulo n et on pose le reste de la réduction de f modulo n égal à 1. Ou bien, après un nombre fini d'opérations élémentaires de réduction modulo n, on tombe sur un produit de la forme

(3)
$$r = a_{x_1}^{y_1} \cdots a_{x_u}^{y_u},$$

où $a_{x_s} \in A^*$, $s = 1, \ldots, u$, $a_{x_s} \neq a_{x_{s+1}}$, $s = 1, \ldots, u-1$, et où $y_1, \ldots, y_u$ sont des nombres de la suite $1, \ldots, n-1$. La réduction modulo n ne peut pas être poursuivie plus longtemps et r est appelé le reste de la réduction de f modulo n. Pour une composition finie donnée f d'éléments de A, le reste r de la réduction modulo n est défini de façon unique. Il est égal a 1, si f est totalement réductible modulo n, et il est $\neq 1$ dans le cas contraire. Soit, par exemple, $A = \{a, b, c\}$. Alors la composition finie $f(a, b, c) = a^2 b^7 c^4 a^{12} c^2 b^5 a^4$ est totalement réductible modulo 3, alors que la composition $g(a, b, c) = a^4 b^7 c^5 a^2 b$ n'est pas totalement réductible modulo 3 et le reste de sa réduction modulo 3 est égal à $abc^2 a^2 b$.

Si une composition finie donnée d'éléments de A est totalement réductible selon deux modules différents k et l, elle est aussi totalement réductible selon le module $d = $ p.g.c.d. de k et l. Mais elle n'est pas forcément réductible selon le module $m = $ p.p.c.m. de k et l. En voici un exemple. La composition $f(a, b) = b^3 ab^2 a^3 ba^2$ est totalement réductible modulo 2 et modulo 3, mais elle n'est pas réductible modulo 6.

4. *Opérations qui n'altèrent pas la réductibilité*

Les opérations suivantes n'altèrent pas la réductibilité totale tout court ou la réductibilité totale modulo n d'une composition finie $f(a_1, \ldots, a_k)$ d'éléments $a_1, \ldots, a_k$ d'un ensemble A de générateurs d'un groupe multiplicatif G:

1°. La multiplication. En multipliant, dans un ordre quelconque, un nombre fini quelconque de telles compositions, on obtient une composition du même type. Donc aussi, par itération, on n'altère pas la réductibilité totale d'une composition donnée d'éléments de A.

2°. La permutation circulaire des facteurs n'altère pas la réductibilité totale d'une composition finie d'éléments de A. C'est ainsi, par exemple, que les compositions suivantes des deux éléments a et b de A: $a^2 b^3 a$, $ab^3 a^2$, $b^3 a^3$, $b^2 a^3 b$, $ba^3 b^2$, $a^3 b^3$ sont toutes totalement réductibles modulo 3.

3°. La transformation par n'importe quel élément de G n'altère pas la réductibilité totale tout court, ni la réductibilité modulo n d'une composition finie d'éléments de A.

4°. L'inversion. Si f est totalement réductible modulo n, f^{-1} l'est également.

5°. La réduction effectuée, compte tenu uniquement des axiomes de groupe multiplicatif, n'altère pas la réductibilité totale modulo n d'une composition finie d'éléments de A.

6°. La réduction partielle modulo n n'altère pas la réductibilité totale modulo n d'une composition finie d'éléments de A.

7°. Le remplacement, dans une composition finie totalement réductible modulo n, $f(a_1, \ldots, a_k)$ d'éléments de A, d'un élément quelconque a_i de A par a_i^m, quel que soit l'entier m, n'altère pas la réductibilité totale modulo n.

8°. Le remplacement, dans une composition finie totalement réductible modulo n, $f(a_1, \ldots, a_k)$, de n'importe quel élément a_i de A par un produit de la forme ga_ih, où g et h sont des compositions finies totalement réductibles modulo n d'éléments de A, n'altère pas la réductibilité totale modulo n de la composition envisagée.

5. *Les différentes classes de groupes multiplicatifs caractérisées par des propriétés communes à toutes les relations caractéristiques de certains ensembles de générateurs de ces groupes*

1. *Les groupes libres.* Un groupe est libre s'il possède (au moins) un ensemble A de générateurs qui ne sont liés que par des relations triviales, c'est-à-dire par des relations de la forme $f(a_1, \ldots, a_k) = 1$, dont le premier membre est une composition finie d'éléments de A, totalement réductible par le seul jeu des axiomes de groupe multiplicatif.

Autrement dit, un groupe est libre s'il possède au moins un ensemble de générateurs A—dits libres—qui ne sont liés que par des relations qui découlent des axiomes de groupe multiplicatif.

2. *Les groupes quasi libres.* Un groupe multiplicatif G est quasi libre s'il possède au moins un ensemble de générateurs A—dits quasi libres— qui ne sont liés que par des relations quasi triviales de la forme $f(a_1, \ldots, a_k) = 1$, où f est une composition finie des éléments $a_1, \ldots, a_k$ de A, de degré nul par rapport à chacun de ces éléments.

3. *Les groupes pseudo-libres.* Un groupe multiplicatif G est pseudo-libre s'il possède au moins un ensemble de générateurs A—dits pseudo-libres— qui ne sont liés que par des relations pseudo-triviales de la forme $f(a_1, \ldots, a_k) = 1$, où f est une composition finie d'éléments $a_1, \ldots, a_k$ de A, de degré nul par rapport a l'ensemble de ces éléments.

Soit à présent n un entier ≥ 2 donné, fixe.

4. *Les groupes libres modulo n.* On dit qu'un groupe multiplicatif G est libre modulo n s'il possède au moins un ensemble A de générateurs—

appelés générateurs libres modulo n—qui ne sont liés que par des relations triviales modulo n de la forme $f(a_1, \ldots, a_k) = 1$, où f est une composition finie d'éléments $a_1, \ldots, a_k$ de A, totalement réductible modulo n.

5. *Les groupes quasi libres modulo n.* Un groupe multiplicatif G est appelé quasi libre modulo n s'il possède au moins un ensemble de générateurs A—dits quasi libres modulo n—dont les éléments ne sont liés que par des relations quasi triviales modulo n, c'est-à-dire par des relations de la forme $f(a_1, \ldots, a_k) = 1$, où f est une composition finie d'éléments $a_1, \ldots, a_k$ de A, de degré $\equiv 0$ (mod n) par rapport à chacun d'eux.

6. *Les groupes pseudo-libres modulo n.* Un groupe multiplicatif G est appelé pseudo-libre modulo n s'il possède au moins un ensemble de générateurs A—dits pseudo-libres modulo n—qui ne sont liés que par des relations de la forme $f(a_1, \ldots, a_k) = 1$ où le premier membre est une composition finie d'éléments $a_1, \ldots, a_k$ de A, de degré $\equiv 0$ (mod n) par rapport a l'énsemble de ces éléments.

Soit $\mathfrak{m}$ un nombre cardinal transfini donné, quelconque.

Soit

E_{gl} l'ensemble de tous les groupes libres de puissance $\leq \mathfrak{m}$;

E_{gql} l'ensemble de tous les groupes quasi libres de puissance $\leq \mathfrak{m}$;

E_{gpl} l'ensemble de tous les groupes pseudo-libres de puissance $\leq \mathfrak{m}$.

Soit n un entier ≥ 2 donné, quelconque, et soit

$E_{gl \,(\text{mod } n)}$ l'ensemble de tous les groupes libres modulo n, de puissance $\leq \mathfrak{m}$;

$E_{gql \,(\text{mod } n)}$ l'ensemble de tous les groupes quasi libres modulo n, de puissance $\leq \mathfrak{m}$;

$E_{gpl \,(\text{mod } n)}$ l'ensemble de tous les groupes pseudo-libres modulo n, de puissance $\leq \mathfrak{m}$.

Tout groupe libre est aussi libre modulo n, quel que soit $n = 2, 3, \ldots$. Tout groupe quasi libre est aussi quasi libre modulo n, $n = 2, 3, \ldots$. Tout groupe pseudo-libre est aussi pseudo-libre modulo n, quel que soit $n \geq 2$. Mais les réciproques de ces trois derniers énoncés sont en défaut.

Tout groupe libre est quasi libre, mais la réciproque n'est pas vraie et il existe une infinité de groupes quasi libres qui ne sont pas libres.

Tout groupe quasi libre est pseudo-libre, mais la réciproque est en défaut.

Tous les groupes libres et quasi libres sont fondamentaux, mais les groupes pseudo-libres ne sont pas nécessairement fondamentaux.

Convenons de désigner par le symbole $E_{gl \, (\text{mod } 0)}$ l'ensemble E_{gl}, par le symbole $E_{gql \, (\text{mod } 0)}$ l'ensemble E_{gql} et par le symbole $E_{gpl \, (\text{mod } 0)}$ l'ensemble E_{gpl}. D'autre part, appelons $E_{gl \, (\text{mod } 1)}$ la réunion de tous les ensembles $E_{gl \, (\text{mod } p)}$, étendue à tous les nombres premiers $p \geq 2$; soit $E_{gql \, (\text{mod } 1)}$ la réunion de tous les $E_{gql \, (\text{mod } p)}$, étendue à tous les nombres premiers $p \geq 2$, et soit $E_{gpl \, (\text{mod } 1)}$ la réunion des ensembles $E_{gpl \, (\text{mod } p)}$ étendue a tous les nombres premiers $p \geq 2$.

Chacun des trois ensembles $T_1 = \{E_{gi \, (\text{mod } n)}\}$, $T_2 = \{E_{gql \, (\text{mod } n)}\}$, $T_3 = \{E_{gpl \, (\text{mod } n)}\}$, $n = 0, 1, 2, 3, \ldots$, peut être muni d'une structure de treillis modulaire et distributif qui possède un élément nul et un élément universel. Voici comment on définit les opérations de treillis union ($\cup$) et inter ($\cap$) pour l'ensemble T_1. Quels que soient les entiers non négatifs k et l, on pose $E_{gl \, (\text{mod } k)} \cup E_{gl \, (\text{mod } l)} = E_{gl \, (\text{mod } d)}$, où d est le p.g.c.d. de k et l, $E_{gl \, (\text{mod } k)} \cap E_{gl \, (\text{mod } l)} = E_{gl \, (\text{mod } m)}$ où m est le p.p.c.m. de k et l. Ces opérations satisfont aux huit axiomes de treillis et aux axiomes supplémentaires des treillis modulaires et distributifs. On ordonne partiellement T_1 en convenant que $E_{gl \, (\text{mod } k)} \leq E_{gl \, (\text{mod } l)}$ si et seulement si $E_{gl \, (\text{mod } k)} \cup E_{gl \, (\text{mod } l)} = E_{gl \, (\text{mod } l)}$, ce qui implique que k est un multiple de l. D'après cette convention, $E_{gl \, (\text{mod } 0)} \leq E_{gl \, (\text{mod } n)}$ quel que soit l'entier $n \geq 0$ et $E_{gl \, (\text{mod } n)} \leq E_{gl \, (\text{mod } 1)}$ quel que soit l'entier $n \geq 0$. Donc $E_{gl \, (\text{mod } 0)}$ est l'élément nul et $E_{gl \, (\text{mod } 1)}$ est l'élément universel du treillis T_1. On traite de façon analogue les ensembles T_2 et T_3. Chacun des trois treillis T_1, T_2, T_3 est filtrant supérieurement et inférieurement.

La réunion des ensembles $E_{gl \, (\text{mod } p)}$, étendue à tous les nombres premiers p, se confond avec la réunion de tous les ensembles $E_{gl \, (\text{mod } n)}$, étendue à tous les entiers $n \geq 0$. D'autre part, l'ensemble E_{gl} fait partie de l'intersection de tous les ensembles $E_{gl \, (\text{mod } n)}$, $n = 2, 3, \ldots$.

Si une composition finie $f(a_1, \ldots, a_k)$ d'éléments d'un ensemble A de générateurs d'un groupe multiplicatif G est totalement réductible modulo n, quel que soit $n = 2, 3, \ldots$, elle est totalement réductible en vertu des seuls axiomes de groupes. La réciproque étant aussi vraie, il s'ensuit que tout groupe libre est aussi libre modulo n, quel que soit $n = 2, 3, \ldots$ et que réciproquement tout groupe multiplicatif qui possède au moins un ensemble de générateurs qui sont simultanément libres modulo n, quel que soit $n = 2, 3, \ldots$, est libre.

19—к.

6. *Les groupes libres modulo n*

La classe des groupes libres modulo n est beaucoup plus vaste que celle des groupes libres et comprend les groupes libres comme cas particulier.

Parmi les groupes libres modulo n, on distingue les groupes libres modulo n élémentaires et les groupes libres modulo n non élémentaires.

Un groupe G libre modulo n est appelé élémentaire s'il possède au moins un ensemble $C = A \cup B$ de générateurs libres modulo n, liés par les seules relations caractéristiques $b^n = 1$, quel que soit $b \in B$, l'un ou l'autre des sous-ensembles A, B de C pouvant être vide. Si $B = \emptyset$, G est libre, il est lié dans le cas contraire. $A \cap B = \emptyset$.

Tout groupe libre modulo n élémentaire est le produit libre des groupes cycliques engendrés par les éléments de l'ensemble C. Il est entièrement caractérisé par un couple ordonné de nombres cardinaux $(\mathfrak{m}, \mathfrak{n})$, dont le premier est la puissance de l'ensemble A et le second celle de B. Cela veut dire que si deux groupes libres modulo n élémentaires, G et G^*, sont engendrés, le premier par l'ensemble $A \cup B$ et le second par l'ensemble $A^* \cup B^*$ de générateurs libres modulo n liés par les seules relations caractéristiques $b^n = 1$, quel que soit $b \in B$, respectivement $b^{*n} = 1$, quel que soit $b^* \in B^*$, et si les ensembles A et A^* sont de même puissance $\mathfrak{m}$ alors que B et B^* sont d'égale puissance $\mathfrak{n}$, les deux groupes G et G^* sont isomorphes. Un groupe libre modulo n élémentaire est donc indépendant de la nature de ses éléments, qui peuvent être de purs symboles, comme les éléments des groupes libres.

Pour tout couple ordonné de nombres cardinaux $(\mathfrak{m}, \mathfrak{n})$, il existe un groupe libre modulo n élémentaire caractérisé par ces deux nombres.

Si n est premier, tout sous-groupe d'ordre > 1 d'un groupe libre modulo n élémentaire est à son tour un groupe libre modulo n élémentaire.

Si un groupe libre modulo n élémentaire n'est pas cyclique d'ordre n et s'il est engendré par l'ensemble $A \cup B$ de générateurs libres modulo n liés par les seules relations caractéristiques $b^n = 1$, quel que soit $b \in B$, tout élément de G qui n'est pas un itéré d'un élément de B, ni le transformé d'un tel itéré par un élément quelconque de G, est d'ordre ∞. Si donc G n'est pas cyclique, il possède une infinité de sous-groupes libres.

Il existe aussi des groupes libres modulo n non élémentaires qui sont les produits libres des groupes cycliques engendrés par les éléments de certains ensembles de leurs générateurs libres modulo n, certains de ces générateurs pouvant être libres et l'ordre de l'un au moins de ces générateurs étant mn, où m désigne un entier > 1, l'ordre de tout générateur de l'ensemble envisagé, pour autant qu'il est fini, étant un multiple de n.

Mais tout groupe libre modulo n n'est pas le produit libre des groupes cycliques engendrés par les éléments d'un de ses ensembles de générateurs libres modulo n. Il existe même une infinité indénombrable de groupes libres modulo n qui ne sont pas décomposables en produit libre de groupes cycliques engendrés par les éléments d'un ensemble de leurs générateurs libres modulo n. En particulier, le groupe libre modulo 2 de transformations des entiers engendré par les deux transformations a et b, dont la première transforme tout nombre entier i en $i+1$ et la seconde fait passer de tout entier négatif $-i$ à $-i-1$, de tout nombre impair positif $2i+1$ à $2i+2$, et de tout nombre pair $2i+2$ au nombre $2i-1$ n'est pas le produit libre des groupes cyclique engendrés par a et b. Chacun des éléments a, b est d'ordre infini et ces deux éléments sont liés par de nombreuses relations triviales modulo 2; on a, par exemple $(a^{2h}b^{2h})^{h+1} = 1$, quel que soit l'entier h. Nous désignerons ce groupe par le symbole $G(a, b)$.

Soit à présent G un groupe libre modulo n quelconque, engendré par un ensemble A de générateurs libres modulo n.

Tout élément c de G peut s'exprimer par une composition finie d'éléments de A et le reste r de la réduction modulo n de toute composition finie d'éléments de A qui représente l'élément a est le même.

On répartit les éléments de G en classes d'équivalence, en prenant dans une même classe C_r tous les éléments de G qui ont le même reste r modulo n. On munit l'ensemble G_C de ces classes C_r d'une structure de groupe multiplicatif en appelant produit de deux classes C_r, C_{r*} l'ensemble $C_r C_{r*} = \{ab, a \in C_r, b \in C_{r*}\}$. L'élément neutre du groupe G_C est la classe C_1 formée de tous les éléments de G qui s'expriment par des compositions finies totalement réductibles modulo n d'éléments de A. Les classes C_r ont un caractère intrinsèque, indépendant de l'ensemble "admis" de générateurs libres modulo n a partir duquel elles ont été définies. L'ensemble des éléments de la classe C_1 est un sous-groupe invariant de G. Le groupe G_C n'est, en général, pas abélien.

Par définition, les éléments de A ne sont liés que par des relations triviales modulo n, de la forme $f(a_1, \ldots, a_k) = 1$, où f est une composition finie totalement réductible modulo n d'éléments de A. Il s'ensuit que f est de degré $\equiv 0 (\bmod\ n)$ par rapport à tout élément de A et par suite que la relation envisagée est aussi quasi triviale modulo n. Cela étant quelle que soit la relation entre éléments de A, il s'ensuit que le groupe G est aussi quasi libre modulo n, que A est un ensemble de générateurs quasi libres modulo n de ce groupe, et la théorie des groupes quasi libres modulo n permet d'affirmer que tout groupe libre modulo n est fondamental, que tout ensemble de ses générateurs libres modulo n est irréductible, que l'on peut répartir les éléments de G en classes d'équivalence en prenant dans une même classe M deux éléments de G dans le cas et ce cas seulement où leurs degrés par rapport à tout élément de A sont congrus $(\bmod\ n)$, on peut munir l'ensemble Γ de ces classes de la structure d'un groupe abélien et à tout sous-groupe γ de Γ correspond un sous-groupe invariant de G, notamment la réunion des classes M qui font partie du groupe γ. En particulier, la classe M dite nulle, formée de tous les éléments de G de degré $\equiv 0 (\bmod\ n)$ par rapport à tout élément de A, classe qui est l'élément nul du groupe Γ, est un sous-groupe invariant de G.

Deux éléments a et b de G sont dits symétriques modulo n si $ab \in C_1$. En particulier, a et a^{-1} sont symétriques modulo n et tout élément b symétrique d'un élément donné a de G est de la forme $b = a^{-1}c$, $c \in C_1$. Deux éléments c et d de G sont dits conjugués modulo n s'il existe un couple a, b d'éléments symétriques modulo n de G, tels que $acb = d$. Deux éléments conjugués modulo n de G ne sont pas forcément du même ordre, il peut même arriver que l'un d'eux soit d'ordre fini et l'autre d'ordre infini. Deux sous-groupes g_1 et g_2 de G sont dits conjugués modulo n s'il existe un couple a, b d'éléments symétriques modulo n de G, tels que $ag_1b = g_2$. Soit g un sous-groupe de G et a, b un couple d'éléments symétriques modulo n de G. La condition nécessaire et suffisante pour que l'ensemble agb soit un sous-groupe de G, c'est que $ba \in g$.

7. *Les sous-groupes invariants modulo n d'un groupe libre modulo n*

Un sous-groupe g de G est dit invariant modulo n si l'on a l'égalité $agb = g$, quel que soit le couple a, b d'éléments de G, symétriques modulo n. Tout groupe libre modulo n possède des sous-groupes invariants modulo n. C_1 est un tel sous-groupe. G est son propre sous-groupe invariant modulo n. Tout sous-groupe invariant modulo n de G contient la classe C_1 qui est le plus petit sous-groupe invariant modulo n de G, et il contient, avec tout élément a d'une classe C_r cette classe entière. A tout sous-groupe invariant g_C de G_C correspond un sous-groupe invariant modulo n de G, notamment la réunion des classes C_r qui font partie du groupe g_C. Réciproquement, à tout sous-groupe g invariant modulo n de G correspond un sous-groupe g_C du groupe G_C, formé de toutes les classes C_r qui contiennent des éléments de g.

Désignons par $\mathfrak{G}$ l'ensemble de tous les sous-groupes invariants modulo n de G. On peut munir cet ensemble $\mathfrak{G}$ d'une structure de treillis, en convenant d'appeler inter de deux éléments g_1 et g_2 de G et de désigner par le symbole $g_1 \cap g_2$ l'intersection de g_1 et de g_2 et d'appeler union de g_1 et de g_2 et de désigner par le symbole $g_1 \cup g_2$ le sous-groupe de G engendré par la reunion des éléments de g_1 et de g_2. Chacun des ensembles $g_1 \cap g_2$ et $g_1 \cup g_2$ est un sous-groupe invariant modulo n de G et l'ensemble $\mathfrak{G}$ où sont définies les deux opérations internes $\cap$ et $\cup$ et qui est partiellement ordonné par la relation "être sous-groupe de" est un treillis filtrant inférieurement et supérieurement, dont C_1 est l'élément nul et G l'élément universel.

On appelle série de composition modulo n de G une suite finie de groupes $G_0 \supset G_1 \supset \cdots \supset G_k$ tels que $G_0 = G$, $G_k = C_1$ et que G_i est un sous-groupe maximal invariant modulo n de G_{i-1}, $i = 1, \ldots, k$. Chacun des groupes G_i contient, avec tout élément a de G tous ses symétriques modulo n. Il existe des groupes libres modulo n qui possèdent des suites infinies normales modulo n, c'est a dire des suites infinies de la forme $G_0 \supset G_1 \supset G_2 \supset \cdots$ où G_i un sous-groupe invariant modulo n de G_{i-1} quel que soit $i \geq 1$ et $G_0 = G$.

8. Les ensembles admis de générateurs libres modulo n de G et les éléments libres modulo n de G

A partir d'un ensemble donné A de générateurs libres modulo n d'un groupe G libre modulo n, on peut déduire d'autre ensembles de générateurs libres modulo n par les procédés suivants.

On peut remplacer dans A tout élément a par son inverse a^{-1}.

Si a est un élément d'ordre fini de A, l'ordre de a est un multiple de n. Soit mn cet ordre (m est un entier ≥ 1). Alors quel que soit l'entier k de la suite $1, 2, \ldots, nm - 1$, premier avec nm, on obtient à partir de A un nouvel ensemble de générateurs libres modulo n de G en remplaçant a par a^k.

Quel que soit l'élément a de A et quelles que soient les compositions finies g et h, totalement réductibles modulo n d'éléments de l'ensemble $A - \{a\}$, on déduit de A un nouvel ensemble de générateurs libres modulo n de G en remplaçant dans A l'élément a par gah.

Par application répétée de ces divers procédés, on déduit, à partir de A, tous les ensembles "admis" de générateurs libres modulo n de G.

Un élément de G est dit libre modulo n s'il appartient à un ensemble au moins de générateurs libres modulo n de G.

Si un élément a de G est libre modulo n et s'il est d'ordre infini, a^{-1} est libre modulo n; et si a est d'ordre fini mn, alors a^k est libre modulo n, quel que soit l'entier k de la suite $1, 2, \ldots, mn - 1$, premier avec mn.

Si un élément a de G est libre modulo n et s'il fait partie d'un ensemble A de générateurs libres modulo n de G, quelles que soient les compositions finies g et h totalement réductibles modulo n d'éléments de l'ensemble $A - \{a\}$, l'élément gah de G est libre modulo n.

Tout élément libre modulo n d'un groupe libre modulo n est soit d'ordre infini, soit d'ordre fini congru à zéro modulo n.

9.   *Le sous-groupe commutateur d'un groupe libre modulo n non cyclique*

Un groupe libre modulo n n'est abélien que s'il est cyclique. Le commutateur de deux éléments distincts a_1 et a_2 de G, faisant partie d'un même ensemble de générateurs libres modulo n, ne saurait être égal a 1, car la relation $a_1 a_2 a_1^{-1} a_2^{-1} = 1$ n'est pas triviale modulo n.

Soit G un groupe libre modulo n élémentaire non cyclique et soit C un ensemble de générateurs libres modulo n de G. Décomposons C en deux sous-ensembles disjoints A et B (C est la réunion de A et de B), de façon que A contienne tous les éléments d'ordre infini et B tous les éléments d'ordre fini de C. Le sous-groupe commutateur de G, G' se compose de tous les éléments de G qui s'expriment par des compositions finies d'éléments de G, de degré nul par rapport à tout élément de A et de degré $\equiv 0 \pmod{n}$ par rapport à tout élément de B.

10.   *Le sous-groupe commutateur modulo n d'un groupe libre modulo n*

Un commutateur modulo n de deux elements a et b de G est un produit de la forme aba^*b^* ou a^* est un symétrique de a et b^* est un symétrique de b modulo n. L'ensemble de tous les commutateurs modulo n des couples d'éléments de G engendre un sous-groupe invariant modulo n de G, appelé le sous-groupe commutateur modulo n de G.

11. Divers résultats concernant les groupes libres modulo n

Soit G un groupe libre modulo n.

Quel que soit l'élément a de G, libre modulo n, il existe un sous-groupe invariant de G qui ne contient pas a.

Un groupe libre modulo n non élémentaire peut posséder des sous-groupes non fondamentaux. Ainsi le groupe $G(a, b)$ possède, entre autres, les sous-groupes[1] non fondamentaux $\mathfrak{S}$ et $\mathfrak{A}$. La classe C_1 d'éléments de ce groupe comprend aussi bien une infinité d'éléments d'ordre fini qu'une infinité d'éléments d'ordre infini. Tout élément d'ordre fini de $G(a, b)$ est representé par une composition finie réduite des éléments a et b, du même degré par rapport à a que par rapport à b, totalement réductible modulo 2, et toute composition de cette nature est un élément d'ordre fini de $G(a, b)$.

Tout groupe libre modulo n est un produit libre modulo n des groupes cycliques engendrés par les éléments de n'importe quel ensemble de générateurs libres modulo n de G.

Pour tout nombre cardinal donné $\mathfrak{m}$ et pour tout entier $n \geq 2$, il existe un groupe libre modulo n dont tout ensemble A de générateurs libres modulo n est de puissance $\mathfrak{m}$.

Les groupes libres modulo n ont été découverts en 1964. Leur étude est beaucoup plus délicate que celle des groupes quasi libres modulo n et ils posent encore bien des problèmes auxquels les jeunes chercheurs ne tarderont pas à apporter la solution.

---

[1] On désigne par $\mathfrak{S}$ [$\mathfrak{A}$] le groupe multiplicatif de toutes les substitutions [de classe paire] d'un nombre fini quelconque d'entiers quelconques.

Proc. Internat. Conf. Theory of Groups, Austral. Nat. Univ. Canberra,
August 1965, p. 279. © Gordon and Breach Science Publishers, Inc. 1967

Abelian groups with endomorphic images of special type

K. M. RANGASWAMY

Problem 86 of Fuchs [1] asks for the characterization of all groups [1] G with the property (P): every endomorphic image of G is a direct summand of G. In a paper which will appear in $J.$ $Algebra$, we give conditions under which a group has (P). We start with characterizing the groups in which every endomorphic image is pure (neat). In particular, these groups belong to the class $\mathscr{C}$ of the groups G with elementary maximal torsion subgroup G_t and divisible G/G_t. However, an example is constructed to show that not every group in $\mathscr{C}$ has (P): this answers negatively a question of Kertész and Szele [2]. We investigate the class $\mathscr{C}$, considering it as a category in which the maps from A to B (A, $B \in \mathscr{C}$) are those homomorphisms from A to B for which the kernel and the image are pure in A and B respectively. $\mathscr{C}$ has enough $\mathscr{C}$-injectives. A group X in $\mathscr{C}$ is $\mathscr{C}$-injective if and only if $X = D \oplus R$ where D is torsion free divisible and $R = \sum^* E_p$ where the summation is taken over an arbitrary set of primes p and each E_p is an elementary p-group. The relative homological algebra for $\mathscr{C}$ is worked out and the consequent results are derived.

References

[1] L. FUCHS, *Abelian groups*, Akadémiai Kiadó, Budapest, 1958.
[2] A. KERTÉSZ and T. SZELE, On abelian groups every multiple of which is a direct summand, *Acta Sci. Math. Szeged* **14** (1952), 157–165.

---

[1] "Group" means abelian group.

Proc. Internat. Conf. Theory of Groups, Austral. Nat. Univ. Canberra,
August 1965, pp. 281–301. © Gordon and Breach Science Publishers, Inc. 1967

Classification of involutions and centralizers of involutions in certain simple groups

RIMHAK REE

Introduction. In this paper we shall classify involutions and compute their centralizers for Chevalley groups of types (G_2), (F_4), (E_i), $i =$ 6, 7, 8, over a finite field of odd characteristic, in the hope of understanding these groups better. The final result is not complete for the type (E_7), mainly due to the fact that, for this type, the centralizer of an involution in the underlying algebraic group is not connected. The main tool is a theorem of Lang [4] on connected algebraic groups defined over a finite field, which enables one to put a given involution in the diagonal form.

The centralizer of an involution is thoroughly discussed only when the involution centralizes a Sylow 2-subgroup of the Chevalley group, since the centralizers of other involutions can be treated in exactly the same way.

In the first section we shall discuss some common features of the types (G_2), (F_4), (E_6), (E_8), and in the subsequent sections we shall deal with particulars of each type separately.

1. Preliminaries

Throughout this paper, we shall follow strictly the notation introduced in [3]. Thus G and G' are the groups defined in [3] using a simple root system Σ and a field K. The field K will be always assumed to be finite, consisting of q elements, q odd. We shall denote by Ω a fixed universal domain containing K. The groups obtained from G, $\mathfrak{H}$, etc., by extending the field K to Ω will be denoted by G_Ω, $\mathfrak{H}_\Omega$, etc. Then G_Ω and $\mathfrak{H}_\Omega$ are connected algebraic groups, and G, $\mathfrak{H}$ are the set of all rational points over K of G_Ω, $\mathfrak{H}_\Omega$ respectively (cf. [5]).

An element of order 2 in a group will be called an involution. Since K is of odd characteristic, any involution in G is semisimple (cf. [2], p. 66 for definition). We shall state a few basic facts and lemmas.

Lemma 1.1. *Any semisimple element in G_Ω is conjugate in G_Ω to an element in $\mathfrak{H}_\Omega$ (cf. [5], p. 312).*

Lemma 1.2. *Two elements in G are conjugate in G if they are conjugate in G_Ω and the centralizer in G_Ω of one of them is a connected algebraic group.*

This lemma can be obtained immediately if one applies a result of Lang ([4], p. 557) to the centralizer. See also [8].

Lemma 1.3. *Two elements h_1, h_2 in $\mathfrak{H}$ are conjugate in G if and only if $\omega(w)^{-1}h_1\omega(w) = h_2$ for some w in the Weyl group W.* (In this case we shall say that h_1 and h_2 are conjugate by W.)

The following proposition is fundamental in this paper, and will be verified for each type of the root systems separately.

Proposition 1.4. *Let the root system Σ be of one of the types (G_2), (F_4), (E_i), $i = 6$, 8, and let $h(\chi)$ be an involution in $\mathfrak{H}$. Let $\Sigma' = \{r \in \Sigma \mid \chi(r) = 1\}$ and $W' = \{w \in W \mid w(\Sigma') = \Sigma'\}$. Then W' is generated by the set $\{w_r \mid r \in \Sigma'\}$.*

In the proof of Proposition 1.4, it will be shown that (a) any element $w \in W'$ induces an inner automorphism[1] $\varphi(w)$ of Σ' and that (b) the kernel of the homomorphism φ of W' onto the Weyl group W^* of Σ' consists of the identity element only. Then $|W'| = |W^*|$. The group generated by $\{w_r \mid r \in \Sigma'\}$ is contained in W' and is isomorphic to W^*. From this Proposition 1.4 follows.

Proposition 1.5. *Let Σ and $h(\chi)$ be as in Proposition 1.4. Then the centralizer of $h(\chi)$ in G_Ω is a connected algebraic group.*

Proof. An element $uh\omega(w)u'$, where $u \in \mathfrak{U}_\Omega$, $h \in \mathfrak{H}_\Omega$, $u' \in (\mathfrak{U}''_w)_\Omega$ in G_Ω centralizes $h(\chi)$ if and only if each of u, $\omega(w)$, u' centralizes it. Write u in the form

$$u = x_{r_1}(t_1)x_{r_2}(t_2)\cdots x_{r_m}(t_m),$$

---

[1] An automorphism w of a root system Σ is a permutation of Σ which is linear with respect to the addition in Σ; w is called inner if it belongs to the Weyl group of Σ, and outer otherwise.

where $0 < r_1 < r_2 < \cdots < r_m$, and $0 \neq t_i \in \Omega$. Then the uniqueness of this expression implies that u centralizes $h(\chi)$ if and only if all $r_1, \ldots, r_m$ belong to Σ'. Similarly for u'. Also, $\omega(w)^{-1} h(\chi) \omega(w) = h(\chi \circ w)$ implies that $\omega(w)$ centralizes $h(\chi)$ if and only if $w \in W'$. Since we can assume that $\omega(w_r) = x_r(1) x_{-r}(-1) x_r(1)$, it follows from Proposition 1.4 that the centralizer $\mathbf{C}(h(\chi))_\Omega$ of $h(\chi)$ in G_Ω is generated by $\{(\mathfrak{X}_r)_\Omega \mid r \in \Sigma'\}$ and $\mathfrak{H}_\Omega$. Each of these groups is connected. Hence $\mathbf{C}(h(\chi))_\Omega$ is connected.

Since any involution $h(\chi)$ in $\mathfrak{H}$ is rational over K, we obtain from the above the following

Theorem 1.6. *If Σ is of one of the types (G_2), (F_4), (E_i), $i = 6$, 8, then any involution in G is conjugate in G to an involution in $\mathfrak{H}$. Two involutions in $\mathfrak{H}$ are conjugate in G if and only if they are conjugate by W, the Weyl group of Σ.*

The above argument also shows the following

Proposition 1.7. *The notation being as above, we have*

$$\mathbf{C}_G(h(\chi)) = \bigcup_{w \in W'} \mathfrak{u}' \mathfrak{H} \omega(w)(\mathfrak{u}' \cap \mathfrak{u}''_w),$$

$$\mathbf{C}_{G'}(h(\chi)) = \bigcup_{w \in W'} \mathfrak{u}' \mathfrak{H}' \omega(w)(\mathfrak{u}' \cap \mathfrak{u}''_w),$$

where $\mathfrak{u}'$ is the group generated by $\{\mathfrak{X}_r \mid r \in \Sigma', r > 0\}$.

The set Σ', being a subsystem of Σ, is the union of some mutually orthogonal simple subsystems $\Sigma'_1, \ldots, \Sigma'_m$. Denote by W'_i the group generated by $\{w_r \mid r \in \Sigma'_i\}$. Then from Proposition 1.4 we have

$$W' = W'_1 \times W'_2 \times \cdots \times W'_m \quad \text{(direct)},$$

and for each $w = w_1 w_2 \cdots w_m$, $w_i \in W'_i$, in W'

$$\mathfrak{u}' \cap \mathfrak{u}''_w = (\mathfrak{u}'_1 \cap \mathfrak{u}''_{w_1}) \times \cdots \times (\mathfrak{u}'_m \cap \mathfrak{u}''_{w_m}),$$

where $\mathfrak{u}'_i$ is the group generated by $\{\mathfrak{X}_r \mid r \in \Sigma'_i, r > 0\}$. Hence[2]

$$\sum_{w \in W'} |\mathfrak{u}' \cap \mathfrak{u}''_w| = \prod_{i=1}^{m} \left(\sum_{w \in W'} |\mathfrak{u}'_i \cap \mathfrak{u}''_{w_i}| \right)$$

$$= \prod_{i=1}^{m} \left(\sum_{w \in W'} q^{n_i(w)} \right),$$

[2] $|S|$ denotes the cardinal number of the set S.

where $n_i(w)$ is the number of positive roots r in Σ_i' such that $w_i(r) < 0$. Hence

Proposition 1.8.   *The notation being as above, we have*

$$|\mathbf{C}_G(h(\chi))| = q^{N'}(q-1)^l \prod_{i=1}^{m} \left(\sum_{w \in W_i'} q^{n_i(w)} \right),$$

where N' is the number of positive roots in Σ' and where l is the rank of Σ.

2.   *Groups of type (G_2)*

The root system Σ of type (G_2) consists of the roots

$$\pm \xi_i, \quad \pm(\xi_i - \xi_j)$$

where i, j range over 1, 2, 3, and are distinct, and where $\xi_1 + \xi_2 + \xi_3 = 0$. The Weyl group W of Σ consists of the transformations $w \colon \xi_i \to e\xi_{\pi(i)}$. where $e = \pm 1$, and where π is an arbitrary permutation of 1, 2, 3.

Let V be the vector space spanned by ξ_1, ξ_2, ξ_3 over the field of two elements,[3] and let V^* be the dual space of V. Then it is clear that there is a one-to-one correspondence between the involutions in $\mathfrak{H}$ and the elements of $V^* - \{0\}$. The action of the Weyl group W on $\mathfrak{H}$ corresponds to the action of W on V^* defined as follows: for $f \in V^*$, $w \in W$, define f^w by $f^w(\xi) = f(w\xi)$, where $\xi \in V$. Let $f_i \in V^*$ be defined by $f_i(\xi_i) = 0$, $f_i(\xi_j) = 1$ $(i \neq j)$. Then $V^* = \{0, f_1, f_2, f_3\}$, and it is clear from the above that W can cause any permutation of f_1, f_2, f_3. Hence
(2.1)   *Any two involutions in $\mathfrak{H}$ are conjugate under W.*
Let $h(\chi)$ be the involution corresponding to f_3. Then

$$\Sigma' = \{r \in \Sigma \mid \chi(r) = 1\}$$

consists of the roots r such that $f_3(r) = 0$. Hence

$$\Sigma' = \{\pm(\xi_1 - \xi_2), \pm \xi_3\} = \Sigma_1' \cup \Sigma_2',$$

where $\Sigma_1' = \{\pm(\xi_1 - \xi_2)\}$, $\Sigma_2' = \{\pm \xi_3\}$. Σ_1' and Σ_2' are mutually orthogonal

---

[3] For brevity, we shall, throughout the paper, use the same notation ξ_i for the corresponding element in V.

root systems of type (A_1). If $W' = \{w \in W \mid w(\Sigma') = \Sigma'\}$ contained an element outside the group generated by $\{w_r \mid r \in \Sigma'\}$, W' would contain an element w such that $w(\xi_1 - \xi_2) = \xi_3$, $w(\xi_3) = \xi_1 - \xi_2$. But this is clearly impossible. Hence

(2.2) *Proposition* 1.4 *holds for the root system of type (G_2) and any involution $h(\chi)$ in $\mathfrak{H}$.*

Hence from Proposition 1.8 we have

$$(2.3) \qquad |\mathbf{C}_G(h(\chi))| = q^6(q^2 - 1)^2.$$

In order to consider the structure of $\mathbf{C}_G(h(\chi))$, let $\mathfrak{L}_1$ (resp. $\mathfrak{L}_2$) be the group generated by $\{\mathfrak{X}_r, \mathfrak{X}_{-r}\}$ where $r = \xi_1 - \xi_2$ (resp. $r = \xi_3$). Since $\mathfrak{L}_1$ and $\mathfrak{L}_2$ share an involution in their centers, it follows that both $\mathfrak{L}_1$ and $\mathfrak{L}_2$ are isomorphic to $SL(2, K)$, that the group $\mathfrak{L}_1\mathfrak{L}_2$ is of index 2 in $\mathbf{C}_G(h(\chi))$, and that $\mathbf{C}_G(h(\chi))$ is generated by $\mathfrak{L}_1\mathfrak{L}_2$ and any element $h(\chi_1)$ such that $\chi_1(\xi_1 - \xi_2)$ is not a square in K^*. (For example, one can take $\chi_1(\xi_1) = z$, $\chi_1(\xi_2) = 1$, where z is a nonsquare in K^*. Then $\chi_1(\xi_1 - \xi_2) = z$, $\chi_1(\xi_3) = z^{-1}$.) From this it can be seen easily that
(2.4) $\mathbf{C}_G(h(\chi))$ *is isomorphic to the group $\mathfrak{C}$ defined as follows: Let $\mathfrak{L} = SL(2, K)$ and let $\mathfrak{L} \times \mathfrak{L}$ be the cartesian product of $\mathfrak{L}$ by itself. Set $\mathfrak{C}' = (\mathfrak{L} \times \mathfrak{L})/\langle(Z, Z)\rangle$, where Z is the central involution of $\mathfrak{L}$. Let $\varphi : \mathfrak{L} \times \mathfrak{L} \to \mathfrak{C}'$ be the canonical homomorphism, and write $X * Y = \varphi(X, Y)$, where $X, Y \in \mathfrak{L}$. Extend $\mathfrak{C}'$ by an element H such that*

$$H^2 = \begin{pmatrix} z & 0 \\ 0 & z^{-1} \end{pmatrix} * \begin{pmatrix} z & 0 \\ 0 & z^{-1} \end{pmatrix},$$

$$H^{-1}\left(\begin{pmatrix} a & b \\ c & d \end{pmatrix} * \begin{pmatrix} a' & b' \\ c' & d' \end{pmatrix}\right) H = \begin{pmatrix} a & z^{-1}b \\ zc & d \end{pmatrix} * \begin{pmatrix} a' & z^{-1}b' \\ zc' & d' \end{pmatrix}$$

(where z is a nonsquare element in K^*). Then $\mathfrak{C} = \langle \mathfrak{C}', H \rangle$.
   Summarizing the above, we have

Theorem 2.5. *The group G has only one class of involutions. The centralizer of an involution in G is isomorphic to the group $\mathfrak{C}$ defined in (2.4).*

3. *Groups of type* (F_4)

The root system Σ of type (F_4) consists of the 48 roots:

$$\xi_i, \quad \xi_i + \xi_j, \quad \tfrac{1}{2}(\xi_i + \xi_j + \xi_k + \xi_l),$$

where: i, j, k, l range over ± 1, ± 2, ± 3, ± 4 with distinct absolute values; $\xi_{-i} = -\xi_i$ for all i. As a system of fundamental roots one can take

$$a_1 = \tfrac{1}{2}(\xi_1 - \xi_2 - \xi_3 - \xi_4), \quad a_2 = \xi_4, \quad a_3 = \xi_3 - \xi_4, \quad a_4 = \xi_2 - \xi_3.$$

Denote by w_i the Weyl reflection associated with a_i, i.e., $w_i = w_r$, where $r = a_i$. Then

$$w_1: \begin{cases} a_1 \rightarrow -a_1 \\ a_2 \rightarrow a_1 + a_2 \\ a_3 \rightarrow a_3 \\ a_4 \rightarrow a_4 \end{cases} \qquad w_2: \begin{cases} a_1 \rightarrow a_1 + a_2 \\ a_2 \rightarrow -a_2 \\ a_3 \rightarrow 2a_2 + a_3 \\ a_4 \rightarrow a_4 \end{cases}$$

$$w_3: \begin{cases} a_1 \rightarrow a_1 \\ a_2 \rightarrow a_2 + a_3 \\ a_3 \rightarrow -a_3 \\ a_4 \rightarrow a_3 + a_4 \end{cases} \qquad w_4: \begin{cases} a_1 \rightarrow a_1 \\ a_2 \rightarrow a_2 \\ a_3 \rightarrow a_3 + a_4 \\ a_4 \rightarrow -a_4 \end{cases}$$

Let V be the vector space spanned by the roots over the field of two elements, V^* the dual space of V, and $f_1, \ldots, f_4$ the dual basis of $a_1, \ldots, a_4$. Defining the action of the Weyl group W on V^* by $f^w(\xi) = f(w\xi)$, where $f \in V^*$, $\xi \in V$, $w \in W$, we have from the above

$$w_1: \begin{cases} f_1 \rightarrow f_1 + f_2 \\ f_2 \rightarrow f_2 \\ f_3 \rightarrow f_3 \\ f_4 \rightarrow f_4 \end{cases} \qquad w_2: \begin{cases} f_1 \rightarrow f_1 \\ f_2 \rightarrow f_1 + f_2 \\ f_3 \rightarrow f_3 \\ f_4 \rightarrow f_4 \end{cases}$$

$$w_3: \begin{cases} f_1 \rightarrow f_1 \\ f_2 \rightarrow f_2 \\ f_3 \rightarrow f_2 + f_3 + f_4 \\ f_4 \rightarrow f_4 \end{cases} \qquad w_4: \begin{cases} f_1 \rightarrow f_1 \\ f_2 \rightarrow f_2 \\ f_3 \rightarrow f_3 \\ f_4 \rightarrow f_3 + f_4 \end{cases}$$

Writing $f \sim g$ for $f, g \in V^*$ if $f^w = g$ for some $w \in W$, we obtain from the above

$$f_1 \sim f_1 + f_2 \sim f_2;$$

$$f_3 \sim f_2 + f_3 + f_4 \sim f_2 + f_4 \sim f_1 + f_2 + f_4 \sim f_1 + f_2 + f_3 + f_4$$

$$\sim f_1 + f_3 + f_4 \sim f_1 + f_4; \; f_1 + f_2 + f_3 + f_4 \sim f_1 + f_3$$

$$\sim f_1 + f_2 + f_3 \sim f_2 + f_3 \sim f_3 + f_4 \sim f_4.$$

Hence we have

(3.1) *The involutions in $\mathfrak{H}$ are divided into two conjugate classes under W. They are represented by $h(\chi_1)$ and $h(\chi_4)$ respectively, where $\chi_i(a_i) = -1$, $\chi_i(a_j) = 1 \; (i \neq j)$.*

The positive roots in the set $\Sigma' = \{r \in \Sigma \mid \chi_1(r) = 1\}$ consist of roots of the form $\sum n_i a_i$, where n_i are integers ≥ 0, and $n_1 \equiv 0 \pmod 2$. Clearly $n_1 = 0$ or $n_1 = 2$. Those with $n_1 = 0$ are the roots ξ_i, $\xi_i \pm \xi_j$, where $i, j = 2, 3, 4$, while those with $n_1 = 2$ are clearly ξ_1, $\xi_1 \pm \xi_i$, $i = 2, 3, 4$. Hence $\Sigma' = \{\xi_i, \xi_i + \xi_j\}$. This shows that Σ' is a simple root system of type (B_4). Hence Σ' does not admit any outer automorphism. Hence $W' = \{w \in W \mid w(\Sigma') = \Sigma'\}$ is generated by $\{w_r \mid r \in \Sigma'\}$.

The positive roots in the set $\Sigma'' = \{r \in \Sigma \mid \chi_4(r) = 1\}$ consist of roots of the form $\sum n_i a_i$, where n_i are integers ≥ 0, and $n_4 = 0 \pmod 2$. If $n_4 \geq 2$, then $a_1 = 2$ and $\sum n_i a_i = \xi_1 + \xi_2$, and hence $n_2 = 4$, $n_3 = 3$. Those roots with $n_4 = 0$ clearly form a root system Σ_1'' of type (C_3). Hence Σ'' is the direct sum $\Sigma_1'' \cup \Sigma_2''$, where $\Sigma_2'' = \{\pm (\xi_1 + \xi_2)\}$ is of type (A_1). Hence Σ'' admits no outer automorphism, and $W'' = \{w \in W \mid w(\Sigma'') = \Sigma''\}$ is generated by $\{w_r \mid r \in \Sigma''\}$. Thus we have

(3.2) *For the root system of type (F_4), (1.4) is verified.*

From the above and Proposition 1.8 we have

(3.3) $|C_G(h(\chi_1))| = q^{16}(q^2 - 1)(q^4 - 1)(q^6 - 1)(q^8 - 1),$

$|C_G(h(\chi_4))| = q^{10}(q^2 - 1)^2(q^4 - 1)(q^6 - 1).$

As a corollary we have

(3.4) *The involution $h(\chi_1)$ centralizes a Sylow 2-subgroup of G while $h(\chi_4)$ does not.*

In order to clarify the structure of $C_G(h(\chi_1))$, let $\mathfrak{g}$ be the simple Lie algebra of type (F_4) over the complex field with Chevalley root

20—K.

vectors X_r, $r \in \Sigma$, and let $\mathfrak{g}'$ be the subalgebra of $\mathfrak{g}$ generated by $\{X_r \mid r \in \Sigma'\}$. Then $\mathfrak{g}'$ is a simple Lie algebra of type (B_4), and $\mathfrak{g}$, regarded as a $\mathfrak{g}'$-module, is the direct sum of the irreducible $\mathfrak{g}'$-modules $\mathfrak{g}'$ and $\mathfrak{g}''$, where $\mathfrak{g}''$ is the space spanned by[4] $\{X_r \mid r \in \Sigma - \Sigma'\}$. The basis $\{X_r \mid r \in \Sigma - \Sigma'\}$ of $\mathfrak{g}''$ is in fact regular in the terminology of [6, p. 491]. The weights of $\mathfrak{g}''$ (with respect to the Cartan subalgebra $\mathfrak{g}' \cap \mathfrak{h}$ of $\mathfrak{g}'$) are not roots of $\mathfrak{g}'$, and hence the additive group generated by them contains all the weights of $\mathfrak{g}'$. From this it follows easily that $\mathbf{C}_G(h(\chi_1))$ is exactly the group $G_K(\mathfrak{g})$ defined in [6] with respect to the algebra $\mathfrak{g}'$. Since $\mathfrak{g}'$ is of type (B_4), we have (cf. [6, p. 493]),

(3.5)   *The group* $\mathbf{C}(h(\chi_1))$ *is isomorphic to the spinor group* $\mathrm{Spin}(9, K)$ *defined by the quadratic form* $\sum_{i=0}^{4} x_i x_{-i}$ *over* K.

Summarizing the above, we have

Theorem 3.6.   *Any involution in G is conjugate in G to an involution in $\mathfrak{H}$. There are exactly two classes of involutions in G. The involutions in one class centralize Sylow 2-subgroups of G, while the involutions in the other do not. G and* $\mathrm{Spin}(9, K)$ *have isomorphic Sylow 2-subgroups.*

4.   *Groups of type* (E_6)

In this case $|G : G'| = 3$ or 1 according as $q \equiv 1 \pmod 3$ or not. The root system Σ of type (E_6) consists of the roots

$$\xi_i - \xi_j, \quad \pm(\xi_i + \xi_j + \xi_k - \xi), \quad \pm\xi,$$

where i, j, k are distinct indices ranging over $1, 2, \ldots, 6$, and where $3\xi = \xi_1 + \xi_2 + \cdots + \xi_6$. As fundamental roots one can take the following:

$$a_i = \xi_i - \xi_{i+1} \quad (1 \leq i \leq 5), \qquad a_6 = \xi_4 + \xi_5 + \xi_6 - \xi.$$

Let w_i be the Weyl reflection associated with the root a_i. Then w_i $(1 \leq i \leq 5)$ interchanges ξ_i and ξ_{i+1} and leaves ξ_j $(j \neq i, i+1)$ fixed. The action of w_6 is as follows:

$$\xi_i \to \xi_i \quad (1 \leq i \leq 3),$$

$$\xi_4 \to \xi - (\xi_5 + \xi_6),$$

$$\xi_5 \to \xi - (\xi_6 + \xi_4),$$

$$\xi_6 \to \xi - (\xi_4 + \xi_5).$$

[4] $S - S'$ denotes the complement of the set S' in the set S.

Let V be the vector space spanned by $\xi_1, \ldots, \xi_6, \xi$ over the field of two elements. It is clear that $(\xi_1, \ldots, \xi_6)$ is a basis of V and that V has a basis consisting of roots. Hence the involutions in $\mathfrak{H}$ are in one-to-one correspondence with the nonzero elements in the dual space V^* of V; the action of W on $\mathfrak{H}$ corresponds to the action of W on V^* defined by $f^w(x) = f(wx)$, where $f \in V^*$, $x \in V$, $w \in W$. Let $(f_1, f_2, \ldots, f_6)$ be the dual basis of $(\xi_1, \ldots, \xi_6)$. Then w_i $(1 \leq i \leq 5)$ interchanges f_i and f_{i+1}, and leaves f_j $(j \neq i, i+1)$ fixed. The operation w_6 transforms f_i $(1 \leq i \leq 3)$ into $f_i + f_4 + f_5 + f_6$ and leaves f_j $(j > 3)$ fixed. Denoting by $\sim$ the conjugacy under W, we have

$$f_1 \sim f_1 + f_4 + f_5 + f_6 \sim f_1 + f_2 + f_3 + f_4 \sim f_1 + f_2 + f_3 + f_5 + f_6;$$

$$f_1 + f_4 \sim f_1 + f_5 + f_6 \sim f_1 + \cdots + f_6.$$

Thus we have proved

(4.1) *The involutions in* $\mathfrak{H}$ *are divided into two conjugate classes under* W. *They are represented by* $h(\chi_1)$ *and* $h(\chi_2)$, *where* χ_1, χ_2 *are defined by*

$$\chi_1(\xi_i) = 1 \quad (1 \leq i \leq 5), \qquad \chi_1(\xi_6) = -1;$$

$$\chi_2(\xi_i) = -1 \quad (1 \leq i \leq 6).$$

We have

$$\Sigma' = \{r \in \Sigma \mid \chi_1(r) = 1\}$$

$$= \{\xi_i - \xi_j, \pm(\xi_i + \xi_j + \xi_6 - \xi) \mid i, j = 1, 2, \ldots, 5; i \neq j\}.$$

This is a root system of type (D_5), and as a system of fundamental roots we have

$$\xi_1 - \xi_2, \ \xi_2 - \xi_3, \ \ldots, \ \xi_4 - \xi_5, \ \xi_4 + \xi_5 + \xi_6 - \xi.$$

Suppose $W' = \{w \in W \mid w(\Sigma') = \Sigma'\}$ contains an element which induces on Σ' an outer automorphism of Σ'. Then W' would contain an element w such that

$$w(\xi_i - \xi_{i+1}) = \xi_i - \xi_{i+1} \quad (1 \leq i \leq 3),$$

$$w(\xi_4 - \xi_5) = \xi_4 + \xi_5 + \xi_6 - \xi,$$

$$w(\xi_4 + \xi_5 + \xi_6 - \xi) = \xi_4 - \xi_5.$$

Consider the root $r = w(\xi_5 - \xi_6)$ in Σ. It can be seen from the above that r is orthogonal to the roots $\xi_i - \xi_j$, $i, j = 1, 2, \ldots, 5$, but such a root does not exist in Σ. Hence every element in W' induces on Σ' an inner automorphism. Let W^* be the Weyl group of Σ'. For $w \in W'$ denote by $\varphi(w)$ the restriction of w to Σ'. Then $\varphi(w) \in W^*$, and φ is surjective. Let $w \in \mathrm{Ker}\ \varphi$. Then one can see easily that the root $w(\xi_5 - \xi_6)$ cannot be but $\xi_5 - \xi_6$. Hence $w = 1$, and $\mathrm{Ker}\ \varphi = \{1\}$. From this it follows easily that W' is generated by $\{w_r \mid r \in \Sigma'\}$.

Consider now $\Sigma'' = \{r \in \Sigma \mid \chi_2(r) = 1\}$. We have

$$\Sigma'' = \{\pm \xi, \xi_i - \xi_j \mid i, j = 1, 2, \ldots, 6;\ i \neq j\}.$$

Hence Σ'' is the direct sum of a root system $\Sigma_1'' = \{\pm \xi\}$ of type (A_1) and a root system $\Sigma_2'' = \{\xi_i - \xi_j\}$ of type (A_5). Let $W'' = \{w \in W \mid w(\Sigma'') = \Sigma''\}$. We shall show that every element in W'' induces on Σ'' an inner automorphism. Suppose the contrary. Then W'' would contain an element w such that

$$w(\xi) = \xi, \qquad w(\xi_i - \xi_{i+1}) = \xi_{6-i} - \xi_{7-i} \quad (1 \le i \le 5).$$

Consider the root $r = w(\xi_4 + \xi_5 + \xi_6 - \xi)$ in Σ. Since any root of the form $\xi_i - \xi_j$ is the image under w of a root of the same form, we have $r = \pm(\xi_i + \xi_j + \xi_k - \xi)$. Since $r - \xi = w(\xi_4 + \xi_5 + \xi_6 - 2\xi) = w(\xi_1 + \xi_2 + \xi_3) \in \Sigma$, we have $r = \xi_i + \xi_j + \xi_k - \xi$. Since $r + (\xi_2 - \xi_3) \in \Sigma$, $r = \xi_3 + \xi_j + \xi_k - \xi$. On the other hand, r is orthogonal to $\xi_1 - \xi_2$, $\xi_2 - \xi_3$, $\xi_4 - \xi_5$, $\xi_5 - \xi_6$. But this is impossible. Hence such an element w does not exist. From this it follows easily that W'' is generated by $\{w_r \mid r \in \Sigma''\}$. Thus,

(4.2)   *For the root system Σ of type (E_6) and any involution $h(\chi)$ in $\mathfrak{H}$,* (1.4) *is verified.*

From the above we have

(4.3) $|\mathbf{C}_G(h(\chi_1))| = q^{12}(q-1)(q^2-1)(q^4-1)(q^6-1)(q^8-1)(q^5-1)$;

$|\mathbf{C}_G(h(\chi_2))| = q^{11}(q^2-1)^2(q^3-1)(q^4-1)(q^5-1)(q^6-1)$.

Hence we obtain

(4.4)   *The involution $h(\chi_1)$ centralizes a Sylow 2-subgroup of G (or G') while $h(\chi_2)$ does not.*

In order to clarify the structure of $\mathbf{C}_G(h(\chi_1))$, we set

$$\mathfrak{L}_1 = \text{the group generated by } \{\mathfrak{X}_r \mid r \in \Sigma'\},$$

$$\mathfrak{L}_2 = \{h(\chi) \in \mathfrak{H} \mid \chi(a_i) = 1 \quad \text{for all } i \neq 5\}.$$

Then $\mathfrak{L}_2$ is a cyclic group of order $q-1$, and centralizes $\mathfrak{L}_1$. The intersection $\mathfrak{L}_1 \cap \mathfrak{L}_2$ is the center of $\mathfrak{L}_1$. We shall compute this center. Let $h(\chi) \in \mathfrak{L}_1 \cap \mathfrak{L}_2$. The condition $h(\chi) \in \mathfrak{L}_1$ is equivalent to saying that χ is of the form

$$\chi = \chi_{a_1, z_1} \chi_{a_2, z_2} \chi_{a_3, z_3} \chi_{a_4, z_4} \chi_{a_6, z_6},$$

where $z_i \in K^*$. The condition $h(\chi) \in \mathfrak{L}_2$ now becomes

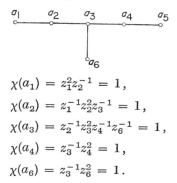

$$\chi(a_1) = z_1^2 z_2^{-1} = 1,$$
$$\chi(a_2) = z_1^{-1} z_2^2 z_3^{-1} = 1,$$
$$\chi(a_3) = z_2^{-1} z_3^2 z_4^{-1} z_6^{-1} = 1,$$
$$\chi(a_4) = z_3^{-1} z_4^2 = 1,$$
$$\chi(a_6) = z_3^{-1} z_6^2 = 1.$$

The above reduces to $z_4^4 = 1$. Since $\chi(a_5) = z_4^{-1}$, we have

(4.5) $|\mathfrak{L}_1 \cap \mathfrak{L}_2| = 4$ or 2 *according as* $q \equiv 1 \pmod 4$ *or not.*

Or, equivalently,

(4.6)   *The center* $\mathbf{Z}(\mathfrak{L}_1)$ *of* $\mathfrak{L}_1$ *is of order* 4 *or* 2 *according as* $q \equiv 1 \pmod 4$ *or not.* $\mathfrak{L}_1$ *is isomorphic to the spinor group* Spin$(10, K)$ *defined by the quadratic form* $\sum_{i=1}^{4} x_i x_{-i}$.

The group $\mathfrak{L}_1 / \mathbf{Z}(\mathfrak{L}_1)$ is isomorphic to the simple Chevalley group of type (D_5) over K (cf. [6], pp. 497–498), and hence by [3], p. 64,

$$|\mathfrak{L}_1 / \mathbf{Z}(\mathfrak{L}_1)| = u^{-1} q^{12} (q^2 - 1)(q^4 - 1)(q^6 - 1)(q^8 - 1)(q^5 - 1),$$

where $u = 4$ or 2 according as $q \equiv 1 \pmod 4$ or not. Hence by (4.6),

$$|\mathfrak{L}_1| = q^{12} (q^2 - 1)(q^4 - 1)(q^6 - 1)(q^8 - 1)(q^5 - 1).$$

Then by (4.3) and (4.5) we have (with the same u as above)

(4.7) $$|\mathbf{C}_G(h(\chi_1)) : \mathfrak{L}_1 \mathfrak{L}_2| = u.$$

By an elementary computation one sees that an element $h(\chi) \in \mathfrak{H}$ belongs to $\mathfrak{L}_1 \mathfrak{L}_2$ if and only if

$$\chi(a_1)^2 \chi(a_3)^2 \chi(a_4) \chi(a_6)^{-1} = \lambda^4$$

for some $\lambda \in K^*$. Hence if we take χ_0 such that $\chi_0(a_i) = 1$ for all $i < 6$, and such that $\chi_0(a_6)$ is a generator of the Sylow 2-subgroup of K^*, then

$$\mathbf{C}_G(h(\chi_1)) = \langle \mathfrak{L}_1 \mathfrak{L}_2, h(\chi_0) \rangle.$$

Summarizing, we have

Theorem 4.8. *Any involution in G (or G') is conjugate in G' to an involution in $\mathfrak{H}$. There are exactly two classes of involutions in G (or G'). The involutions in one class centralize Sylow 2-subgroups of G (or G') while the involutions in the other do not. The centralizer in G of an involution in the former class is a certain extension of the spinor group $\mathrm{Spin}(10, K)$ by an abelian group and can be described precisely (as above).*

5. Groups of type (E_7)

In this case, $|G : G'| = 2$. The root system Σ of type (E_7) consists of the roots

$$\xi_i - \xi_j, \quad \pm(\xi_i + \xi_j + \xi_k - \xi), \quad \pm(\xi_i - \xi),$$

where i, j, k range over distinct values of $1, 2, \ldots, 7$, and where $3\xi = \xi_1 + \cdots + \xi_7$. As a system of fundamental roots one can take

$$a_i = \xi_i - \xi_{i+1} \quad (1 \le i \le 6), \qquad a_7 = \xi_5 + \xi_6 + \xi_7 - \xi.$$

Let w_i be the Weyl reflection associated with a_i. Then w_i $(1 \le i \le 6)$ interchanges ξ_i and ξ_{i+1} and leaves ξ_j $(j \ne i, i+1)$ fixed. The reflection w_7 transforms ξ_i $(i \le 4)$ into itself, and ξ_i $(i \ge 5)$ into $\xi_i - a_7$.

Let P (resp. P') be the additive group generated by the ξ, $\xi_1, \ldots, \xi_7$ (resp. by the roots). (It can be seen easily that P is in fact the additive group generated by the weights.) An involution $h(\chi) \in \mathfrak{H}$ is given by $\chi \in \mathrm{Hom}(P', \{\pm 1\})$, where $\{\pm 1\} \subseteq K^*$. Let Ω^* be the multiplicative group of the algebraically closed extension Ω of K, extend χ to a homomorphism $P \to \Omega^*$, and set $\chi(\xi_i) = x_i$, $\chi(\xi) = x$, $\chi(a_i) = \epsilon_i = \pm 1$. Then

$$\epsilon_i = x_i x_{i+1}^{-1} \quad (1 \le i \le 6), \qquad \epsilon_7 = x_5 x_6 x_7 x^{-1}, \quad x^3 = x_1 x_2 \cdots x_7.$$

From this one derives

$$(5.1) \qquad x = x_1 \epsilon_2 \epsilon_4 \epsilon_6, \qquad x_1^2 = \epsilon_1 \epsilon_3 \epsilon_7.$$

Hence $x_1, x_2, \ldots, x_7$ determine $\epsilon_1, \epsilon_2, \ldots, \epsilon_7$ and hence χ uniquely. We shall write

$$\chi = (x_1, x_2, \ldots, x_7) = (-x_1, -x_2, \ldots, -x_7).$$

If $i \leq 6$, then we obtain corresponding expressions for $\chi^{w_i} = \chi \circ w_i$ by interchanging x_i and x_{i+1} in the above expressions. Let

$$\chi^{w_7} = (x_1', \ldots, x_7') = (-x_1', \ldots, -x_7').$$

Then

$$x_i' = \begin{cases} x_i & (1 \leq i \leq 4), \\ x_1 x_2 x_3 x_4^{-1} x_i & (5 \leq i \leq 7). \end{cases}$$

If $\epsilon_1 \epsilon_3 \epsilon_7 = 1$, then by (5.1) we have $x_i = \pm 1$ for all i. Expressing by $\sim$ the conjugacy under W, we have

$$(-1, 1, 1, 1, 1, 1, 1) \sim (-1, 1, 1, 1, -1, -1, -1)$$

$$\sim (-1, -1, -1, 1, 1, 1, -1)$$

$$\sim (-1, -1, -1, 1, -1, -1, 1).$$

Remembering $(x_1, \ldots, x_7) = (-x_1, \ldots, -x_7)$, we obtain

(5.2)   *Any two elements* $\neq 1$ *in* $\mathrm{Hom}(P', \{\pm 1\})$ *such that* $\epsilon_1 \epsilon_3 \epsilon_7 = 1$ *are conjugate under* W.

If $\epsilon_1 \epsilon_3 \epsilon_7 = -1$, then $x_i = \pm \rho$ for all i, where ρ is a primitive fourth root of unity in Ω^*. Then

$$(\rho, \rho, \rho, \rho, \rho, \rho, \rho) \sim (\rho, \rho, \rho, \rho, -\rho, -\rho, -\rho)$$

$$\sim (\rho, \rho, -\rho, -\rho, -\rho, \rho, \rho) \sim (\rho, \rho, -\rho, -\rho, \rho, -\rho, -\rho);$$

$$(\rho, \rho, \rho, \rho, \rho, \rho, -\rho) \sim (\rho, \rho, \rho, \rho, -\rho, -\rho, \rho).$$

Hence we obtain

(5.3)   *The elements in* $\mathrm{Hom}(P', \{\pm 1\})$ *such that* $\epsilon_1 \epsilon_3 \epsilon_7 = -1$ *are divided into two conjugacy classes under* W. *They are represented by*

$$(\rho, \rho, \rho, \rho, \rho, \rho, \rho) \quad and \quad (\rho, \rho, \rho, \rho, \rho, \rho, -\rho)$$

respectively.

Since $\rho \in K$ if and only if $q \equiv 1 \pmod 4$, we have

(5.4)   *If* $q \equiv 1 \pmod 4$, *then all the involutions in* $\mathfrak{H}$ *are in* $\mathfrak{H}' \subseteq G'$, *and they are divided into three conjugacy classes under* W. *If* $q \equiv 3 \pmod 4$ *then, among the involutions* $h(\chi)$ *in* $\mathfrak{H}$, *only those for which* $\chi(a_1)\chi(a_3)\chi(a_7) = 1$ *belong to* $\mathfrak{H}'$, *and they form a single conjugacy class under* W.

Define χ_1, χ_2, $\chi_3 \in \mathrm{Hom}(P', \{\pm 1\})$ by

$$\chi_1(\xi_i) = 1 \quad (i < 7), \qquad \chi_1(\xi_7) = -1;$$

$$\chi_2(\xi_i) = \rho \quad (1 \le i \le 7);$$

$$\chi_3(\xi_i) = \rho \quad (i < 7), \qquad \chi_3(\xi_7) = -\rho.$$

Then $h(\chi_1)$, $h(\chi_2)$, and $h(\chi_3)$ represent the three conjugacy classes mentioned above. Let $\Sigma' = \{r \in \Sigma \mid \chi_1(r) = 1\}$, $W' = \{w \in W \mid w(\Sigma') = \Sigma'\}$, and define Σ'', W'', Σ''', W''' similarly for $h(\chi_2)$ and $h(\chi_3)$.

We have $\Sigma' = \Sigma_1' \cup \Sigma_2'$, where

$$\Sigma_1' = \{\xi_i - \xi_j, \pm(\xi_i + \xi_j + \xi_7 - \xi) \mid 1 \le i, j < 7, i \ne j\},$$

$$\Sigma_2' = \{\pm(\xi_7 - \xi)\}.$$

Hence Σ_1' is a simple root system of type (D_6), while Σ_2' is a root system of type (A_1). We shall show that W' contains no element which induces an outer automorphism of Σ'. Assume the contrary. Then W' would contain an element w such that

$$w(\xi_7 - \xi) = \xi_7 - \xi, \qquad w(\xi_i - \xi_j) = \xi_i - \xi_j \quad (1 \le i, j \le 5),$$

$$w(\xi_5 - \xi_6) = \xi_5 + \xi_6 + \xi_7 - \xi, \qquad w(\xi_5 + \xi_6 + \xi_7 - \xi) = \xi_5 - \xi_6.$$

Consider the root $r = w(\xi_6 - \xi_7) \in \Sigma$. Then r is orthogonal to $\xi_1 - \xi_2$, $\xi_2 - \xi_3, \ldots, \xi_5 - \xi_6$. It is clear that no such root r exists in Σ. Thus every element in W' induces on Σ' an inner automorphism of Σ'. From this it follows easily that W' is generated by $\{w_r \mid r \in \Sigma'\}$.

We have $\Sigma'' = \{\xi_i - \xi_j, \pm(\xi_i - \xi) \mid i, j = 1, 2, \ldots, 7; i \ne j\}$. Hence Σ'' is a simple root system of type (A_7). Let

$$a_1^* = \xi_7 - \xi, \qquad a_i^* = \xi_{8-i} - \xi_{9-i} \quad (2 \le i \le 6),$$

$$a_7^* = \xi_5 + \xi_6 + \xi_7 - \xi.$$

Then $a_1^*, a_2^*, \ldots, a_7^*$ can be regarded as a system of fundamental roots of Σ. Hence there is an automorphism w_0 of Σ such that $w_0(a_i) = a_i^*$. Since Σ admits no outer automorphism, $w_0 \in W$. We have $w_0 \in W''$ and $w_0^2 = 1$. Denote by W_0'' the subgroup of W'' generated by $\{w_r \mid r \in \Sigma''\}$. Then the group generated by w_0 and W_0'' is the automorphism group of Σ''. From this it follows that

(5.5) $$W'' = \langle W_0'', w_0 \rangle = W_0'' \cup W_0'' w_0.$$

We have

$$\Sigma''' = \{\xi_i - \xi_j, \ \pm(\xi_i + \xi_j + \xi_k - \xi), \ \pm(\xi_7 - \xi) \mid i, j, k = 1, 2, \ldots, 6\}.$$

This is a simple root system of type (E_6). Suppose W''' contains an element which induces on Σ''' an outer automorphism of Σ'''. Then W''' would contain an element w such that

$$w(\xi_4 + \xi_5 + \xi_6 - \xi) = \xi_4 + \xi_5 + \xi_6 - \xi,$$

$$w(\xi_i - \xi_{i+1}) = \xi_{6-i} - \xi_{7-i} \quad (1 \le i \le 5).$$

Consider the root $r = w(\xi_5 + \xi_6 + \xi_7 - \xi)$. Then r is orthogonal to $\xi_1 - \xi_2$, $\xi_3 - \xi_4$, $\xi_4 - \xi_5$, $\xi_5 - \xi_6$. Moreover, $r + (\xi_2 - \xi_3)$ and $r - (\xi_4 + \xi_5 + \xi_6 - \xi)$ are in Σ. It can be seen easily that such a root r does not exist in Σ. Hence every element in W''' induces on Σ''' an inner automorphism $\varphi(w)$ of Σ'''. Let W^* be the Weyl group of Σ'''. Then $\varphi \colon W''' \to W^*$ is an epimorphism. Let $w \in \mathrm{Ker}\,\varphi$, and consider the root

$$r = w(\xi_5 + \xi_6 + \xi_7 - \xi).$$

Clearly r is orthogonal to $\xi_1 - \xi_2$, $\xi_2 - \xi_3$, $\xi_3 - \xi_4$, $\xi_5 - \xi_6$; moreover, $r + (\xi_4 - \xi_5)$, $r + (\xi_4 - \xi_6)$, and $r - (\xi_4 + \xi_5 + \xi_6 - \xi)$ are in Σ. From this it follows that $r = \xi_5 + \xi_6 + \xi_7 - \xi$. Hence $w = 1$, $\mathrm{Ker}\,\varphi = 1$, and φ is an isomorphism. Now it is easy to see that W''' is generated by $\{w_r \mid r \in \Sigma'''\}$.

From the above and [3], p. 64, we have

(5.6) $|\mathbf{C}_G(h(\chi_1))| = q^{31}(q^2-1)^2(q^4-1)(q^6-1)(q^8-1)(q^{10}-1)(q^6-1);$

$|\mathbf{C}_G(h(\chi_2))| = 2q^{28}(q^2-1)(q^3-1)(q^4-1)(q^5-1)$
$$\times (q^6-1)(q^7-1)(q^8-1);$$

$|\mathbf{C}_G(h(\chi_3))| = q^{36}(q-1)(q^2-1)(q^5-1)(q^6-1)$
$$\times (q^8-1)(q^9-1)(q^{12}-1).$$

Among the involutions $h(\chi_i)$, $i=1$, 2, 3, only $h(\chi_1)$ centralizes a Sylow 2-subgroup of G'.

(5.7) *The group $\mathbf{C}_G(h(\chi_1))_\Omega$ and $\mathbf{C}_G(h(\chi_3))_\Omega$ are connected algebraic groups, while $\mathbf{C}_G(h(\chi_2))_\Omega$ has two components, represented respectively by 1 and $\omega(w_0)$.*

We shall classify involutions in G. Let a be an involution in G. Then there exists an element x in G_Ω such that $x^{-1}ax = h \in \mathfrak{H}$, and we can assume that h is one of $h(\chi_i)$, $i = 1, 2, 3$. If $i \neq 2$, then $\mathbf{C}(h(\chi_i))_\Omega$ is a connected algebraic group by (5.7), and hence a and h are conjugate in G. Suppose $h = h(\chi_2)$. Using the notation introduced in [4] we have $(x^{(q)})^{-1}ax^{(q)} = h$. Hence $x^{-1}x^{(q)} \in \mathbf{C}(h(\chi_2))_\Omega$. If $x^{-1}x^{(q)} \in \mathbf{C}_0$, the connected component of $\mathbf{C}(h(\chi_2))_\Omega$, then, by [4], $x^{-1}x^{(q)} = x_1^{-1}x_1^{(q)}$ with some $x_1 \in \mathbf{C}_0$. Then $x_1 x^{-1} = (x_1 x^{-1})^{(q)}$, $x_1 x^{-1} \in G$, and $a = xhx^{-1} = xx_1^{-1}hx_1 x^{-1}$. Hence a is conjugate in G to h. If $x^{-1}x^{(q)} \in \mathbf{C}_0\omega_0$, where $\omega_0 = \omega(w_0)$, then $x^{-1}x^{(q)} = x_1^{-1}\omega_0 x_1^{(q)}$ for some $x_1 \in \mathbf{C}_0$. (This can be seen by an argument entirely similar to the one in [4].) Set $y = xx_1^{-1}$. Then $y^{-1}y^{(q)} = \omega_0$ and $a = yhy^{-1}$. Conversely, if y is any element in G_Ω such that $y^{-1}y^{(q)} = \omega_0$, then $yhy^{-1} = yh(\chi_2)y^{-1}$ is an involution in G, since $(yhy^{-1})^{(q)} = y^{(q)}h(y^{(q)})^{-1} = y\omega_0 h\omega_0^{-1}y^{-1} = yhy^{-1}$. Any two such involutions in G are conjugate in G to each other, since $y^{-1}y^{(q)} = y_1^{-1}y_1^{(q)} = \omega_0$ imply $y_1 y^{-1} \in G$. We shall now show that the involution yhy^{-1}, where $y^{-1}y^{(q)} = \omega_0$, is not conjugate in G to $h = h(\chi_2)$. Suppose $yhy^{-1} = zhz^{-1}$ with $z \in G$. Set $y_1 = z^{-1}y$. Then $y_1^{-1}y_1^{(q)} = \omega_0$ and $y_1 \in \mathbf{C}(h(\chi_2))$. If $y_0 \in \mathbf{C}_0$, then $y_1^{(q)} \in \mathbf{C}_0$ and hence $\omega_0 \in \mathbf{C}_0$, a contradiction. If $y_1 = \omega_0 z_1$ with $z_1 \in \mathbf{C}_0$, then $z_1^{-1}z_1^{(q)} = \omega_0$, again a contradiction. Thus we have proved

(5.8) *The group G has exactly four classes of involutions. They are represented by $h(\chi_i)$, $i = 1, 2, 3$, and $yh(\chi_2)y^{-1}$, where y is an element in G_Ω such that $y^{-1}y^{(q)} = \omega_0$.*

Now we shall consider the involutions in G'. First assume $q \equiv 1 \pmod 4$. Let $\bar{G}_\Omega$ be a simply connected covering of G_Ω, and let $\bar{G}$ be the set of rational points over K in $\bar{G}_\Omega$. Then, by [6], there is an epimorphism $\varphi: \bar{G} \to G'$. Let a be any involution in G' and choose $\bar{a} \in \bar{G}$ such that $\varphi(\bar{a}) = a$. Then $\bar{a}^4 = 1$. Since $\bar{G}$ is simply connected and $\bar{a}$ semisimple, the centralizer of $\bar{a}$ in $\bar{G}_\Omega$ is connected (cf. [8]), and the assumption $q \equiv 1 \pmod 4$ implies that the conjugates of $\bar{a}$ in $\mathfrak{H}$ are rational over K. Hence $\bar{a}$ is conjugate in $\bar{G}$ to an element in $\bar{\mathfrak{H}}$. From this it follows that a is conjugate in G' to an involution in $\mathfrak{H}'$. Then by (5.6) we have

(5.9) *If $q \equiv 1 \pmod 4$, then G' has exactly three classes of involutions, they are represented by $h(\chi_i)$, $i = 1, 2, 3$.*

If $q \equiv 3 \pmod 4$, then the above argument cannot be applied. The author has been unable to decide whether or not, in this case, the involution $yh(\chi_2)y^{-1}$ belongs to G'. By (5.4) one can assert only the following

(5.10) *If* $q \equiv 3 \pmod 4$, *then* G' *has exactly* ν *classes of involutions, where* $\nu = 2$ *or* 1 *according as* $yh(\chi_2)y^{-1}$, *defined above, belongs to* G' *or not.*

The centralizer of $yh(\chi_2)y^{-1}$ in G is

$$\{yxy^{-1} \mid x \in \mathbf{C}(h(\chi_2))_\Omega,\ \omega_0 x^{(q)} \omega_0^{-1} = x\}.$$

From this one can show without difficulty that $yh(\chi_2)y^{-1}$ cannot centralize a Sylow 2-subgroup of G'.

We shall consider the groups $\mathbf{C}_G(h(\chi_1))$ and $\mathbf{C}_{G'}(h(\chi_1))$. Let $\mathfrak{L}_i$ be the group generated by $\{\mathfrak{X}_r \mid r \in \Sigma_i'\}$. Then $\mathfrak{L}_1$ and $\mathfrak{L}_2$ centralize each other. The center $\mathbf{Z}(\mathfrak{L}_1)$ of $\mathfrak{L}_1$ consists of $h(\chi)$, where

$$\chi = \chi_{a_1, z_1} \chi_{a_2, z_2} \chi_{a_3, z_3} \chi_{a_4, z_4} \chi_{a_5, z_5} \chi_{a_7, z_7},$$

such that $\chi(a_i) = 1$ $(1 \le i \le 5)$, $\chi(a_7) = 1$. In terms of z_i this becomes $z_1^2 z_2^{-1} = z_1^{-1} z_2^2 z_3^{-1} = z_2^{-1} z_3^2 z_4^{-1} = z_3^{-1} z_4^2 z_5^{-1} z_7^{-1} = z_4^{-1} z_5^2 = z_4^{-1} z_7^2 = 1$. There are four solutions for $(z_1, \ldots, z_5, z_7)$, but they give only two elements $h(\chi) = 1$ and $h(\chi_1)$. Thus

(5.11) $\mathfrak{L}_1$ *and* $\mathfrak{L}_2$ *centralize each other.* $\mathfrak{L}_1 \cap \mathfrak{L}_2 = \{1, h(\chi_1)\}$ *and this is the center for* $\mathfrak{L}_1$ *as well as for* $\mathfrak{L}_2$.

The group $\mathfrak{L}_2$ is clearly isomorphic to $SL(2, K)$.

The group $\mathfrak{L}_1$ can be clarified as follows. Let $\mathfrak{g}$ be the simple Lie algebra of type (E_7) over the complex field with Chevalley root vector $\{X_r \mid r \in \Sigma\}$. Let $\mathfrak{g}_i'$ be the subalgebra generated by the $\{X_r \mid r \in \Sigma_i'\}$, $i = 1, 2$, and let $\mathfrak{g}_3'$ (resp. $\mathfrak{g}_4'$) be the subspace of $\mathfrak{g}$ spanned by

$$\{X_r \mid r = \xi_i + \xi_j + \xi_k - \xi,\ \xi_i - \xi_7,\ \xi - \xi_i\}$$

(resp. by $\{X_r \mid r = \xi - (\xi_i + \xi_j + \xi_k),\ \xi_7 - \xi_i,\ \xi_i - \xi\}$), where i, j, k range over distinct values of $1, 2, \ldots, 6$. Then $\mathfrak{g}$ is the direct sum

$$\mathfrak{g} = \mathfrak{g}_1' + \mathfrak{g}_2' + \mathfrak{g}_3' + \mathfrak{g}_4' \quad \text{(direct)},$$

of $\mathfrak{g}_1'$-submodules. $\mathfrak{g}_1'$ acts on $\mathfrak{g}_2'$ trivially (i.e., $[\mathfrak{g}_1', \mathfrak{g}_2'] = 0$), but $\mathfrak{g}_3'$ and $\mathfrak{g}_4'$ are irreducible $\mathfrak{g}_1'$-modules of dimension 32. The root vectors X_r form a regular basis of $\mathfrak{g}_3'$ and $\mathfrak{g}_4'$ in the sense of [6], p. 491. The group $\mathfrak{L}_1$ is the group $G_k'(V)$ defined in [6], p. 495, where $V = \mathfrak{g}_3'$. This is in fact the spinor group Spin(12, K).

Using (5.11) above, and the fact that the factor group of $\mathfrak{L}_1$ over its center is isomorphic to the simple Chevalley group of type (D_6), one sees easily that $\mathfrak{L}_1\mathfrak{L}_2$ is a subgroup of index 4 of $\mathbf{C}_G(h(\chi_1))$, and consequently, a subgroup of index 2 of $\mathbf{C}_{G'}(h(\chi_1))$. (Note that $\mathfrak{L}_1\mathfrak{L}_2 \subset G'$.)

Let $h(\chi) \in H$. Then the condition $h(\chi) \in \mathfrak{L}_1\mathfrak{L}_2$ is equivalent to saying that χ is of the form

$$\chi = \left(\prod_{i=1}^{5} \chi_{a_i, z_i}\right)\chi_{a_7, z_7}\chi_{a, z},$$

where $a = \xi_7 - \xi$, and $z_i, z \in K^*$. This condition can be seen easily to be equivalent to saying that $\chi(a_1)\chi(a_3)\chi(a_7)$ and $\chi(a_5)\chi(a_7)$ are squares in K^*.

On the other hand, the condition $h(\chi) \in G'$ is equivalent to saying that χ can be extended to a homomorphism $P \to K^*$. This condition turns out to be equivalent to saying that $\chi(a_1)\chi(a_3)\chi(a_7)$ is a square in K^*.

Let ζ be a nonsquare element in K^* and define χ_4 and χ_5 by $\chi_4(a_1) = \zeta$, $\chi_4(a_i) = 1$ $(i \neq 1)$ and $\chi_5(a_5) = \zeta$, $\chi_5(a_i) = 1$ $(i \neq 5)$. Then from the above we have

(5.12) $\mathbf{C}_G(h(\chi_1)) = \langle \mathfrak{L}_1\mathfrak{L}_2, h(\chi_4), h(\chi_5)\rangle,$

$\mathbf{C}_{G'}(h(\chi_1)) = \langle \mathfrak{L}_1\mathfrak{L}_2, h(\chi_5)\rangle.$

Summarizing, we have

Theorem 5.13.   *G has exactly four classes of involutions of which only three have representatives in $\mathfrak{H}$. If $q \equiv 1 \pmod 4$, then G' has exactly three classes of involutions, all having representatives in $\mathfrak{H}'$. If $q \equiv 3 \pmod 4$, then G' has exactly ν classes of involutions, where $1 \leq \nu \leq 2$. Of these only one class has representatives in $\mathfrak{H}'$. The centralizer of an involution which centralizes a Sylow 2-subgroup of G is a certain extension by an abelian group of the central product of $\mathrm{Spin}(12, K)$ and $SL(2, K)$ and can be described precisely (as in (5.12)).*

6.   Groups of type (E_8)

For these groups $G = G'$. The root system Σ of type (E_8) consists of the roots

$$\xi_i - \xi_j, \quad \pm(\xi_i + \xi_j + \xi_k),$$

where i, j, k range over distinct values of $1, 2, \ldots, 9$, and where $\xi_1 + \xi_2 + \cdots + \xi_9 = 0$. As a system of fundamental roots one can take

$$a_i = \xi_i - \xi_{i+1} \quad (1 \leq i \leq 7), \qquad a_8 = \xi_6 + \xi_7 + \xi_8.$$

Let $w_i = w_r$, where $r = a_i$. Then w_i $(i \leq 7)$ interchanges ξ_i and ξ_{i+1} and leaves ξ_j $(j \neq i, i+1)$ fixed. The operation w_8 transforms ξ_i $(i \leq 5)$ into $\xi_i + \frac{1}{3} a_8$, and ξ_i $(i \geq 6)$ into $\xi_i - \frac{2}{3} a_8$.

Let V be the vector space spanned by $\xi_1, \ldots, \xi_9$ over the field of two elements. It is easy to see that $(a_1, a_2, \ldots, a_8)$ is a basis of V. The action of the Weyl group W on V is obvious, and the action of W on V^*, the dual space of V is defined by $f^w(x) = f(wx)$, where $f \in V^*$, $x \in V$, $w \in W$. Let $(f_1, \ldots, f_8)$ be the dual basis of $(\xi_1, \ldots, \xi_8)$. Then w_i $(i \leq 7)$ interchanges f_i and f_{i+1} and leaves f_j $(j \neq i, i+1)$ fixed. And, w_8 leaves f_i $(i \leq 5)$ fixed and transforms f_j $(j \geq 6)$ into $(f_1 + \cdots + f_5) + f_j$. Expressing by $\sim$ the conjugacy under W, and remembering that W can cause any permutation of $f_1, \ldots, f_8$, we have

$$f_6 \sim f_1 + \cdots + f_6 \sim f_3 + f_4 + \cdots + f_8$$

$$\sim f_1 + f_2 + f_6 + f_7 + f_8 \sim f_2 + f_3 + \cdots + f_6 \sim f_1 + f_6;$$

$$f_6 + f_7 + f_8 \sim f_1 + f_2 + \cdots + f_8 \sim f_4 + f_5 + f_6$$

$$\sim f_1 + f_2 + f_3 + f_6 \sim f_5 + f_6 + f_7 + f_8$$

$$\sim f_1 + f_2 + f_3 + f_4 + f_6 + f_7 + f_8.$$

Hence we have

(6.1)   *The involutions in $\mathfrak{H}$ are divided into two conjugacy classes by W; they are represented by $h(\chi_1)$ and $h(\chi_2)$, where*

$$\chi_1(\xi_i) = -1 \ (1 \leq i \leq 8); \qquad \chi_2(\xi_i) = 1 \ (1 \leq i \leq 7), \ \chi_2(\xi_8) = -1.$$

Set $\Sigma' = \{r \in \Sigma \mid \chi_1(r) = 1\}$, $\Sigma'' = \{r \in \Sigma \mid \chi_2(r) = 1\}$,   and $W' = \{w \in W \mid w(\Sigma') = \Sigma'\}$, $W'' = \{w \in W \mid w(\Sigma'') = \Sigma''\}$. Then we have

$$\Sigma' = \{\xi_i - \xi_j, \ \pm(\xi_i + \xi_j + \xi_9)\} \mid i, j = 1, 2, \ldots, 8; \ i \neq j\}.$$

This is a root system of type (D_8). Suppose W' contains an element which induces on Σ' an outer automorphism of Σ'. Then W' would contain an element w such that

$$w(\xi_i - \xi_{i+1}) = \xi_i - \xi_{i+1} \quad (1 \leq i \leq 7),$$

$$w(\xi_7 - \xi_8) = \xi_7 + \xi_8 + \xi_9, \quad w(\xi_7 + \xi_8 + \xi_9) = \xi_7 - \xi_8.$$

Consider the root $r = w(\xi_8 - \xi_9)$. Then r is orthogonal to all $\xi_i - \xi_j$, where $1 \leq i \leq 8$, and $r + (\xi_7 + \xi_8 + \xi_9) \in \Sigma$. It can be seen easily that no such root exists in Σ. Hence every element in W' induces on Σ' an inner automorphism of Σ'. From this it follows easily that W' is generated by $\{w_r \mid r \in \Sigma'\}$.

The system Σ'' is a direct sum of Σ_1'' and Σ_2'', where

$$\Sigma_1'' = \{\xi_i - \xi_j,\ \pm(\xi_i + \xi_j + \xi_k),\ \pm(\xi_i + \xi_8 + \xi_9)\},$$

i, j, k ranging over distinct values of $1, 2, \ldots, 7$, and where

$$\Sigma_2'' = \{\pm(\xi_8 - \xi_9)\}.$$

Σ_1'' is a root system of type (E_7), and Σ_2'' is a root system of type (A_1), so there cannot be any outer automorphism of Σ''. Hence every element in W'' induces on Σ'' an inner automorphism of Σ''. From this it follows easily that W'' is generated by $\{w_r \mid r \in \Sigma''\}$. Thus

(6.2)   *For the root system of type (E_8) and any involution in $\mathfrak{H}$, (1.4) is verified.*

Consulting [3], p. 64, we obtain

(6.3) $$|\mathbf{C}_G(h(\chi_1))| = q^{56}(q^8 - 1) \prod_{i=1}^{7} (q^{2i} - 1),$$

$$|\mathbf{C}_G(h(\chi_2))| = q^{64}(q^2 - 1)^2(q^6 - 1)(q^8 - 1)(q^{10} - 1)(q^{12} - 1)(q^{14} - 1)(q^{18} - 1)$$

The involution $h(\chi_1)$ centralizes a Sylow 2-subgroup of G, while $h(\chi_2)$ does not.

Denote by $\mathfrak{L}$ the group generated by $\{\mathfrak{X}_r \mid r \in \Sigma'\}$. The center of $\mathfrak{L}$ consists of $h(\chi)$, where χ are of the form

(*) $$\chi = \chi_{a,z} \prod_{i=1}^{7} \chi_{a_i, z_i},$$

$a = \xi_7 + \xi_8 + \xi_9$, $z_i \in K^*$, and such that $\chi(a) = \chi(a_i) = 1$ for all $i = 1, 2, \ldots, 7$. Written in terms of z, z_i, the above becomes $z_1^2 z_2^{-1} = z_1^{-1} z_2^2 z_3^{-1} = z_2^{-1} z_3^2 z_4^{-1} = z_3^{-1} z_4^2 z_5^{-1} = z_4^{-1} z_5^2 z_6^{-1} = z_5^{-1} z_6^2 z_7^{-1} z^{-1} = z_6^{-1} z_7^2 = z_6^{-1} z^2 = 1$. These equations imply $z_1 = \pm 1$, $z = \pm 1$. Since $\chi(a_8) = z z_6^{-1}$, it follows that the center of $\mathfrak{L}$ is $\{1, h(\chi_1)\}$. Since the quotient group of $\mathfrak{L}$ over its center is the simple Chevalley group of type (D_8) over K, we conclude that $\mathfrak{L}$ is of index 2 in $\mathbf{C}_G(h(\chi_1))$. Also it can be seen easily that χ can be written in the form (*), above, if and only if $\chi(a_1)\chi(a_3)\chi(a_5)\chi(a_7)$ is a square in K^*. Hence let ζ be a nonsquare element in K^*, and define χ_0 by $\chi_0(a_i) = 1$ $(i < 7)$, $\chi_0(a_7) = \zeta$. Then

$$(6.4) \qquad \mathbf{C}_G(h(\chi_1)) = \langle \mathfrak{L}, h(\chi_0) \rangle.$$

The group $\mathfrak{L}$ can be clarified as follows. Let $\mathfrak{g}$ be the simple Lie algebra of type (E_8) over the complex field with Chevalley root vectors $\{X_r \mid r \in \Sigma\}$. Let $\mathfrak{g}'$ be the subalgebra generated by $\{X_r \mid r \in \Sigma'\}$. Let $\mathfrak{g}_1'$ be the subspace of $\mathfrak{g}$ spanned by $\{X_r \mid r \in \Sigma - \Sigma'\}$. Then $\mathfrak{g} = \mathfrak{g}' + \mathfrak{g}_1'$ (direct), and it can be seen easily that $\mathfrak{g}_1'$ is an irreducible $\mathfrak{g}'$-module of dimension 128. Hence $\mathfrak{g}_1'$ gives the spin representation of $\mathfrak{g}'$. Hence by [6], pp. 497–498, $\mathbf{C}_G(h(\chi_1))$ is the group $G_K'(V)$, where $V = \mathfrak{g}_1'$. Summarizing the above,

Theorem 6.5. *Any involution in G is conjugate in G to an involution in $\mathfrak{H}$. There are exactly two conjugacy classes of involutions in G. The involutions in one class centralize Sylow 2-subgroups of G, while the involutions in the other do not. The centralizer of an involution in the former class is a certain extension of the spinor group $\mathrm{Spin}(16, K)$ by a group of order 2, and can be described precisely (as in (6.4)).*

References

[1] C. CHEVALLEY, *Theory of Lie groups*, I, Princeton Univ. Press, Princeton, 1946.

[2] C. CHEVALLEY, *Théorie des groupes de Lie, II, Groupes algébriques*, Actualités Sci. Ind. No. 1152, Hermann, Paris, 1951.

[3] C. CHEVALLEY, Sur certains groupes simples, *Tohoku Math. J.* (2) **7** (1955), 14–66.

[4] SERGE LANG, Algebraic groups over finite fields, *Amer. J. Math.* **78** (1956), 555–563.

[5] TAKASHI ONO, Sur les groupes de Chevalley, *J. Math. Soc. Japan* **10** (1958), 307–313.

[6] RIMHAK REE, Construction of certain semi-simple groups, *Canad. J. Math.* **16** (1964), 490–508.

[7] SÉMINAIRE C. CHEVALLEY, 1956–1958, *Classification des groupes de Lie algébriques*, vols. 1, 2, Ecole Normale Supérieure, Paris, 1958.

[8] ROBERT STEINBERG, Representations of algebraic groups, *Nagoya J. Math.* **22** (1963), 33–56.

Proc. Internat. Conf. Theory of Groups, Austral. Nat. Univ. Canberra,
August 1965, pp. 303–305. © Gordon and Breach Science Publishers, Inc. 1967

Remarks on system normalizers
and Carter subgroups

JOHN S. ROSE*

In investigations into the abnormal structure of a finite soluble group, a natural problem is posed by the relationship between system normalizers and Carter subgroups. At present this eludes a satisfactory general solution. For an account of results already obtained, reference may be made to papers [1] of J. L. Alperin and [2] of R. W. Carter.

Some extensions of results of Carter on A-groups, and methods of proof different from his, are reported here. The approach adopted depends on a property of A-groups which follows from

Theorem 1. *Suppose that G is a finite soluble group, with abelian Sylow p-subgroups for some prime factor p of $|G|$. Let D be a system normalizer of G, and D_p the Sylow p-subgroup of D. Then D_p is a Sylow p-subgroup of a normal subgroup of G.*

Corollary 1. *Suppose that G is an A-group and D a system normalizer of G. Then each Sylow subgroup of D is also a Sylow subgroup of some normal subgroup of G.*

This property is perhaps rather unexpected, since D is contained in no proper normal subgroup of G. The relevance of Corollary 1 to the problem of relating system normalizers and Carter subgroups is seen by elementary considerations of *pronormality*. The definition and basic properties of pronormal subgroups are due to Professor P. Hall.

Definition. A subgroup H is pronormal in a group G if any two conjugates of H in G are already conjugate in their join.

* The author wishes to acknowledge financial support from The Australian National University, NATO, and The University of Newcastle upon Tyne, which enabled him to attend the International Conference on the Theory of Groups.

This isolates a useful property of Sylow subgroups in a finite group. In fact, one has

Lemma 1. *Sylow subgroups of normal subgroups are pronormal.*

Thus Corollary 1 implies that if G is an A-group, then each Sylow subgroup of a system normalizer of G is pronormal in G. From this fact, the following deduction can be made.

Corollary 2. *If G is an A-group, then the system normalizers of G are pronormal in G.*

It seems that this property of an A-group underlies Carter's results on system normalizers and Carter subgroups, for the following precise analogues of Theorems 6, 9, and 10 of [2] can be proved. (For any soluble group X, $l(X)$ denotes the nilpotent length of X.)

Theorem 2. *Suppose that G is a finite soluble group such that G and all its subgroups have pronormal system normalizers. Define $D_0 = 1$, $B_0 = G$; and, inductively, for each positive integer i, $D_i = a$ system normalizer of B_{i-1}, $B_i = N_G(D_i)$, the normalizer in G of D_i. Then*

(i) $D_{i+1} \geq D_i$, $B_{i+1} \leq B_i$, *for all i.*

(ii) *There is a Carter subgroup C of G such that $D_i \leq C \leq B_i$ for all i.*

(iii) *If $D_{i+1} = D_i$, then $D_i = C$. If $B_{i+1} = B_i$, then $B_i = C$.*

(iv) *For any Carter subgroup C of G, there is a uniquely determined sequence D_0, B_0, D_1, B_1, ... with C as its limit.*

(v) *If $l(B_i) \geq 3$ for any particular i, then $l(B_{i+1}) \leq l(B_i) - 2$.*

(vi) *If $l(G) \leq 2n + 1$, then $B_n = C$. If $l(G) \leq 2n$, then $D_n = C$.*

In establishing his results, Carter made extensive use of four invariants, which he associated with each subgroup of a soluble group. These are not used in the proof of Theorem 2, which relies on simple properties of pronormal subgroups, rather than on more special features of A-groups. In particular, the following facts are needed.

Lemma 2. *If H is a pronormal subgroup of a group G, then the normalizer $N_G(H)$ of H in G is abnormal in G. Moreover, every subgroup of G in which H is subnormal is contained in $N_G(H)$, so that $N_G(H)$ is the subnormalizer of H in G.*

Lemma 3. *No two distinct conjugates of a pronormal subgroup are permutable.*

It follows at once from Corollary 2 and Lemma 2 that if D is a system normalizer of an A-group G, then $N_G(D)$ is the subnormalizer of D in G. This was also proved in [2] (Theorem 5), by different means, and used in the proofs of the subsequent theorems.

Details of the results described here will be given in [3].

References

[1] J. L. ALPERIN, System normalizers and Carter subgroups, *J. Algebra* **1** (1964), 355–366.

[2] R. W. CARTER, Nilpotent self-normalizing subgroups and system normalizers, *Proc. London Math. Soc.* (3) **12** (1962), 535–563.

[3] J. S. ROSE, Finite soluble groups with pronormal system normalizers, *Proc. London Math. Soc.* [to appear].

Proc. Internat. Conf. Theory of Groups, Austral. Nat. Univ. Canberra, August 1965, pp. 307–314. © Gordon and Breach Science Publishers, Inc. 1967

Group-theoretical interpretation of projective incidence theorems

H. SCHWERDTFEGER

1. Certain groups can be provided with the structure of a projective plane where the group elements are the points and the normalizers and their cosets are the straight lines. An example of such a group is the one-dimensional affine group $\mathfrak{G}_F$ over a field F, that is the group of the transformations

$$x \rightarrow ax + \alpha, \quad \text{represented by} \quad A = (a, \alpha)$$

where $a \neq 0$ and α are elements of the field F which should have its characteristic different from 2. If $B = (b, \beta)$ we have $AB = (ab, a\beta + \alpha)$, the unit element $I = (1, 0)$ and the inverse $A^{-1} = (a^{-1}, -a^{-1}\alpha)$.

The group plane then is the cartesian a, α-plane from which the line $a = 0$ has been removed. It can also be considered as a projective plane from which two straight lines $\mathfrak{L}_0$, $\mathfrak{L}_\infty$ have been removed. As explained in [2], any two lines whose intersection lies on $\mathfrak{L}_0$ are left cosets of the normalizer whose line passes through this intersection; we call them left parallel. Any two lines intersecting at a point of $\mathfrak{L}_\infty$ are right cosets of the normalizer passing through this point; we call them right parallel. The line through the unit element and the point $\mathcal{U} = \mathfrak{L}_0 \cap \mathfrak{L}_\infty$ is the normal subgroup $\mathfrak{H} \triangleleft \mathfrak{G}_F$, consisting of all elements $H = (1, \eta)$ $(\eta \in F)$. In this case both kinds of cosets (parallelisms) coincide.

2. It will be shown now that an apparently wider class of abstract groups can be considered as such a modified projective plane. A group $\mathfrak{G}$ has been termed a T_1-group (cf. [1]) if it contains a subgroup $\mathfrak{H}$ such that for every $A, T \in \mathfrak{G}$, $A \notin \mathfrak{H}$ there is a unique $H \in \mathfrak{H}$ so that

$$TAT^{-1} = HAH^{-1}.$$

Every conjugate TAT^{-1} of A is thus obtainable by transformation with a uniquely defined element $H \in \mathfrak{H}$. It is not difficult to verify that $\mathfrak{G}_F$ is a T_1-group.

For every T_1-group one can readily prove the following facts:

$$(2.1) \qquad\qquad \mathfrak{H} \vartriangleleft \mathfrak{G}.$$

Indeed for every $A \notin \mathfrak{H}$ also all the conjugates $TAT^{-1} = HAH^{-1} \notin \mathfrak{H}$. Hence the set $\mathfrak{G} - \mathfrak{H}$ is a system of complete classes of conjugate elements; so is therefore the subgroup $\mathfrak{H}$ itself and thus $\mathfrak{H} \vartriangleleft \mathfrak{G}$.

(2.2) Every commutator of elements of $\mathfrak{G}$ is an element of $\mathfrak{H}$.

In fact, let $C = TAT^{-1}A^{-1}$. If $A \in \mathfrak{H}$ the statement is evident. If $A \notin \mathfrak{H}$ we have $C = H(AH^{-1}A^{-1}) \in \mathfrak{H}$ for some $H \in \mathfrak{H}$.

Hence we conclude that $TAT^{-1} = CA$ which implies that all conjugates of A are elements of the coset $\mathfrak{H}A$:

$$\mathfrak{C}_A \subseteq \mathfrak{H}A \quad \text{if} \quad A \notin \mathfrak{H}.$$

If $\mathfrak{H}$ is finite, then the number of elements in $\mathfrak{H}A$ equals the order $|\mathfrak{H}|$. According to the definition of a T_1-group this is also the number of conjugate elements of A; thus $|\mathfrak{C}_A| = |\mathfrak{H}|$ and therefore

$$(2.3) \qquad\qquad \mathfrak{C}_A = \mathfrak{H}A, \qquad (A \notin \mathfrak{H}).$$

This might not be the case for all infinite T_1-groups. If however a T_1-group has this property, it will be called a T_1'-group. Every group $\mathfrak{G}_F$ is readily seen to be a T_1'-group.

From (2.3) follows that in every T_1'-group all the elements of $\mathfrak{H}$ are commutators of the form $HAH^{-1}A^{-1}$ where $H \in \mathfrak{H}$ and A a fixed element outside $\mathfrak{H}$. In particular it is seen that $\mathfrak{H}$ then is the commutator group of $\mathfrak{G}$.

3. Let $\mathfrak{N}_A$ denote the normalizer of an element $A \notin \mathfrak{H}$, i.e., the system of all $N \in \mathfrak{G}$ so that $NAN^{-1} = A$. The unique element of $\mathfrak{H}$ which transforms A into itself is the unit element I. Hence

$$(3.1) \qquad\qquad \mathfrak{N}_A \cap \mathfrak{H} = I.$$

Now let $B \notin \mathfrak{N}_A$. Also the intersection

$$(3.2) \qquad\qquad (B\mathfrak{N}_A) \cap \mathfrak{H} = H$$

is a single element of $\mathfrak{H}$. For if $T \in B\mathfrak{N}_A$, i.e., $T = BN$ for an element $N \in \mathfrak{N}_A$, then $TAT^{-1} = BAB^{-1}$ and B can be replaced by a unique $H \in \mathfrak{H}$ so that $BAB^{-1} = HAH^{-1}$.

Similarly

(3.3)
$$(\mathfrak{N}_A B) \cap \mathfrak{H} = H'$$

is a single element of $\mathfrak{H}$. The proof is reduced to the one of (3.2) by the remark that

$$\mathfrak{N}_A B = B(B^{-1}\mathfrak{N}_A B) = B\mathfrak{N}_{B^{-1}AB}.$$

Further let $B \notin \mathfrak{H}$ and consider the intersection $\mathfrak{N}_A \cap \mathfrak{H}B$. Its elements are of the form $N = HB$ ($H \in \mathfrak{H}$). With regard to (3.3) there is an element $N \in \mathfrak{N}_A$ and an element $H \in \mathfrak{H}$ so that $NB^{-1} = H$. Moreover $HB \in \mathfrak{N}_A$ is unique since the relation $HBA(HB)^{-1} = A$ is equivalent to $BAB^{-1} = H^{-1}AH$ where $H \in \mathfrak{H}$ is unique; hence

(3.4)
$$\mathfrak{N}_A \cap (\mathfrak{H}B) = N$$

is a single element of $\mathfrak{N}_A$.

Similarly we show that for any $D \notin \mathfrak{N}_A$,

(3.5)
$$(D\mathfrak{N}_A) \cap (\mathfrak{H}B)$$

is a single element. Determine $N \in \mathfrak{N}_A$ and $H \in \mathfrak{H}$ so that $DN = HB$. This is possible since for all $N \in \mathfrak{N}_A$

$$A = NAN^{-1} = D^{-1}BHA(D^{-1}BH)^{-1}$$

and so there must exist a unique $H \in \mathfrak{H}$ for which

$$B^{-1}DA(B^{-1}D)^{-1} = HAH^{-1}.$$

Thus the intersection (3.5) is the element HB.

4. Now we observe that the elements of $\mathfrak{H}$ form a complete system of representatives mod $\mathfrak{N}_A$ ($A \notin \mathfrak{H}$) in $\mathfrak{G}$ so that

(4.1)
$$\mathfrak{G} = \mathfrak{H} \cdot \mathfrak{N}_A = \mathfrak{N}_A \cdot \mathfrak{H} = \bigcup_{H \in \mathfrak{H}} H\mathfrak{N}_A.$$

Indeed $H\mathfrak{N}_A = H'\mathfrak{N}_A$ implies $H^{-1}H' \in \mathfrak{N}_A$ so that by (3.1) $H = H'$. But also

(4.2)
$$\mathfrak{G} = \bigcup_{N \in \mathfrak{N}_A} N\mathfrak{H} \quad \text{for any fixed } A \notin \mathfrak{H}$$

is the decomposition of $\mathfrak{G}$ into cosets mod $\mathfrak{H}$. Indeed $N\mathfrak{H} = N'\mathfrak{H}$ by (3.1) again implies $N' = N$.

Further recalling that $\mathfrak{H}$ is the commutator group of $\mathfrak{G}$ we conclude that the factor group

(4.3) $\mathfrak{G}/\mathfrak{H} \simeq \mathfrak{N}_A$ *is abelian* whatever $A \notin \mathfrak{H}$.

Hence for any $N_A \in \mathfrak{N}_A \ (N_A \neq I)$

(4.4) $\mathfrak{N}_{N_A} = \mathfrak{N}_A$.

By (2.3) all the elements of $\mathfrak{H}A$ are the conjugates of A in $\mathfrak{G}$. Thus for distinct $H \in \mathfrak{H}$, or for distinct $H' \in \mathfrak{H}$ respectively, the normalizers

$$H\mathfrak{N}_A H^{-1} = \mathfrak{N}_{HAH^{-1}} = \mathfrak{N}_{AH'}$$

are distinct conjugates of $\mathfrak{N}_A$. For if $A, B \notin \mathfrak{H}$ and $\mathfrak{N}_A = \mathfrak{N}_B$ then $B \in \mathfrak{N}_A$ and if $B = HAH^{-1} = AH'$ it follows that $A^{-1}B \in \mathfrak{H}$, i.e., $A = B$.

Conversely all $\mathfrak{N}_A, \mathfrak{N}_B$ are conjugate subgroups in $\mathfrak{G}$. Indeed $\mathfrak{N}_A$ has by (3.4) a common element N with each coset $\mathfrak{H}B$ and each element of $\mathfrak{H}B$ is a certain conjugate HBH^{-1} of B and so a certain $\mathfrak{N}_{HBH^{-1}} = H\mathfrak{N}_B H^{-1}$ must coincide with a given $\mathfrak{N}_A$.

Finally it follows that

(4.5) $\mathfrak{N}_A \cap \mathfrak{N}_B = I$, if $\mathfrak{N}_A \neq \mathfrak{N}_B$.

For if $N_A \in \mathfrak{N}_A$ and $N_A \in \mathfrak{N}_B$ then $B \in \mathfrak{N}_{N_A} = \mathfrak{N}_A$ whence $\mathfrak{N}_A = \mathfrak{N}_B$.

5. It remains to investigate the intersection of different cosets of normalizers $\mathfrak{N}_A, \mathfrak{N}_B \ (A, B \notin \mathfrak{H})$.

Evidently all distinct $C\mathfrak{N}_A \ (C \in \mathfrak{G})$, i.e., all $H\mathfrak{N}_A \ (H \in \mathfrak{H})$, have no common elements in $\mathfrak{G}$. They will be called right-parallel (to each other). Similarly all distinct $\mathfrak{N}_A C$, i.e., all $\mathfrak{N}_A H$, are disjoint and will be called left-parallel.

Further we note that every left coset $\mathfrak{N}_A C = C \cdot C^{-1}\mathfrak{N}_A C$ is right coset of some conjugate normalizer, namely $\mathfrak{N}_{C^{-1}AC}$. Thus it is sufficient to study the case of two right cosets $C\mathfrak{N}_A, D\mathfrak{N}_B$.

Their intersection $\mathfrak{D} = (C\mathfrak{N}_A) \cap (D\mathfrak{N}_B)$ cannot contain more than one element. For if $\mathfrak{D} = (F, F', F'', \ldots)$ so that $\mathfrak{D} = (F\mathfrak{N}_A) \cap (F\mathfrak{N}_B)$, then

$$F^{-1}\mathfrak{D} = (I, F^{-1}F', F^{-1}F'', \ldots) = F^{-1}(F\mathfrak{N}_A \cap F\mathfrak{N}_B)$$
$$= \mathfrak{N}_A \cap \mathfrak{N}_B = I,$$

and in fact $\mathfrak{D} = F$.

Further we may assume that one of the cosets is a normalizer itself, for

$$C^{-1}\mathfrak{D} = (C^{-1}D\mathfrak{N}_B) \cap \mathfrak{N}_A$$

which is empty if $\mathfrak{D} = \emptyset$. Thus it remains to prove

(5.1) If for some $E \in \mathfrak{G}$ the intersection $\mathfrak{N}_A \cap (E\mathfrak{N}_B) = \emptyset$ then either $\mathfrak{N}_A = \mathfrak{N}_B$ or $E\mathfrak{N}_B$ is a left coset of $\mathfrak{N}_A$.

For the proof we consider another decomposition of $\mathfrak{G}$ into cosets mod $\mathfrak{N}_B$. This contains all $N_A\mathfrak{N}_B$ where $N_A \in \mathfrak{N}_A$ and all these are distinct, thus disjoint: In fact $N_A\mathfrak{N}_B = N'_A\mathfrak{N}_B$ implies that

$$N_A^{-1}N'_A \in \mathfrak{N}_B,$$

and thus by (4.5) $N'_A = N_A$. But the complete decomposition of $\mathfrak{G}$ (mod $\mathfrak{N}_B$) may contain other terms which must be of the form $H\mathfrak{N}_B$ where $H \in \mathfrak{H}$ and no $N_A \in H\mathfrak{N}_B$ (cf. (4.1)).

If now $E\mathfrak{N}_B = N_A\mathfrak{N}_B$ then $\mathfrak{N}_A \cap (E\mathfrak{N}_B) = N_A$. If, however, $E\mathfrak{N}_B = H\mathfrak{N}_B$ (no $N_A \in H\mathfrak{N}_B$) then

$$(E\mathfrak{N}_B) \cap \mathfrak{N}_A = \emptyset$$

and we conclude that $E\mathfrak{N}_B = H\mathfrak{N}_B$ must be a coset of $\mathfrak{N}_A$, either left or right. In the first case $H\mathfrak{N}_B = \mathfrak{N}_A H'$ ($H' \in \mathfrak{H}$) so that for a certain element $N_A \in \mathfrak{N}_A$ we have $H = N_A H'$; since $H, H' \in \mathfrak{H}$ this implies $N_A = I$ and thus $H' = H$ is the uniquely defined element of $\mathfrak{H}$ which transforms $\mathfrak{N}_B$ into $\mathfrak{N}_A = H\mathfrak{N}_B H^{-1}$. Thus $E\mathfrak{N}_B$ is left coset of $\mathfrak{N}_A$ and unique.

In the second case $H\mathfrak{N}_B = H'\mathfrak{N}_A$ which again implies that $H' = H$ and therefore $\mathfrak{N}_B = \mathfrak{N}_A$.

6. Now we adjoin to the group ideal elements which will be denoted by capital script letters $\mathscr{A}, \mathscr{B}, \ldots$. The symbol $\mathscr{A}'$ will be defined by the system of all left cosets of a normalizer $\mathfrak{N}_A$ in the same way as a point at infinity is defined in plane by a pencil of parallel lines. We shall write

$$\mathscr{A}' = \bigcap_{H \in \mathfrak{H}} \mathfrak{N}_A H.$$

Similarly

$$\mathscr{A} = \bigcap_{H \in \mathfrak{H}} H\mathfrak{N}_A.$$

The system of all $\mathscr{A}'$ (all $\mathscr{A}$) may be considered as a "straight line at infinity" $\mathfrak{L}_0$ ($\mathfrak{L}_\infty$ resp.). Both together constitute "the absolute" of $\mathfrak{G}$.

Their intersection $\mathscr{U}$ must be defined by a subgroup for which left and right parallelism coincide; this is the normal subgroup $\mathfrak{H}$. So we say that all cosets of $\mathfrak{H}$ pass through $\mathscr{U} = \mathfrak{L}_0 \cap \mathfrak{L}_\infty$. Then we have $\mathfrak{N}_A \cap \mathfrak{L}_0 = \mathscr{A}'$, $\mathfrak{N}_A \cap \mathfrak{L}_\infty = \mathscr{A}$, $\mathfrak{H} \cap \mathfrak{L}_0 = \mathfrak{H} \cap \mathfrak{L}_\infty = \mathscr{U}$.

Now we call *straight lines in* $\mathfrak{G}$:

1. all normalizers $\mathfrak{N}_A$ and their cosets, left and right, completed by their intersections $\mathscr{A}$, $\mathscr{A}'$ with $\mathfrak{L}_\infty$, $\mathfrak{L}_0$;
2. the normal subgroup $\mathfrak{H}$ and its cosets, completed by the point $\mathscr{U} = \mathfrak{H} \cap \mathfrak{L}_0 = \mathfrak{H} \cap \mathfrak{L}_\infty$;
3. the ideal (infinite) lines $\mathfrak{L}_0$, $\mathfrak{L}_\infty$.

All these together with the group elements and the ideal elements as points define a projective plane. In view of the facts demonstrated in Sections 2–5 the following fundamental axioms are satisfied.

(a) Any two straight lines have one and only one point in common.

(b) Any two points have one and only one straight line in common.

Following the scheme developed in the paper [2] for the group $\mathfrak{G}_F$ it is now possible to give a geometrical interpretation to all group operations in $\mathfrak{G}$ and conversely: to give a group theoretical interpretation to geometrical incidence relations in the group plane obtained from $\mathfrak{G}$ by adjoining the two "infinite" lines $\mathfrak{L}_0$ and $\mathfrak{L}_\infty$ to $\mathfrak{G}$.

7. The basic problems which have been dealt with in [2] are the following.

I. To find the line through two given points A, B (not both in $\mathfrak{H}$). The line is given by the coset $A\mathfrak{N}_{A^{-1}B}$ ($= \mathfrak{N}_{AB^{-1}}B = \cdots$).

II. To find the product AB for any two points A, B. There are four different cases, according to the situation of A and B with respect to $\mathfrak{H}$ (cf. [2], p. 687). These constructions also lead to a projective method of finding the inverse A^{-1} for any given $A \in \mathfrak{G}$.

The set of all elements T which transform a given normalizer $\mathfrak{N}_A$ into another given normalizer $\mathfrak{N}_B$ is the (only) coset which is right-parallel to $\mathfrak{N}_A$ and left-parallel to $\mathfrak{N}_B$. So if

$$\mathscr{A} = \mathfrak{N}_A \cap \mathfrak{L}_\infty, \qquad \mathscr{B}' = \mathfrak{N}_A \cap \mathfrak{L}_0,$$

$$\mathscr{A}' = \mathfrak{N}_B \cap \mathfrak{L}_\infty, \qquad \mathscr{B} = \mathfrak{N}_B \cap \mathfrak{L}_0,$$

then the line $\overline{\mathscr{A}\mathscr{B}}$ carries the coset:

$$\overline{\mathscr{A}\mathscr{B}} = T\mathfrak{N}_A = \mathfrak{N}_B T.$$

Correspondingly

$$\overline{\mathscr{A}'\mathscr{B}'} = \mathfrak{N}_A S = S \mathfrak{N}_B.$$

The intersection $\overline{\mathscr{A}\mathscr{B}} \cap \overline{\mathscr{A}'\mathscr{B}'}$ is therefore an element W such that $W \mathfrak{N}_A W^{-1} = \mathfrak{N}_B$ and $W \mathfrak{N}_B W^{-1} = \mathfrak{N}_A$. Thus

$$W^{-1} = W.$$

If $\mathfrak{N}_C$ is a third normalizer and

$$\mathscr{C} = \mathfrak{N}_C \cap \mathfrak{L}_\infty, \qquad \mathscr{C}' = \mathfrak{N}_C \cap \mathfrak{L}_0$$

we find

$$\overline{\mathscr{A}\mathscr{C}} \cap \overline{\mathscr{A}'\mathscr{C}'} = V \quad \text{and again} \quad V^{-1} = V.$$

Moreover the two triangles $\mathscr{A}\mathscr{B}\mathscr{C}$ and $\mathscr{A}'\mathscr{B}'\mathscr{C}'$ are seen to be perspective with the centrum I. Hence the three points

$$\mathscr{U} = \mathfrak{L}_0 \cap \mathfrak{L}_\infty = \overline{\mathscr{B}\mathscr{C}} \cap \overline{\mathscr{B}'\mathscr{C}'}, \qquad V = \overline{\mathscr{A}\mathscr{C}} \cap \overline{\mathscr{A}'\mathscr{C}'},$$

$$W = \overline{\mathscr{A}\mathscr{B}} \cap \overline{\mathscr{A}'\mathscr{B}'},$$

are collinear if and only if Desargues' theorem is true in the $\mathfrak{G}$-plane. So we see that this theorem is equivalent with the statement: All elements of order 2 in $\mathfrak{G}$ form a single coset of $\mathfrak{H}$ in $\mathfrak{G}$, or a complete class of conjugate elements.

8. In (4.3) it was stated that all normalizers $\mathfrak{N}_A$ ($A \notin \mathfrak{H}$) in the T_1'-group $\mathfrak{G}$ are abelian. Can the same be said for the normal subgroup $\mathfrak{H} \lhd \mathfrak{G}$? It will be seen that this depends on the validity of Pappus' theorem in the $\mathfrak{G}$-plane.

Let $\mathscr{A}, \mathscr{A}', \mathscr{B}, \mathscr{B}', \mathscr{C}, \mathscr{C}'$ be the vertices of a Pappus hexagon where $\mathscr{A}, \mathscr{B}, \mathscr{C}$ lie on $\mathfrak{L}_0$ and $\mathscr{A}', \mathscr{B}', \mathscr{C}'$ on $\mathfrak{L}_\infty$. We may assume that

$$\overline{\mathscr{A}'\mathscr{B}} \cap \overline{\mathscr{C}\mathscr{C}'} = I$$

so that $\overline{\mathscr{A}'\mathscr{B}}$ and $\overline{\mathscr{C}\mathscr{C}'}$ carry normalizers $\mathfrak{N}$ and $\mathfrak{N}'$ respectively. Further put

$$\overline{\mathscr{A}\mathscr{A}'} \cap \overline{\mathscr{C}\mathscr{B}'} = A, \qquad \overline{\mathscr{B}\mathscr{B}'} \cap \overline{\mathscr{A}\mathscr{C}'} = B.$$

Pappus' theorem states that the three points I, A, B are collinear. For this it is sufficient that $B \in \mathfrak{N}_A$, if $A \notin \mathfrak{H}$. If however $A \in \mathfrak{H}$, then also $B \in \mathfrak{H}$.

For the proof we follow the sides of the hexagon in the group $\mathfrak{G}$, beginning at the vertex B

$$\overline{\mathscr{B}\mathscr{A}'} = \mathfrak{N}, \quad \overline{\mathscr{A}'\mathscr{A}} = A\mathfrak{N},$$

$$\overline{\mathscr{A}\mathscr{C}'} = A\mathfrak{N}A^{-1}\cdot B,$$

$$\overline{\mathscr{C}'\mathscr{C}} = B^{-1}A\mathfrak{N}A^{-1}B = B^{-1}A\mathfrak{N}(B^{-1}A)^{-1} = \mathfrak{N}',$$

$$\overline{\mathscr{C}\mathscr{B}'} = \mathfrak{N}'A = B^{-1}A\mathfrak{N}(B^{-1}A)^{-1}A,$$

$$\overline{\mathscr{B}'\mathscr{B}} = BA^{-1}\mathfrak{N}'A = BA^{-1}B^{-1}A\mathfrak{N}(B^{-1}A)^{-1}A = \mathfrak{N}B.$$

Hence if

$$[B, A^{-1}] = BA^{-1}B^{-1}A = C$$

it follows that

$$C\mathfrak{N}C^{-1} = \mathfrak{N}$$

and since C as a commutator is an element of $\mathfrak{H}$, it follows that $C = I$ which is equivalent to

$$BA = AB.$$

Thus from the validity of Pappus' theorem in the $\mathfrak{G}$-plane follows the commutativity of $\mathfrak{H}$.

References

[1] H. SCHWERDTFEGER, Über eine spezielle Klasse Frobeniusscher Gruppen, *Arch. Math.* **13** (1962), 283–289.
[2] H. SCHWERDTFEGER, Projective geometry in the one-dimensional affine group, *Canad. J. Math.* **16** (1964), 683–700.

Proc. Internat. Conf. Theory of Groups, Austral. Nat. Univ. Canberra, August 1965, pp. 315–319. © Gordon and Breach Science Publishers, Inc. 1967

On the Galois cohomology of linear algebraic groups

ROBERT STEINBERG

Our purpose is to discuss the theorem **A** below, some consequences, and some relatives. To do this we must start with some preliminaries.

Recall that a linear algebraic group G is a subgroup of some $GL_n(K)$ ($n \geq 1$, K algebraically closed field) which is the complete set of zeros of a set of polynomials in the matrix entries. The word algebraic will be omitted. For example, SL_n (the subgroup of elements of determinant 1), $O_n(K)$ (resp. $Sp_n(K)$) (the subgroup fixing a nonsingular symmetric (resp. skew) bilinear form), the subgroup of diagonal or superdiagonal elements of GL_n or SL_n, all are linear groups. An isomorphism of linear groups is required to be not only an isomorphism of the corresponding abstract groups but also birational (in terms of the matrix entries). We will use the Zariski topology, in which the closed subsets of G are the algebraic subsets (i.e., complete sets of zeros of polynomials in the matrix entries). In this topology all of the groups mentioned above are connected except for O_n which has two components, SO_n being the identity component. Recall that G is (defined) over the subfield k of K if the polynomial ideal which defines G has a basis of polynomials with coefficients in k. For example, $SL_n(K)$ is over any subfield of K while $O_n(K)$ is over any subfield which contains the coefficients of the bilinear form which defines the group.

Henceforth k will be a perfect field, K its algebraic closure, and Γ the Galois group of K over k. Let G be a linear algebraic group over k. Consider all maps $\gamma \to x_\gamma$ from Γ to G which

(1) *are cocycles:* $x_{\gamma\delta} = x_\gamma \gamma(x_\delta)$ $(\gamma, \delta \in \Gamma)$,

(2) *are continuous:* $x_\gamma = 1$ *for all γ in a subgroup of Γ corresponding to a finite extension of k.*

Here one uses the finite topology on Γ and the discrete topology on G. The group G acts on the continuous cocycles: $x_\gamma \cdot a = a^{-1} x_{\gamma} \gamma(a)$ $(a \in G, \gamma \in \Gamma)$. Then, by definition, $H^1(k, G)$ is the set of orbits under this action. There is always at least one orbit, that containing the cocycle which is identically 1. If this is the only orbit, we write $H^1(k, G) = 0$.

There are several possible interpretations of H^1, but the one which is most important in the present context may be described as follows. Let S be some algebraic structure over k whose group of automorphisms over K is G. For example, S may be a quadratic form, or a central simple algebra, or a Lie algebra (G is the orthogonal group in the first case). Let S' be a second structure over k which is isomorphic over K to S. Let $\varphi: S \to S'$ be an isomorphism. Form

$$x_\gamma = \varphi^{-1}\gamma(\varphi) \quad (\gamma \in \Gamma, \gamma(\varphi) = \gamma \circ \varphi \circ \gamma^{-1}).$$

Then x is a cocycle, continuous because φ is over some finite extension of k. Here x depends on the choice of φ, but if a different isomorphism $\varphi \circ a$ ($a \in \operatorname{Aut} S = G$) is chosen then x is replaced by an equivalent cocycle. Thus it turns out that $H^1(k, G)$ classifies the structures S' over k which are isomorphic over K to S, modulo isomorphism over k.

Let us consider some examples. If $G = GL_1$, GL_n, SL_n, or Sp_n, then $H^1(k, G) = 0$. The first result is due to Hilbert (Theorem 90), the second to Speiser, the third follows easily from the first two, and the fourth is a consequence of the fact that any two nonsingular skew bilinear n-forms with coefficients in k are isomorphic over k. If $G = O_n$ then the field k plays a role: if k is the field of real (resp. rational) numbers, then the cardinality of $H^1(k, G)$ is $[n/2] + 1$ (resp. infinite).

Using the notation dim k for the cohomological dimension of Γ (topologized as above), we can now state our central theorem.

A. *If k is a perfect field such that* (I) dim $k \leq 1$, *then* (II) $H^1(k, G) = 0$ *for every connected linear group over k.*

Remarks. (a) Equivalent to dim $k \leq 1$ is: there are no noncommutative finite dimensional division algebras over k. Some cases in which dim $k \leq 1$ are as follows: if k is finite; if k is C_1 (i.e., for every n every homogeneous polynomial of degree n in $n + 1$ variables over k has a nontrivial zero); if k is complete relative to a discrete valuation and has an algebraically closed residue class field.

(b) There is a converse to **A**: if k is a perfect field then (II) implies (I).

(c) In general the assumption that k is perfect cannot be dropped in **A**, but if it is assumed that G is semisimple (i.e., has no nontrivial connected solvable normal subgroup), it appears from recent work of Grothendieck (for which there is a simplified version due to Borel and Springer [7]), that it can be.

(d) Further details about the examples and cohomological concepts introduced above may be found in [5]. The proof of **A** in some special cases may be found in [4] and in general in [8].

We will now indicate, very briefly, some of the ideas involved in the proof. The case k finite follows almost immediately from the theorem of Lang [2] that if k is a finite field of q elements, if σ is the homomorphism which replaces all matrix entries of all elements of G by their qth powers, and if x is any element of G, then there exists a in G such that $x = a\sigma(a)^{-1}$ (if $\gamma \to x_\gamma$ is any continuous cocycle, one applies Lang's result with $x = x_\sigma$). In the general case we consider the result **A*** obtained from **A** by adding the assumption:

(*) *G contains a Borel* (i.e., *a maximal connected solvable*) *subgroup over* k.

Assume **A*** has been proved. Let G be as in **A**. Introducing a suitable group G^* which satisfies (*) and using **A*** applied to G^*, one can make a favorable comparison of G and G^* to show that G also satisfies (*), whence **A*** applied a second time, this time to G, yields **A** (see [8], p. 80). Thus we are reduced to proving **A***. Recall that an algebraic group is called a torus if it is isomorphic (over K) to the direct product of a finite number of copies of GL_1. Now in case G is a torus, Serre ([4], Proposition 3.1.2) has proved **A***. (This is the only place in our proof of **A** where the assumption dim $k \leq 1$ is used.) Thus in the general case **A*** follows from:

B. *If k is a perfect field and G is a connected linear group over k which satisfies* (*), *then every element of $H^1(k, G)$ can be reduced to a torus over* k.

The proof of **B** depends on the following result.

C. *Assume that G is as in **B** and also that G is semisimple and simply connected. Then every conjugacy class X of semisimple elements which is defined over k contains an element x over* k.

Recall that G is simply connected if every projective rational representation of G is a projection of a linear one, while an element of G is semisimple if it is diagonalizable (over K). (The groups SL_n and Sp_n are simply connected, while SO_n is doubly connected having as its simply connected covering the group Spin_n; all these groups are semisimple.) For the deduction of **B** from **C** see [8], p. 78. Finally, to prove **C** we construct a cross-section S, defined over k, of the collection of semisimple conjugacy classes ([8], 7.16, 9.4, 9.7); then the element $x = X \cap S$ fulfils the requirements of **C**. In this last step one must go rather deeply into the structure of linear groups; in the other steps one only skims the surface.

From the argument used to show that **A*** implies **A**, we see that the following result has also been proved.

D. *Let k be a perfect field. If* $\dim k \le 1$, *then every connected linear group over k satisfies* (*).

A consequence of **D** is that the simple linear groups over k can be classified without much further work (see [4]), in terms of the corresponding classification over K; we recall that over K the simple groups are roughly in one-one correspondence with the simple Lie groups over the complex field (the classical groups together with the five exceptional groups) (see [3]). The converse of D turns out to be true ([6], p. 129).

We close by mentioning two other theorems related to the above results. These concern groups over fields of $\dim \le 2$.

E. *If k is a p-adic field and G is a semisimple simply connected group over k, then* $H^1(k, G) = 0$.

F. *Same as* **E** *except that k is a complete field relative to a discrete valuation and the residue class field has $\dim \le 1$.*

The first result is due to M. Kneser [1], whose proof uses **B** extensively. At the recently concluded 1965 A.M.S. Summer Institute in Algebraic Groups (see [9]), Bruhat, Tits, and Springer have indicated a proof of the more general result **F** by a simpler method which makes direct use of **A**.

References

[1] M. Kneser, Galois-Kohomologie halbeinfacher algebraischer Gruppen über p-adischen Körpern I, II, *Math. Z.* **88** (1965), 40–47; ibid. **89** (1965), 250–272.
[2] Serge Lang, Algebraic groups over finite fields, *Amer. J. Math.* **78** (1956), 555–563.

[3] SÉMINAIRE C. CHEVALLEY, 1956–1958, *Classification des groupes de Lie algébriques, I, II*, Ecole Normale Supérieure, Paris, 1958.

[4] JEAN-PIERRE SERRE, Cohomologie galoisienne des groupes algébriques linéaires, *Colloque sur la théorie des groupes algébriques, Bruxelles*, 1962, pp. 53–68; Gauthier-Villars, Paris, 1962

[5] JEAN-PIERRE SERRE, *Cohomologie galoisienne*, Springer, Berlin–Göttingen–Heidelberg, 1964.

[6] T. A. SPRINGER, Quelques résultats sur la cohomologie galoisienne, *Colloque sur la théorie des groupes algébriques, Bruxelles*, 1962, pp. 129–135; Gauthier-Villars, Paris, 1962.

[7] T. A. SPRINGER and A. BOREL, Rationality properties of linear algebraic groups, *Proc. Summer Inst. Algebraic Groups*, 1965, I-C; Amer. Math. Soc., Providence, R.I., 1965.

[8] R. STEINBERG, Regular elements of semisimple, algebraic groups, *Inst. Hautes Études Sci. Publ. Math.* No. 25 (1965), 49–80.

[9] J. TITS, Simple groups over local fields, *Proc. Summer Inst. Algebraic Groups*, 1965, I-G; Amer. Math. Soc., Providence, R.I., 1965.

Proc. Internat. Conf. Theory of Groups, Austral. Nat. Univ. Canberra, August 1965, p. 321. © Gordon and Breach Science Publishers, Inc. 1967

On the class of certain nilpotent groups

A. G. R. STEWART

P. Hall, in his paper "Some sufficient conditions for a group to be nilpotent" (*Illinois J. Math.*, 1958), has proved the following:

Let N be a normal subgroup of a group G. If N is nilpotent of class c, and G/N' is nilpotent of class d where N' denotes the derived group of N, then G is nilpotent of class at most

$$\binom{c+1}{2}d - \binom{c}{2}.$$

This bound can be improved to $cd + (c-1)(d-1)$. The new bound is, in general, best possible. If further restrictions are placed on G, then a better bound can be obtained: for example, if G is metabelian, the bound is cd.

Proc. Internat. Conf. Theory of Groups, Austral. Nat. Univ. Canberra,
August 1965, pp. 323–346. © Gordon and Breach Science Publishers, Inc, 1967

Metabelian groups with two generators

G. SZEKERES

1. Ideals in $Z[x]$

The purpose of this paper is to determine all finite metabelian groups generated by two elements; the classification will also include infinite groups provided that the commutator subgroup is finite. Metabelian groups with two generators have many applications, particularly in the construction of various group-theoretical examples and counter-examples; a complete classification will no doubt extend their applicability.

Previous determinations were confined to certain special subclasses such as when the commutator subgroup is cyclic, or when the (abelian) commutator subgroup is of prime exponent p (see [4] for a list of known cases). A complete classification has not been attempted before.

The groups will be described in terms of Young diagrams and certain numerical invariants which are uniquely (or almost uniquely) associated with the groups. There is a residual isomorphism problem which cannot be resolved in a neat general fashion. Fortunately the redundancy is not very severe in relation to the great diversity of group structures obtained in the classification.

A metabelian group with two generators is structurally equivalent to (i) a quotient module $A = Z[x, y]/N$ over $Z[x, y]$ where Z is the domain of rational integers and N an ideal in $Z[x, y]$, and (ii) a fairly simply constituted factor system in A. Considerable portion of the work will be taken up with the enumeration of ideals in $Z[x, y]$. If the commutator subgroup is finite then we only have to consider ideals whose norm $|N|$ (i.e., the order of $Z[x, y]/N$) is finite. A further condition is that x and y should have inverses modulo N (Section 5).

For ideals of $Z[x]$ there exists a complete enumeration theory due to Kronecker and Hensel [1]; the version presented here is due to the author [3]. We shall only consider ideals with finite norm, that is ones which have no proper principal ideal factor. Denote by $\mathscr{I}_0[x]$ the set of these ideals and by $\mathscr{I}[x]$ the set of all ideals of $Z[x]$. Members of $\mathscr{I}_0[x]$ always contain a nonzero constant; the smallest positive integer contained in $M \in \mathscr{I}_0[x]$ will be called the *exponent* of M.

Given a rational prime $p > 1$, let p^k be the largest power of p which goes into the exponent, and $M_p = \mathrm{Id}(p^k, M)$ the p-primary component of M. Here $\mathrm{Id}(S)$ for any subset $S \subset Z[x]$ denotes the ideal generated by S. M is completely determined by its p-primary components and in fact

$$M = \bigcap_p M_p.$$

Hence it is quite sufficient (and for our purposes more convenient) to consider p-primary ideals only, that is ideals whose exponent is a power of p. The set of all p-primary ideals of $Z[x]$ will be denoted by $\mathscr{I}_p[x]$.

To obtain an arbitrary member of $\mathscr{I}_p[x]$, take a set of integers $e_\mu \geq 0$, $\mu = 1, 2, \ldots$, and $b_{\mu v}$, $0 \leq v < \mu$, satisfying the conditions

$$(1.1) \qquad 0 < b_{\mu v} < p^{e_\mu} = b_\mu \qquad (0 \leq v < \mu),$$

$$(1.2) \qquad e_\mu = 0 \quad (\text{hence } b_{\mu v} = 0) \quad \text{for } \mu > m,$$

for some $m > 0$. Define the polynomials $g_\mu(x)$, $\mu = 0, 1, 2, \ldots$ recursively by

$$(1.3) \qquad g_0(x) = \prod_{\mu=1}^{\infty} b_\mu = \pi(e_1 + \cdots + e_m),$$

$$(1.4) \qquad b_\mu g_\mu(x) = x g_{\mu-1}(x) + \sum_{v=0}^{\mu-1} b_{\mu v} g_v(x) \qquad (\mu = 1, 2, \ldots)$$

(hence $g_\mu(x) = x^{\mu-m} g_m(x)$ for $\mu \geq m$) where for convenience we write $\pi(a)$ for p^a.

Theorem 1. *Every ideal $M \in \mathscr{I}_p[x]$ has exactly one ideal basis of the form $(g_0(x), g_1(x), \ldots, g_\mu(x), \ldots)$ where the $g_\mu(x)$ are defined as in (1.3), (1.4).*

Actually $(g_0(x), g_1(x), \ldots, g_m(x))$ is already an ideal basis. We shall refer to it as the *canonical basis* of M, and to the numbers e_μ (or b_μ), $b_{\mu v}$ as the *canonical invariants* of M. The basis polynomials $g_\mu(x)$ have the following properties:

(i) $g_\mu(x)$ is divisible by $B_\mu = \prod_{v > \mu} b_v = \pi(e_{\mu+1} + \cdots + e_m)$.

(ii) $g_\mu(x)$ is of degree μ and has B_μ for its leading coefficient.

(iii) $g_\mu(x)$ is minimal in the sense that

$$f(x) = a_\mu x^\mu + a_{\mu-1} x^{\mu-1} + \cdots + a_0 \in M$$

implies $B_\mu | a_\mu$.

(iv) Every $f(x) \in M$ of degree $\leq \mu$ is in the vectorspace

$$V(g_0(x), \ldots, g_\mu(x))$$

over the integers spanned by $g_0, \ldots, g_\mu$, i.e.,

(1.5) $$f(x) = \sum_{v=0}^{\mu} c_v g_v(x), \qquad c_v \in Z,$$

and this representation is unique.

(v) Every $f(x) \in Z[x]$ has a unique representation

(1.6) $$f(x) \equiv \sum_{\mu=0}^{m-1} d_\mu x^\mu \pmod{M} \quad \text{with} \quad 0 \leq d_\mu < B_\mu.$$

Hence

(1.7) $$|M| = \prod_{\mu=0}^{m-1} B_\mu.$$

From (iv) it follows that every $f(x) \in M$ of degree μ is divisible by B_μ. From (i) and (ii) it follows that if we set

(1.8) $$g_\mu(x) = B_\mu h_\mu(x) \qquad (\mu \geq 0),$$

then

(1.9) $$h_\mu(x) = x^\mu + a_{\mu,\mu-1} x^{\mu-1} + \cdots + a_{\mu 0} \in Z[x].$$

Thus

(1.10) $$M = \mathrm{Id}(B_\mu h_\mu(x); \mu \geq 0)$$

with

(1.11) $$B_{\mu-1} = b_\mu B_\mu \qquad (\mu = 1, 2, \ldots)$$

and

(1.12) $$h_0(x) = 1,$$

(1.13) $$h_\mu(x) = x h_{\mu-1}(x) + \sum_{v=0}^{\mu-1} b_{\mu v} \frac{B_0}{B_{\mu-1}} h_v(x), \qquad 0 \leq b_{\mu v} < b_\mu.$$

In applications to metabelian groups it is necessary to know that x has an inverse modulo M. We shall now examine the effect of this condition upon the system of invariants. We denote by $\mathscr{J}_p[x]$ the set of those ideals of $\mathscr{I}_p[x]$ which have this property.

Lemma 1. *M belongs to $\mathscr{J}_p[x]$ if and only if it belongs to $\mathscr{I}_p[x]$ and contains a polynomial*

$$(1.14) \qquad f(x) = a_\mu x^\mu + \cdots + a_0$$

with $a_0 = 1$.

Write $\delta(f(x)) = a_0$ for any $f(x)$ of the form (1.14). Generally for any subset $S \subset Z[x]$ denote by $\delta(S)$ the g.c.d. of all $\delta(f(x))$, $f(x) \in S$. The condition of Lemma 1 is equivalent to $\delta(M) = 1$.

Suppose $\delta(M) = 1$ and $f(x) = 1 - x\varphi(x) \in M$, $\varphi(x) \in Z[x]$. Then $\varphi(x) \equiv x^{-1} \pmod{M}$. Conversely, $\varphi(x) \equiv x^{-1} \pmod{M}$ implies $x\varphi(x) \equiv 1$, $1 - x\varphi(x) \in M$.

Lemma 2. *If $M \in \mathscr{J}_p[x]$ then it contains $1 + x + \cdots + x^{r-1}$ for some $r > 0$.*

For, among the polynomials $1, 1+x, \ldots, 1+x+\cdots+x^{|M|}$ at least two are congruent each other modulo M, hence M contains certainly $x^s + x^{s+1} + \cdots + x^t$ for some $0 \le s \le t \le |M|$. But then because of the existence of $x^{-1} \bmod M$ it also contains $1 + x + \cdots + x^{t-s}$. The proof shows incidentally that r can always be determined so that $r \le |M| + 1$.

To express the assumption $M \in \mathscr{J}_p[x]$ in terms of the invariants of M, write

$$(1.15) \qquad \delta(h_\mu(x)) = d_\mu \in Z, \qquad \mu \ge 0.$$

From (1.12) and (1.13) we get

$$(1.16) \qquad d_0 = 1,$$

$$(1.17) \qquad d_\mu = \sum_{v=0}^{\mu-1} b_{\mu v} \frac{B_v}{B_{\mu-1}} d_v \quad \text{for } \mu > 0.$$

Here $d_\mu = 0$ if $e_\mu = 0$ since then $b_{\mu v} = 0$ for every $v < \mu$. If $e_\mu > 0$ and v is the smallest index $< \mu$ such that $e_\lambda = 0$ for $v < \lambda < \mu$ then $B_{\mu-1} = B_v$ and $B_{v-1}/B_{\mu-1}$ is divisible by p if $v > 0$, hence by (1.17)

$$(1.18) \qquad d_\mu \equiv b_{\mu v} d_v \pmod{p}.$$

If k is the largest value of μ for which $e_\mu > 0$, we must have $(d_k, p) = 1$ since otherwise every $B_\mu d_\mu$ is divisible by p and $\delta(M)$ is divisible by p, by (1.10). But by (1.18), $(d_\mu, p) = 1$ if and only if $(b_{\mu v}, p) = 1$ for all pairs of indices μ, v for which $e_\mu > 0$, $e_v > 0$, and $e_\lambda = 0$ for $v < \lambda < \mu$. Thus we have

Theorem 2. $M \in \mathscr{J}_p[x]$ *if and only if the canonical invariants of* M *have the property that*

(1.19) $$(b_{\mu v}, p) = 1$$

for every $\mu > v \geq 0$ *for which*

(1.20) $$e_\mu > 0, \quad e_v > 0 \quad and \quad e_\lambda = 0 \quad for \quad v < \lambda < \mu.$$
$$(e_0 > 0 \ \text{by convention}).$$

The validity of Theorem 1 rests on the fact that Z is a principal ideal domain; we can take for instance $K[y]$ instead of Z as the domain of coefficients, where K is a field. Thus with trivial modifications Theorem 1 also yields an enumeration of ideals in $K[x, y]$. If the domain of coefficients R is merely a unique factorization domain, as in the case of $Z[x, y]$ or $K[x, y, z]$, a general method of enumeration for the ideals of $R[x]$ is not known.

One of the aims of this paper is to enumerate all ideals of $Z[x, y]$ which have a finite norm $|N|$ and which have the property that both x and y possess an inverse modulo N. We conclude this section with some notations and remarks concerning these ideals.

We denote by $\mathscr{I}[x, y]$ the set of all ideals of $Z[x, y]$ and by $\mathscr{I}_0[x, y]$ the subset of ideals with finite norm. As in the case of $\mathscr{I}_0[x]$, ideals of $\mathscr{I}_0[x, y]$ can be represented uniquely as the intersection of their primary components, i.e., ideals whose exponent ($=$ the smallest positive integer contained in the ideal) is a power of a prime number p. We denote by $\mathscr{I}_p[x, y]$ the set of all p-primary ideals of $\mathscr{I}_0[x, y]$, and by $\mathscr{J}_p[x, y]$ the subset of all $N \in \mathscr{I}_p[x, y]$ which have the property that x and y have inverses modulo N.

Lemma 3. *Let* $N \in \mathscr{I}_p[x, y]$ *and suppose that* $x\varphi(x, y) \equiv 1 \pmod{N}$ *for some* $\varphi(x, y) \in Z[x, y]$. *Then* $x^r \equiv 1 \pmod{N}$ *for some* $r > 0$.

Since $|N|$ is finite,

(1.21) $$x^s \equiv x^t \pmod{N}$$

for suitable $0 \leq s < t \leq |N|$. The lemma is obtained by multiplying congruence (1.21) with $(\varphi(x, y))^s$ and setting $r = t - s$.

As a corollary we obtain: if $N \in \mathscr{J}_p[x, y]$ and $M = N \cap Z[x]$ then $M \in \mathscr{J}_p[x]$.

2. Elementary extensions

Let $N \in \mathscr{J}_p[x, y]$ and let us regard elements of N as polynomials of y with coefficients in $Z[x]$. The leading coefficients $f_\alpha(x) \in Z[x]$ of all

$$F(x, y) = f_\alpha(x)y^\alpha + \cdots + f_0(x) \in N$$

form a p-primary $Z[x]$-ideal M_α,

(2.1) $M_\alpha \subseteq M_{\alpha+1}$ for $\alpha \geq 0$.

By the corollary of Lemma 3, $M_0 \in \mathscr{J}_p[x]$, hence $M_\alpha \in \mathscr{J}_p[x]$ for every $\alpha \geq 0$.

Since the ascending chain condition holds in $\mathscr{I}[x]$, $\exists n$; $M_\alpha = M_n$ for $\alpha \geq n$. If N is to have finite norm, we must in fact have $M_n = \mathrm{Id}(1)$. Our first task will be the enumeration of all chains of ideals

(2.2) $M_0 \subseteq M_1 \subseteq \cdots \subseteq M_n = \mathrm{Id}(1)$

whose members are in $\mathscr{J}_p[x]$.

To describe a chain (2.2) it is clearly sufficient to specify the canonical bases and invariants of the M_α. This on the other hand requires the construction of canonical bases of ideals M_α which satisfy condition (2.1). It will be helpful to associate a Young diagram Y with every $M \in \mathscr{J}_p[x]$.

Let

(2.3) $M = \mathrm{Id}(g_\mu(x); \mu \geq 0)$

be the canonical basis of M,

(2.4) $g_0(x) = B_0 = \prod_{\mu > 0} b_\mu = \pi\left(\sum_{\mu > 0} e_\mu\right),$

(2.5) $b_\mu g_\mu(x) = x g_{\mu-1}(x) + \sum_{v=0}^{\mu-1} b_{\mu v} g_v(x),$

(2.6) $0 \leq b_{\mu v} < b_\mu = \pi(e_\mu).$

Set

(2.7) $$B_\mu = \pi(E_\mu), \qquad E_\mu = \sum_{v > \mu} e_v,$$

so that

(2.8) $$E_{\mu-1} \geq E_\mu \quad (\mu \geq 1), \qquad E_m = 0$$

for suitable $m > 0$. The numbers E_μ thus form the columns of a Young diagram

(2.9) $$Y = [E_0, E_1, \ldots, E_\mu, \ldots] = [E_\mu]$$

(see [2], p. 36); this will be called the *diagram* of M. Conversely, given Y we say that $M \in \mathcal{S}_p[x]$ belongs to the diagram Y if the canonical invariants of M are given by

(2.10) $$e_\mu = E_{\mu-1} - E_\mu \quad (\mu > 0).$$

The zeros at the tail section of (2.9) will usually be omitted; in particular the diagram with $E_0 = 0$ (i.e., the diagram of $\mathrm{Id}(1)$) will be denoted [ ].

With the notation

(2.11) $$|Y| = \sum_{\mu \geq 0} E_\mu$$

we have, by (1.7),

(2.12) $$|M| = \pi(|Y|).$$

Now suppose that

(2.13) $$M \supseteq M^* \in \mathcal{S}_p[x]$$

with the canonical basis

(2.3*) $$M^* = \mathrm{Id}(g_\mu^*(x); \mu \geq 0).$$

We say that M^* is a subideal of M, or M an *extension* of M^*, if (2.13) holds. The $g_\mu^*(x)$ satisfy equations and inequalities (2.4*), (2.5*), (2.6*) analogous to (2.4), (2.5), (2.6), with corresponding starred invariants $b_\mu^* = \pi(e_\mu^*)$, $b_{\mu v}^*$; in particular

(2.6*) $$0 \leq b_{\mu v}^* < b_\mu^*.$$

$Y^* = [E_\mu^*]$, $E_\mu^* = \sum_{v > \mu} e_v^*$ is the diagram of M^*.

Because of (2.13) we have

$$(2.14) \qquad g_\mu^*(x) = q_\mu g_\mu(x) + \sum_{v=0}^{\mu-1} t_{\mu v} g_v(x), \qquad \mu \geq 0,$$

where $t_{\mu v} \in Z$ and

$$(2.15) \qquad\qquad q_\mu = B_\mu^*/B_\mu = \pi(d_\mu),$$

$$(2.16) \qquad\qquad d_\mu = E_\mu^* - E_\mu \geq 0 \qquad (\mu \geq 0).$$

Y^* is said to be an extension of Y, or Y a subdiagram of Y^*, in notation $Y^* \supseteq Y$, if (2.16) holds. Note that $M^* \subseteq M$ implies $Y^* \supseteq Y$ for the associated diagrams. Hence subdiagram corresponds to ideal divisor.

From (2.15) we have for $\mu > 0$

$$b_\mu^* q_\mu B_\mu = b_\mu^* B_\mu^* = B_{\mu-1}^* = q_{\mu-1} B_{\mu-1} = q_{\mu-1} b_\mu B_\mu,$$

$$(2.17) \qquad\qquad b_\mu^* q_\mu = q_{\mu-1} b_\mu,$$

i.e.,

$$(2.18) \qquad\qquad e_\mu^* + d_\mu = d_{\mu-1} + e_\mu, \qquad (\mu > 0).$$

Since $B_\mu^* = \pi(E_\mu^*)$ is the highest power of p that goes into the coefficients of $g_\mu^*(x)$, it follows from (2.14) that

$$(2.19) \qquad t_{\mu v} \equiv 0 \pmod{\pi_{\mu v}}, \qquad \pi_{\mu v} = \text{Max}\{1, B_\mu^*/B_v\}.$$

Now substitute for $g_\mu^*(x)$ from (2.14) into (2.5*), and then for $b_\mu^* q_\mu g_\mu(x) = q_{\mu-1} b_\mu g_\mu(x)$ and $xg_{v-1}(x)$ $(v < \mu)$ from (2.5). We obtain, by collecting the coefficients of $g_v(x)$ and setting it equal to 0,

$$(2.20) \qquad b_{\mu v}^* q_v = b_{\mu v} q_{\mu-1} + b_\mu^* t_{\mu v} - t_{\mu-1,v-1} b_v + a_{\mu v}$$

for $0 \leq v < \mu$ where

$$(2.21) \qquad a_{\mu v} = \sum_{v < \lambda < \mu} (t_{\mu-1,\lambda-1} b_{\lambda v} - b_{\mu \lambda}^* t_{\lambda v}).$$

In order to determine all pairs of ideals $M^* \subseteq M$ which belong to given diagrams $Y^* \supseteq Y$ we must find all integral solutions $b_{\mu v}$, $b_{\mu v}^*$, $t_{\mu v}$ of the system (2.20), (2.21), subject to (2.6), (2.6*). These will be called reduced solutions. We first deal with a case in which a complete explicit solution can be found.

Given a diagram $Y = [E_\mu]$, let j, k $(0 \leq j < k)$ be such that

(2.22) $$e_j > 0, \qquad e_\mu = 0 \quad \text{for } j < \mu < k.$$

We say that Y^* is an elementary extension of Y if and only if

(2.23) $$d_\mu = E_\mu^* - E_\mu = 0 \quad \text{for } 0 \leq \mu < j \text{ and } \mu \geq k,$$

(2.24) $$d_\mu = 1, \qquad q_\mu = p \quad \text{for } j \leq \mu < k.$$

Hence

(2.25) $$e_j^* = e_j - 1, \qquad e_k^* = e_k + 1,$$

(2.26) $$e_\mu^* = 0 \quad \text{for } j < \mu < k,$$

by (2.18). An extension $M \supset M^*$ will be called *elementary, of type* (j, k), if their diagrams have this property.

Given the elementary pair of diagrams $Y^* \supset Y$ we shall now determine all ideal pairs $M^* \subset M$ which belong to these diagrams. To solve equations (2.20) we proceed from lower to higher values of μ and for fixed μ from higher to lower values of v. Several cases will be distinguished.

(a) $\mu < j$. All $t_{\mu v}$ are 0, $q_{\mu - 1} = q_v = 1$ and $b_{\mu v}^* = b_{\mu v}$ for every $v < \mu$.

(b) $\mu = j$. We have $a_{jv} = 0$, $t_{j-1, v-1} = 0$ by (2.21) and the previous result (a). Equation (2.20) becomes

(2.27) $$b_{jv}^* = b_{jv} + b_j^* t_{jv}.$$

Suppose that b_{jv} is arbitrarily given in the interval $0 \leq b_{jv} < b_j = p b_j^*$; then t_{jv} and b_{jv}^* can uniquely be determined from (2.27) so that $0 \leq b_{jv}^* < b_j^*$.

(c) $j < \mu < k$. We have $b_\mu = b_\mu^* = 1$, $b_{\mu v} = b_{\mu v}^* = 0$ from (2.22) and (2.26), and so (2.20) gives

(2.28) $$-t_{\mu v} = t_{\mu - 1, v - 1} b_v + a_{\mu v}.$$

$t_{\mu v}$ is uniquely determined since $a_{\mu v}$ only involves previously determined quantities.

(d) $\mu = k$. The equation to be solved is

(2.29) $$b_{kv}^* q_v = p b_{kv} + b_k^* t_{kv} - t_{k-1, v-1} b_v + a_{kv}.$$

All terms on the right-hand side are divisible by q_v. For $v < j$ this is trivial since then $q_v = 1$; for $j \leq v < k$ it can be inferred either from

(2.19) (which must hold for all solutions of the system) or directly from (2.28) and (2.21). We thus have for $j \leq v < k$ since $b_k^* = pb_k$ by (2.25),

$$(2.30) \qquad\qquad b_{kv}^* = b_{kv} + b_k t_{kv} + a_{kv}'$$

where $a_{kv}' \in Z$ is given.

Set for $j \leq v < k$

$$(2.31) \qquad\qquad b_{kv}^* = b_k c_v + b_{kv}', \qquad 0 \leq b_{kv}' < b_k,$$

$$(2.32) \qquad\qquad\qquad 0 \leq c_v < p.$$

By (2.30) and (2.31) we must have $b_{kv}' \equiv b_{kv} + a_{kv}' \pmod{b_k}$, hence b_{kv}' is uniquely determined provided that b_{kv} is given. Furthermore

$$(2.33) \qquad\qquad\qquad t_{kv} = c_v + a_{kv}'',$$

where a_{kv}'' is given. This shows that c_v can be arbitrarily prescribed in the interval (2.32); alternatively, we may prescribe a residue class modulo p for t_{kv}.

(e) $\mu > k$. Then $q_{\mu-1} = 1$, $b_\mu = b_\mu^*$ and

$$(2.34) \qquad\qquad b_{\mu v}^* q_v = b_{\mu v} + b_\mu t_{\mu v} - t_{\mu-1, v-1} b_v + a_{\mu v}.$$

A solution exists only if the right-hand side is divisible by q_v, i.e., if either $v < j$, or $v \geq k$, or if $j \leq v < k$ and $b_{\mu v}$ falls in a certain residue class modulo p, determined by $a_{\mu v} - t_{\mu-1, v-1} b_v$. There is no such restriction on $b_{\mu v}^*$ which can be prescribed freely in the interval (2.6*); both $b_{\mu v}$ and $t_{\mu v}$ are then uniquely determined from (2.34). We have thus proved

Lemma 4. *Let there be given* (i) *a Young diagram* $Y = [E_\mu]$, (ii) *an elementary extension* $Y^* = [E_\mu^*]$ *of* Y, *of type* (j, k), *and* (iii) *a set of integers* $b_{\mu v}$ $(0 \leq v < \mu \leq k)$, c_v $(j \leq v < k)$, $b_{\mu v}^*$ $(0 \leq v < \mu, \mu > k)$, *subject to*

$$(2.35) \quad 0 \leq b_{\mu v} < \pi(e_\mu), \qquad 0 \leq c_v < p, \qquad 0 \leq b_{\mu v}^* < \pi(e_\mu^*).$$

Then the equations (2.20) *have precisely one system of reduced solutions which takes the given values* $b_{\mu v}$, $b_{\mu v}^*$ *for the indices specified above and for which* c_v *is the integer part of* b_{kv}^*/b_k *for* $j \leq v < k$.

Theorem 3. *Given* Y, *an elementary extension* Y^* *of type* (j, k) *and a set of integers* $b_{\mu v}$, $b_{\mu v}^*$, c_v *as specified in Lemma 3, there exists exactly one pair of ideals* $M^* \subset M \in \mathcal{I}_p[x]$ *with the given diagrams and the given* *canonical invariants* $b_{\mu v}$, $b^*_{\mu v}$ *such that* c_v *is the integer part of* b^*_{kv}/b_k *for* $j \leq v < k$.

An immediate consequence of Theorem 3 is

Theorem 4. *Given Y and an elementary extension Y^*, the number of distinct pairs $M \supset M^*$ with these diagrams is equal to $\pi(|Y^*|)$.*

The result is obtained by counting the number of possible systems of invariants $b_{\mu v}$, $b^*_{\mu v}$, and c_v in (2.35). The count yields

$$\sum_{0 \leq v < k} \pi\left(\sum_{v < \mu \leq k} e_\mu + \sum_{\mu > k} e^*_\mu + d_v\right) + \sum_{v > k} \pi\left(\sum_{\mu > v} e^*_\mu\right)$$

$$= \sum_{0 \leq v < k} \pi(F_v + d_v) + \sum_{v > k} \pi(E^*_v)$$

$$= \pi\left(\sum_{v \geq 0} E^*_v\right) = \pi(|Y^*|).$$

Theorem 4 is also valid for certain nonelementary extensions; for instance it holds for all extensions of $Y = [\ ]$. In fact $|Y^*|$ is equal to the number of distinct ideals with diagram Y^* as seen immediately by counting the number of distinct systems of canonical invariants $b^*_{\mu v}$ which belong to the b^*_μ determined by Y^*.

The following example shows that the theorem is not true for arbitrary extensions $Y^* \supset Y$. Take $Y = [1]$, $Y^* = [2, 1]$ with $p = 2$. There are two ideals with diagram Y, namely $M_1 = (2, x)$, $M_2 = (2, x+1)$; and eight ideals with diagram Y^*, namely

$$M^*_1 = (4, 2x, x^2), \qquad\qquad M^*_2 = (4, 2x, x^2 + x),$$

$$M^*_3 = (4, 2x, x^2 + 2), \qquad\quad M^*_4 = (4, 2x, x^2 + x + 2),$$

$$M^*_5 = (4, 2x + 2, x^2 + x), \qquad M^*_6 = (4, 2x + 2, x^2 + 2x + 1),$$

$$M^*_7 = (4, 2x + 2, x^2 + x + 2), \qquad M^*_8 = (4, 2x + 2, x^2 + 2x + 3).$$

There are 12 pairs $M \supset M^*$ instead of 8:

M_1 contains M^*_i for $i = 1, 2, 3, 4, 5, 7$, and
M_2 contains M^*_i for $i = 2, 4, 5, 6, 7, 8$.

The problem of arbitrary (nonelementary) extensions will be considered in the next section.

3. Ascending chains

To describe a chain

(3.1) $M_0 \subseteq M_1 \subseteq \cdots \subseteq M_n = \mathrm{Id}(1), \qquad M_\alpha \in \mathscr{J}_p[x]$

it will be necessary to extend the results of Section 2 to arbitrary extensions $M^* \subseteq M$. We first show that every extension $M^* \subset M$ has a resolution into a chain of maximal elementary extensions

(3.2) $M = M^{(0)} \supset M^{(1)} \supset \cdots \supset M^{(\varDelta)} = M^*.$

Let

(3.3) $Y = [E_\mu], \qquad Y^* = [E_\mu^*], \qquad Y \subset Y^*$

be the Young diagrams of M and M^*. With the notations (2.10), (2.16), define j, k $(0 \le j < k)$ by

(3.4) $d_\mu = 0 \quad \text{for } \mu \ge k, \qquad d_{k-1} > 0,$

(3.5) $e_\mu = 0 \quad \text{for } j < \mu < k, \qquad e_j > 0.$

We say that Y^* is an extension of type (j, k) of Y; the definition agrees with the earlier one in the elementary case.

Construct now the following diagram $Y^{(1)} = [E_\mu^{(1)}]$:

(3.6) $E_\lambda^{(1)} = E_\lambda \qquad \text{for } 0 \le \lambda < j,$

(3.7) $E_\mu^{(1)} = E_\mu + 1 \quad \text{for } j \le \mu < k,$

(3.8) $E_v^{(1)} = E_v \qquad \text{for } v \ge k.$

Clearly $Y^{(1)}$ is an elementary extension of Y, of type (j, k) and $Y^{(1)} \subseteq Y^*$.

Set $j^{(1)} = j$, $k^{(1)} = k$ and suppose that Y^* is a proper extension of $Y^{(1)}$, of type $(j^{(2)}, k^{(2)})$. It is seen from the construction that

(3.9) $j^{(2)} \le j^{(1)}, \qquad k^{(2)} \le k^{(1)}.$

By repeating the construction we obtain a finite chain of diagrams

(3.10) $Y = Y^{(0)} \subset Y^{(1)} \subset \cdots \subset Y^{(\varDelta)} = Y^*,$

each $Y^{(i)}$ being an elementary extension of $Y^{(i-1)}$, of type $(j^{(i)}, k^{(i)})$. The chain (3.10) is uniquely associated with $Y \supset Y^*$. Since $E_{j^{(i+1)}}^{(i+1)} = E_{j^{(i+1)}}^{(i)} + 1 \ge E_{j^{(i)}}^{(i)} + 1$, we have for the length of the chain

(3.11) $\varDelta \le E_0.$

Returning to the ideals $M \supset M^*$, consider

(3.12) $\quad M^{(1)} = \mathrm{Id}(g_\lambda(x), pg_\mu(x), g_v^*(x); 0 \le \lambda < j \le \mu < k \le v)$

where (j, k) is the type of Y^* over Y. We show that

(3.13) $\qquad\qquad\qquad M \supset M^{(1)} \supseteq M^*.$

The first (proper) inclusion follows from the fact that $g_\mu(x) \notin M^{(1)}$ for $j \le \mu < k$; for, all polynomials of degree $\mu < k$ in $\mathrm{Id}(g_v^*(x); v \ge k)$ are divisible by $\pi(E_\mu^*) \ge \pi(E_{k-1}^*) \ge \pi(E_{k-1}^{(1)}) = \pi(E_{k-1} + 1)$ by (3.7) and property (i) of Theorem 1. Similarly all members of

$$\mathrm{Id}(g_\lambda(x), pg_\mu(x); 0 \le \lambda < j \le \mu < k)$$

are divisible by $\pi(E_{k-1} + 1)$ since $E_{j-1} = e_j + E_j > E_j = E_{k-1}$ by (3.5) whereas the leading coefficient of $g_\mu(x)$ is $\pi(E_\mu) = \pi(E_{k-1})$ for $j \le \mu < k$.

The second inclusion in (3.13) follows from (2.14) and (2.19), since $p | q_\mu$ for $j \le \mu < k$ by (3.7) and $E_\mu^* - E_v \ge E_\mu^{(1)} - E_v = 1$, hence $p | t_{\mu v}$, for $j \le v < \mu < k$ by (3.5) and (3.7). It follows that

$$g_v^*(x) \in \mathrm{Id}(g_\lambda(x), pg_\mu(x); 0 \le \lambda < j \le \mu < k) \quad \text{for } j \le v < k.$$

The proof shows that $M^{(1)}$ is an elementary subideal of M, of type (j, k), and its diagram is $Y^{(1)}$. It is maximal in the sense that it is not properly contained in any elementary extension of M.

By repeating the construction we obtain a unique chain of ideals

(3.14) $\qquad\qquad M = M^{(0)} \supset M^{(1)} \supset \cdots \supset M^{(4)} = M^*$

with the property that each $M^{(i)}$ is elementary in $M^{(i-1)}$, of type $(j^{(i)}, k^{(i)})$. This is the indicated elementary resolution of M over M^*.

To characterize an elementary resolution by invariants, it will be necessary to modify the specifications of Theorem 3 so that the invariants of M^* shall not occur explicitly among the freely prescribable quantities. For only then can we ensure that the invariants of $M^{(i)}$ are available, free from uncontrollable restrictions, for the next link $M^{(i)} \supset M^{(i+1)}$ of the chain.

Henceforth we will assume that all ideals are in $\mathscr{J}_p[x]$, that is, condition (1.19) of Theorem 2 holds for the canonical invariants of M and the corresponding starred condition (1.19*) for M^*. Actually the condition on M^* is sufficient; its implication is that we retain only

23—K.

those determinations of $b^*_{\mu v}$ which satisfy (1.19*). No such determinations exist unless the $b_{\mu v}$ themselves satisfy (1.19).

To obtain the required specification, we take another look at equation (2.34) which was responsible for the appearance of the invariants of M^* in Lemma 4. If $j \le v < k$ then clearly equation (2.34) cannot be solved for $b^*_{\mu v}$ unless

$$(3.15) \qquad b_{\mu v} + b_\mu t_{\mu v} - t_{\mu-1, v-1} b_v + a_{\mu v} \equiv 0 \pmod{p}.$$

If $b_\mu = 1$ then there is no problem; both $b_{\mu v}$ and $b^*_{\mu v}$ are necessarily 0 and $t_{\mu v}$ is uniquely determined from

$$t_{\mu v} = t_{\mu-1, v-1} b_v - a_{\mu v}.$$

If $e_\mu > 0$, $b_\mu \equiv 0 \pmod{p}$ then the congruence to be satisfied is

$$(3.16) \qquad a_{\mu v} + b_{\mu v} \equiv t_{\mu-1, v-1} b_v \pmod{p}.$$

Now the expression (2.21) for $a_{\mu v}$ involves $t_{k v}$ which can be selected freely modulo p according to (2.33). Therefore (3.16) can be regarded as a congruence for $t_{k v}$ provided that its coefficient in $a_{\mu v}$ is not 0. It is at this stage that we utilize the assumption $M^* \in \mathscr{J}_p[x]$. Let

$$\{k = \mu_0 < \mu_1 < \cdots < \mu_r\}$$

be the set of indices $\mu_i \ge k$ for which $e^*_\mu > 0$; then condition (1.19*) gives

$$(3.17) \qquad (b^*_{\mu_i, \mu_i - 1}, p) = 1.$$

Suppose that for some $i \ge 1$ we have already satisfied equation (2.34) with $\mu = \mu_{i-1}$, and that the congruence class of $t_{\mu_{i-1}, v}$ modulo p is freely available in the process. Consider equation (2.34) for $\mu = \mu_i$. The expression (2.21) for $a_{\mu_i, v}$, $j \le v < k$ contains the term $-b^*_{\mu_i, \mu_i - 1} t_{\mu_i - 1, v}$. If $\mu_{i-1} < \mu_i - 1$ then this is the only term containing $t_{\mu_i - 1, v}$ and by (3.17) it can be determined mod p so that $a_{\mu_i, v}$ should satisfy congruence (3.16) for any given value of $b_{\mu_i, v}$.

If $\mu_{i-1} = \mu_i - 1$ then the expression for $a_{\mu_i, v}$ also contains the term $t_{\mu_i, v} b_{v+1, v}$ and it is possible for the coefficient of $t_{\mu_i, v}$ to be divisible by p, namely if

$$(3.18) \qquad b^*_{\mu_i, \mu_i - 1} \equiv b_{v+1, v} \pmod{p}.$$

Actually (3.18) can only occur for $v = k-1$ since for $j \le v < k-1$, $b_{v+1,v} = 0$. Therefore congruence (3.16) can certainly be satisfied for $j \le v < k-1$ and $\mu = \mu_i$, by a suitable determination of $t_{\mu_i,v}$ mod p, and we can set

(3.19) $$b_{\mu_i,v} + a_{\mu_i,v} - t_{\mu_{i-1},v-1}b_v = pb'_{iv}, \quad b'_{iv} \in Z.$$

Determine $\tau_{iv} \in Z$ so that

(3.20) $$0 \le \pi(e_{\mu_i}-1)\tau_{iv} + b'_{iv} < \pi(e_{\mu_i}-1),$$

and set

(3.21) $$t_{\mu_i,v} = \tau_{iv} + c_{iv},$$

(3.22) $$0 \le c_{iv} < p.$$

Then (2.34) and (3.19) give

(3.23) $$b^*_{\mu_i,v} = \pi(e_{\mu_i}-1)\tau_{iv} + b'_{iv} + \pi(e_{\mu_i}-1)c_{iv}$$

and to every c_{iv} in the interval (3.22) there corresponds, by (3.20), a $b^*_{\mu_i,v}$ in the required interval. From (3.21) it follows that $t_{\mu_i,v}$ is freely available modulo p for the next step of the process. This step will fix the value of $t_{\mu_i,v}$ and so each $b^*_{\mu_i,v}$ is uniquely determined, except the last one, $b^*_{\mu_r,v}$ for which $c_{rv} = \mathrm{Int}(b^*_{\mu_r,v}/\pi(e_{\mu_r}-1))$ can be arbitrarily prescribed in the interval $0 \le c_{rv} < p$. Here $\mathrm{Int}(\beta)$ denotes the integer part of β.

The conclusion is still valid for $v = k-1$ provided that

(3.24) $$b^*_{\mu_i,\mu_{i-1}} \not\equiv b_{k,k-1} \pmod{p}$$

whenever $\mu_i = \mu_{i-1}+1$. Since $b^*_{\mu_i+1,\mu_i} \equiv b_{\mu_i+1,\mu_i} \pmod{p}$ for $i \ge 1$ and $b^*_{k+1,k} \equiv b_{k+1,k} - t_{k,k-1} \pmod{p}$ by (2.34) we have to examine the cases when

(i) $\mu_{i+1} = \mu_i + 1$, $i > 1$,

(3.25) $$b_{\mu_i+1,\mu_i} \equiv b_{k,k-1} \pmod{p}$$

and (ii) $\mu_1 = k+1$,

(3.26) $$b_{k+1,k} \equiv b_{k,k-1} + t_{k,k-1} \pmod{p}.$$

In the first case there is no solution unless

(3.27) $\qquad b_{\mu_i+1,k-1} \equiv -a_{\mu_i+1,k-1}+t_{\mu_i,k-2}b_{k-1}$ (mod p).

If $b_{\mu_i+1,k-1}$ happens to have a value which satisfies (3.27) then $t_{\mu_i,k-1}$ is undetermined modulo p and $c_{i,k-1}=\mathrm{Int}(b^*_{\mu_i,k-1}/\pi(e_{\mu_i}-1))$ can be prescribed arbitrarily.

In the second case there is no solution unless

(3.28) $\qquad b_{k+1,k-1} \equiv t_{k,k-2}b_{k-1}$ (mod p);

$t_{k,k-1}$ is then uniquely determined from (3.26) and this particular value of $t_{k,k-1}$ modulo p may or may not be admissible depending on whether (3.26) is true or not. We have thus proved

Lemma 5.   *Given a Young diagram* $Y=[E_\mu]$, *an elementary extension* $Y^*=[E^*_\mu]$ *of type* (j, k), *and a set of integers* $b_{\mu v}$ $(0 \leq v < \mu)$, c_v $(j \leq v < k)$ *subject to*

(3.29) $\qquad\qquad 0 \leq b_{\mu v} < \pi(e_\mu), \qquad 0 \leq c_v < p,$

the equations (2.20) *have at most one system of reduced solutions* $b^*_{\mu v}$, $t_{\mu v}$ *satisfying the condition*

(3.30) $\qquad\qquad \mathrm{Int}(b^*_{\kappa v}/\pi(e_\kappa-1)) = c_v,$

where κ *is the largest index with* $e^*_\kappa > 0$; *except that if* $e_\mu > 0$, $e_{\mu+1} > 0$ *for some* $\mu > k$ *and*

(3.31) $\qquad\qquad b_{\mu+1,\mu} \equiv b_{k,k-1}$ (mod p)

then there is no solution unless $b_{\mu+1,k-1}$ *falls in a certain residue class modulo* p, *given by*

(3.27) $\qquad\qquad b_{\mu+1,k-1} \equiv -a_{\mu+1,k-1}+t_{\mu,k-2}b_{k-1}$ (mod p).

In that case the value of

(3.32) $\qquad\qquad c^*_\mu = \mathrm{Int}(b^*_{\mu,k-1}/\pi(e_\mu-1))$

can also be prescribed arbitrarily in the interval $0 \leq c_\mu < p$.

The statement "at most" refers to the fact that not all systems of solutions are acceptable but only those which satisfy the condition (1.19*) of Theorem 3. The numbers c_v, c^*_μ will be called the *parameters* of the solution.

Although the parameters c_μ^* can be prescribed freely in the interval $0 < c_\mu^* < p$, they do not contribute to the total number of solutions (for given Y, Y^*). For, whenever c_μ^* appears, an equivalent number of $b_{\mu v}$-systems will have an inadmissible value for $b_{\mu+1,k-1}$ and hence must be rejected.

Theorem 5. *Given $M \in \mathscr{J}_p[x]$ with diagram Y, an elementary extension Y^* of Y, a set of parameters $0 \le c_v < p$ $(j \le v < k)$ and $0 \le c_\mu^* < p$ for the values μ specified in Lemma 5, there is at most one subideal $M^* \in \mathscr{J}_p[x]$ with diagram Y^* and satisfying (3.30) and (3.32).*

An arbitrary subideal $M^* \in \mathscr{J}_p[x]$ of a given M is obtained as follows: We take its elementary resolution (3.14) and specify the parameters $c_v^{(i)}$, $c_\mu^{*(i)}$ of each link $M^{(i-1)} \supset M^{(i)}$. M^* is completely determined by M and the parameters, and is available for further resolutions. An ascending chain (3.1) can therefore be specified by prescribing the parameters $0 \le c_{\alpha v}^{(i)} < p$, $0 \le c_{\alpha \mu}^{*(i)} < p$ of each link $M_\alpha^{(i-1)} \supset M_\alpha^{(i)}$ in the elementary chain resolution

$$(3.33) \qquad M_\alpha = M_\alpha^{(0)} \supset M_\alpha^{(1)} \supset \cdots \supset M_\alpha^{(\Delta_\alpha)} = M_{\alpha-1}$$

from M_α to $M_{\alpha-1}$. In the case of $M_\alpha = M_{\alpha-1}$ we set $\Delta_\alpha = 0$. The first member of the composite chain (in descending order) is $M_n = \mathrm{Id}(1)$, hence needs no specification.

Once we have the canonical bases of the M_α, we can forget about the parameters $c_{\alpha v}^{(i)}$, $c_{\alpha \mu}^{*(i)}$ which were needed merely for the systematic construction of these bases. In future applications of ascending chains we shall indeed only use the canonical bases and invariants of the M_α.

4. Ideals in $Z[x, y]$

We are now in a position to enumerate the ideals of $\mathscr{J}_p[x, y]$. Let $N \in \mathscr{J}_p[x, y]$ and denote by $M_\alpha \in \mathscr{J}_p[x]$ the ideal formed by the leading coefficients $f_\alpha(x) \in Z[x]$ of polynomials

$$(4.1) \qquad F(x, y) = f_\alpha(x) y^\alpha + \cdots + f_0(x) \in N.$$

We say, $F(x, y)$ is of degree (α, μ) if $f_\alpha(x)$ in (4.1) is not 0 and is of x-degree μ. Degrees are well-ordered lexicographically:

$$(4.2) \qquad (\alpha, \mu) < (\beta, v) \text{ if and only if } \alpha < \beta \text{ or } \alpha = \beta. \quad \mu < v.$$

Thus the degree of $F(x, y) = \sum_\alpha \sum_\mu c_{\alpha\mu} x^\mu y^\alpha$ is the degree of its leading term $c_{\beta\nu} x^\nu y^\beta$ with highest (β, ν). Also

$$(4.3) \qquad \deg(F(x, y) + G(x, y)) \leq \mathrm{Max}\{\deg F, \deg G\}.$$

Let

$$(4.4) \qquad M_\alpha = \mathrm{Id}(g_{\alpha\mu}(x); \mu \geq 0)$$

be the canonical basis of M_α, described by the invariants $e_{\alpha\mu}$, $b_{\alpha\mu\nu}$;

$$(4.5) \qquad g_{\alpha 0}(x) = \pi\left(\sum_{\nu > 0} e_{\alpha\nu}\right) = B_{\alpha 0},$$

$$(4.6) \qquad b_{\alpha\mu} g_{\alpha\mu}(x) = x g_{\alpha,\mu-1}(x) + \sum_{\nu=0}^{\mu-1} b_{\alpha\mu\nu} g_{\alpha\nu}(x),$$

$$(4.7) \qquad 0 \leq b_{\alpha\mu\nu} < b_{\alpha\mu} = \pi(e_{\alpha\mu}).$$

The numbers $b_{\alpha\mu\nu}$ are obtained from the parameters of the chain

$$(4.8) \qquad M_0 \subseteq M_1 \subseteq \cdots \subseteq M_n = \mathrm{Id}(1)$$

as described in Section 3. The parameters themselves need not concern us any further; we shall only use the canonical invariants $b_{\alpha\mu}$, $b_{\alpha\mu\nu}$ and the quantities $q_{\alpha\mu}$, $t_{\alpha\mu\nu}$ determined from

$$(4.9) \qquad g_{\alpha-1,\mu}(x) = q_{\alpha\mu} g_{\alpha\mu}(x) - \sum_{\nu=0}^{\mu-1} t_{\alpha\mu\nu} g_{\alpha\nu}(x).$$

Corresponding to each $g_{\alpha\mu}(x)$ we have a polynomial

$$(4.10) \quad G_{\alpha\mu}(x, y) = g_{\alpha\mu}(x) y^\alpha + \sum_{\beta=0}^{\alpha-1} f_{\alpha\beta\mu}(x) y^\beta \in N, \qquad \alpha \geq 0, \quad \mu \geq 0.$$

Lemma 6. *Every* $F(x, y) \in N$, $\deg F = (\alpha, \mu)$, *is contained in the vectorspace*

$$V_{\alpha\mu} = V(G_{\beta\nu}(x, y); (\beta, \nu) \leq (\alpha, \mu))$$

over Z,

$$(4.11) \quad F(x, y) = \sum_{\nu=0}^{\mu} c_{\alpha\nu} G_{\alpha\nu}(x, y) + \sum_{\beta=0}^{\alpha-1} \sum_{\nu \geq 0} c_{\beta\nu} G_{\beta\nu}(x, y), \qquad c_{\alpha\nu}, c_{\beta\nu} \in Z,$$

where in the last sum there are only a finite number of nonzero terms. The representation (4.11) *is unique.*

For suppose we had for some $F(x, y) \in N$

$$(4.12) \qquad F(x, y) = a_{\alpha\mu} x^\mu y^\alpha + \cdots \notin V_{\alpha\mu}.$$

We may assume that $F(x, y)$ has lowest possible degree (α, μ). Then $a_{\alpha\mu} = c B_{\alpha\mu}$ for some $c \in Z$ by the definition of $g_{\alpha\mu}(x)$, hence

$$\deg(F(x, y) - cG_{\alpha\mu}(x, y)) < \deg F(x, y)$$

and $F(x, y) - cG_{\alpha\mu}(x, y) \in V_{\beta v}$ for some $(\beta, v) < (\alpha, \mu)$, by the minimum property of deg F. Hence $F(x, y) \in V_{\alpha\mu}$, contrary to (4.12). Uniqueness of the representation is obvious.

Consider now

$$F(x, y) = yG_{\alpha-1,\mu}(x, y) - q_{\alpha\mu}G_{\alpha\mu}(x, y) + \sum_{v=0}^{\mu-1} t_{\alpha\mu v}G_{\alpha v}(x, y).$$

From (4.9) and (4.10) it follows that $F(x, y)$ has y-degree $\alpha - 1$, hence by Lemma 6, it is in $V_{\alpha-1, v}$ for suitable v. Hence

$$
\begin{aligned}
(4.13) \quad q_{\alpha\mu}G_{\alpha\mu}(x, y) = {} & yG_{\alpha-1,\mu}(x, y) + \sum_{v=0}^{\mu-1} t_{\alpha\mu v}G_{\alpha v}(x, y) \\
& + \sum_{\beta=0}^{\alpha-1} \sum_{v \geq 0} t_{\alpha\mu;\beta v}G_{\beta v}(x, y).
\end{aligned}
$$

Since each term in the sum must be divisible by $B_{\alpha-1,\mu} = \pi(E_{\alpha-1,\mu})$, and $B_{\beta v} = \pi(E_{\beta,v})$ is the highest power of p that goes into $G_{\beta v}$, we must have

$$(4.14) \qquad t_{\alpha\mu;\beta v} = \pi_{\alpha\mu;\beta v} b_{\alpha\mu;\beta v}, \quad b_{\alpha\mu;\beta v} \in Z$$

where

$$(4.15) \qquad \pi_{\alpha\mu;\beta v} = \text{Max}\{1, \pi(E_{\alpha-1,\mu} - E_{\beta v})\}.$$

If we replace $G_{\alpha\mu}$ by a suitable combination

$$G_{\alpha\mu} - \sum_{\beta=0}^{\alpha-1} \sum_{v>0} c_{\beta v}G_{\beta v}, \quad c_{\beta v} \in Z,$$

we can achieve that $0 \leq t_{\alpha\mu;\beta v} < q_{\alpha\mu} = \pi(E_{\alpha-1,\mu} - E_{\alpha\mu})$. Hence by (4.14 $b_{\alpha\mu;\beta v} = 0$ if $E_{\beta v} < E_{\alpha\mu}$, and we can write (4.13) in the form

$$
\begin{aligned}
(4.16) \quad q_{\alpha\mu}G_{\alpha\mu}(x, y) = {} & yG_{\alpha-1,\mu}(x, y) + \sum_{v=0}^{\mu-1} t_{\alpha\mu v}G_{\alpha v}(x, y) \\
& + \sum_{\beta=0}^{\alpha-1} \sum_{E_{\beta v} \geq E_{\alpha\mu}} b_{\alpha\mu;\beta v}\pi_{\alpha\mu;\beta v}G_{\beta v}(x, y)
\end{aligned}
$$

with

$$(4.17) \qquad 0 \leq b_{\alpha\mu;\beta v} < \text{Min}\{\pi(E_{\alpha-1,\mu} - E_{\alpha\mu}), \pi(E_{\beta v} - E_{\alpha\mu})\}.$$

$G_{\alpha\mu}(x, y)$ is uniquely fixed by these conditions. Thus N has exactly one ideal basis of the form (4.16), (4.17) where $\pi_{\alpha\mu;\beta\upsilon}$ is given by (4.15).

Conversely, given the chain (4.8) in $\mathscr{I}_p[x]$ and the invariants $b_{\alpha\mu;\beta\upsilon}$ freely in the range (4.17) we can uniquely determine a set of $G_{\alpha\mu}(x, y)$ from (4.16); for each term on the right will be divisible by $q_{\alpha\mu}$, as seen readily by induction on α. The $G_{\alpha\mu}(x, y)$ so obtained are evidently minimal in

$$N = \mathrm{Id}(G_{\alpha\mu}(x, y); \alpha \geq 0, \mu \geq 0).$$

Theorem 6. *Every ideal $N \in \mathscr{I}_p[x, y]$ has a uniquely determined ideal basis formed by minimal polynomials $G_{\alpha\mu}(x, y)$ of degree (α, μ), $0 \leq \alpha \leq n$, $\mu \geq 0$. These polynomials are obtained from* (i) *an ascending chain of ideals in $\mathscr{I}_p[x]$*

$$M_0 \subseteq M_1 \subseteq \cdots \subseteq M_n = \mathrm{Id}(1)$$

and (ii) *invariants $b_{\alpha\mu;\beta\upsilon}$. These can be prescribed freely in the interval* (4.17).

The norm of N is given by

(4.18)
$$|N| = \pi\left(\sum_{\alpha=0}^{n} \sum_{\mu=0}^{m} E_{\alpha\mu}\right).$$

The last statement follows from the remark that all residue classes mod N have a unique representative

$$\sum_{\alpha=0}^{n} \sum_{\mu=0}^{m} c_{\alpha\mu}x^{\mu}y^{\alpha} \quad \text{with } 0 \leq c_{\alpha\mu} < B_{\alpha\mu} = \pi(E_{\alpha\mu}).$$

If elements of $Z[x, y]$ are regarded as polynomials in x, with coefficients in $Z[y]$, one obtains a dual chain

(4.8*) $M_0^* \subseteq M_1^* \subseteq \cdots \subseteq M_m^* = \mathrm{Id}(1)$, $M_\beta^* \in \mathscr{I}_p[y]$,

a dual set of invariants $b_{\beta\upsilon;\alpha\mu}^*$ and a dual canonical basis $G_{\beta\upsilon}^*(x, y)$ for N. In particular $N \in \mathscr{I}_p[x, y]$ if and only if the chain M_α is in $\mathscr{I}_p[x]$ and the chain M_β^* is in $\mathscr{I}_p[y]$.

5. Metabelian groups with two generators

Let a, b be the generators of the metabelian group G, $K = G'$ abelian and finite. aK, bK are generators of G/K, of orders r and s respectively. ($r = 0$ if the order is infinite.) If $r > 0$, $s > 0$, we may assume that $r|s$.

Let α, β be the inner automorphisms induced by a and b in K,

(5.1) $$d^\alpha = a^{-1}da, \qquad d^\beta = b^{-1}db \qquad (d \in K).$$

Clearly α, β are independent of the representatives a, b of aK, bK and

(5.2) $$\alpha^r = \epsilon, \qquad \beta^s = \epsilon$$

(ϵ the identity automorphism of K). Let

$$R = Z[\alpha, \beta] = \left\{ \sum_{i=0}^{m} \sum_{j=0}^{n} c_{ij}\alpha^i\beta^j; \; m \geq 0, \; n \geq 0, \; c_{ij} \in Z \right\},$$

where $\alpha^0 = \beta^0 = \epsilon$. K is an R-group by the usual definition

(5.3) $$d^{\rho\sigma} = (d^\rho)^\sigma, \qquad d^{(\rho+\sigma)} = d^\rho d^\sigma, \qquad d^\epsilon = d \quad \text{for } d \in K, \quad \rho, \sigma \in R,$$

and

(5.4) $$d^{m\rho} = (d^\rho)^m \quad \text{for } m \in Z.$$

For convenience of notation we rewrite K additively. Define a monomorphism ψ from K onto a left R-module $A = K^\psi$ such that

(5.5) $$d_1^\psi + d_2^\psi = (d_1 d_2)^\psi$$

(5.6) $$\rho d^\psi = (d^\rho)^\psi$$

for every d_1, d_2, $d \in K$, $\rho \in R$. Then

$$(\rho\sigma)d^\psi = (d^{\rho\sigma})^\psi = \sigma(d^\rho)^\psi = \sigma(\rho d^\psi),$$
$$(\rho+\sigma)d^\psi = (d^{\rho+\sigma})^\psi = (d^\rho d^\sigma)^\psi = \rho d^\psi + \sigma d^\psi,$$

so that the left R-module A carries an antirepresentation of R. This of course is also a direct representation as R is commutative.

Lemma 7. *A is monogenic and is generated by c^ψ where*

(5.7) $$c = a^{-1}b^{-1}ab.$$

For let B be the R-module generated by c^ψ, $H = B^{\psi*}$ where $\psi\psi* = \epsilon$. Since B admits α and β, $H^\alpha = H$, $H^\beta = H$ hence $a^{-1}Ha = H$, $b^{-1}Hb = H$, also $c \in H$. Therefore G/H is abelian, generated by aH, bH, hence $H \supseteq K$; but $H \subseteq K$ is obvious, so $K = H$.

Let $\varphi: Z[x, y] \to R$ be the homomorphism

$$F(x, y)^\varphi = F(\alpha, \beta).$$

The structure of A is completely determined by the ideal N of all $F \in Z[x, y]$ such that F^φ acts trivially on A, i.e.,

(5.8) $F(\alpha, \beta) \cdot c^\psi = 0.$

Since K is finite, N must contain a nonzero constant h, the exponent of K; also

(5.9) $x^r - 1 \in N, \qquad y^s - 1 \in N$

by (5.2). If $r = 0$ then x must have an inverse mod N; therefore $N \in \mathscr{I}_p[x, y]$ irrespective whether r and s are 0 or not.

The complete description of G thus requires

(a) a specification of $N \in \mathscr{I}_p[x, y]$ and

(b) a specification of the elements a^r, b^s in K.

The first point (a) has already been settled in Section 4. If A_p is the p-primary component of A and N_p the ideal of all $F \in Z[x, y]$ such that $F(\alpha, \beta)$ acts trivially on A_p, then N_p is just the p-primary component of N. Hence Theorem 6 applies and N_p is completely specified by its parameters and invariants, or for practical purposes more conveniently by its canonical basis polynomials $G_{\alpha\mu}(x, y)$.

Concerning point (b): The elements a^r, $b^s \in K$ represent the factor-system of G. Let

(5.10) $(a^r)^\psi = \rho c^\psi, \qquad (b^s)^\psi = -\sigma c^\psi,$

where

(5.11) $\rho = \varphi(\alpha, \beta), \qquad \sigma = \psi(\alpha, \beta)$

are suitable elements of $R = Z[\alpha, \beta]$.

Since a commutes with a^r, we must have $\alpha\rho = \rho$, i.e.,

(5.12) $(x - 1)\varphi(x, y) \in N_p$

for each p. Furthermore since $b^{-1}ab = ac$ by (5.7), we have for $r > 0$:

$$b^{-1}a^r b = (ac)^r = aca^{-1}a^2ca^{-2}\cdots a^r ca^{-r}a^r,$$

hence by (5.10), $\beta\rho c^\psi = (b^{-1}a^r b)^\psi = (1 + \alpha + \cdots + \alpha^{r-1} + \rho)c^\psi,$

(5.13) $(y - 1)\varphi(x, y) - (1 + x + \cdots + x^{r-1}) \in N_p.$

Similarly we get for $\psi(x, y)$

(5.14)
$$(y-1)\psi(x, y) \in N_p,$$

(5.15)
$$(x-1)\psi(x, y) - (1+y+\cdots+y^{s-1}) \in N_p$$

if $s > 0$.

Equations (5.12)–(5.15) represent the Schreier conditions for the factorsystem. Each p-primary component of a^r and b^s is represented by polynomials $\varphi(x, y)$, $\psi(x, y)$ mod N_p which satisfy these conditions. If $r = 0$ then $\varphi(x, y) \equiv 0 \pmod{N_p}$ for each p and similarly if $s = 0$ then $\psi(x, y) \equiv 0 \pmod{N_p}$.

φ and ψ are independent of the representatives a, b of the cosets aK, bK. For suppose we replace a by ad, $d \in K$ where $d^\psi = \Lambda(\alpha, \beta)c^\psi$. Then a simple computation shows that c^ψ is replaced by $(c^\psi)^* = (1 + (\beta - 1)\Lambda(\alpha, \beta))c^\psi$, and $\varphi(x, y)$ by $\varphi^*(x, y)$ where

$$\varphi(x, y) = \varphi^*(x, y)\{1 + (y-1)\Lambda(x, y)\} - (1 + x + \cdots + x^{r-1})\Lambda(x, y).$$

But $(y-1)\varphi^*(x, y) \equiv (1 + x + \cdots + x^{r-1}) \bmod N_p$ by (5.13), hence $\varphi(x, y) \equiv \varphi^*(x, y) \bmod N_p$.

Equations (5.12)–(5.15) do not always have a solution. A necessary (though not sufficient) condition for the existence of a solution is

(5.16)
$$(1 + x + \cdots + x^{r-1})(1 + y + \cdots + y^{s-1}) \in N_p$$

for every p, as seen from (5.13) or (5.15). For the existence of a splitting factorsystem it is necessary (though not sufficient) that

(5.17) $1 + x + \cdots + x^{r-1} \equiv 0, \qquad 1 + y + \cdots + y^{s-1} \equiv 0 \pmod{N_p};$

the conditions are vacuous if $r = 0$ or $s = 0$. From Theorem 2 it is seen that for given $N \in \mathscr{J}_p[x, y]$ there always exist values $r = r_0$, $s = s_0$ for which (5.17) is satisfied. In that case both φ and ψ are in the vectorspace determined by the equations

(5.18) $(x-1)\varphi(x, y) \equiv 0, \qquad (y-1)\varphi(x, y) \equiv 0 \pmod{N_p}.$

The final result can be summarized as follows. To obtain an arbitrary metabelian group with two generators and finite commutator subgroups:

(a) Select a set of p-primary ideals $N_p \in \mathscr{J}_p[x, y]$, one for each p in a finite set $\mathscr{P}$ of prime numbers.

(b) Determine the smallest pair of integers $r_0 \geq 0$, $s_0 > 0$ such that $x^{r_0} - 1 \in N_p$, $y^{s_0} - 1 \in N_p$ for each $p \in \mathscr{P}$. Such integers always exist by

Theorem 2. Take suitable multiples $r = mr_0$, $s = ns_0$ such that if $m > 0$ then $r \mid s$ and if $m = 0$ then $n = 0$; furthermore $1 + x + \cdots + x^{r-1} \in N_p$, if necessary.

(c) Take solutions $\varphi_p(x, y)$, $\psi_p(x, y)$ of equations (5.12)–(5.15). At least one such solution (namely $\varphi_p = \psi_p = 0$) exists if (5.17) is satisfied.

The corresponding G consists of elements

$$(5.19) \qquad a^m b^n \prod_{p \in P} c_p^{F_p(\alpha, \beta)}, \qquad 0 \le m < r, \quad 0 \le n < s,$$

where $F_p(x, y)$ is modulo N_p and the following generating relations hold:

(G1) $$c_p c_q = c_q c_p \quad \text{all } p, q \in P \, ;$$

(G2) $$a^{-1} c_p a = c_p^\alpha, \qquad b^{-1} c_p b = c_p^\beta \, ;$$

(G3) $$a^{-1} b^{-1} a b = \prod_{p \in P} c_p \, ;$$

(G4) $$a^r = \prod_{p \in P} c_p^{\varphi_p(\alpha, \beta)}, \qquad b^r = \prod_{p \in P} c_p^{\psi_p(\alpha, \beta)} .$$

G is uniquely determined by these relations.

It would be desirable to find conditions under which two of these groups are isomorphic. Alternatively, and more hopefully, one might try to impose conditions upon N which will select one particular set of generators of G. Both N and the factorsystem are independent of the representatives of the cosets aK, bK, but replacement of aK, bK by new generating cosets will in general affect both N and the factorsystem. It seems to be difficult to formulate explicit conditions which will make this selection unique.

References

[1] L. Kronecker and K. Hensel, *Vorlesungen über Zahlentheorie*, Leipzig, 1901.
[2] G. de B. Robinson, *Representation theory of the symmetric group*, Univ. of Toronto Press, Toronto, 1961.
[3] G. Szekeres, A canonical basis for the ideals of a polynomial domain, *Amer. Math. Monthly* **59** (1952), 379–386.
[4] G. Szekeres, On finite metabelian p-groups with two generators, *Acta Sci. Math. Szeged* **21** (1960), 270–291.

Proc. Internat. Conf. Theory of Groups, Austral. Nat. Univ. Canberra,
August 1965, pp. 347–355. © Gordon and Breach Science Publishers, Inc. 1967

A generalized character theory
on finite groups

OLAF TAMASCHKE

I. A survey of the theory

The principal notion of this theory is that of an S-ring (i.e., a Schur-ring) on a finite group introduced by Wielandt ([8], 23.1). It arose from investigations of finite permutation groups having a regular subgroup. There it yields remarkable results in a rather elementary way without using character theory ([6], [7], [8]) though some connections with representation theory have already been pointed out in [8], Chapter V.

Here we exploit extensively the semisimplicity of S-rings from which there results a theory which can be considered as a generalization of character theory, and which in fact includes character theory as a special case.

Let G be a finite group, and let Γ be its group algebra over the field $\mathbf{C}$ of complex numbers. A subalgebra T of Γ is said to be an S-ring on G ([8], 23.1) if there exists a decomposition

$$G = \mathscr{T}_1 \cup \cdots \cup \mathscr{T}_t$$

of G into nonempty, trivially intersecting subsets $\mathscr{T}_i$ with the properties:

(1) *The elements $\tau_i = \sum_{g \in \mathscr{T}_i} g \; (i = 1, \ldots, t)$ of Γ form a $\mathbf{C}$-basis of T.*

(2) *For every $\mathscr{T}_i$ there exists a $\mathscr{T}_j$ consisting exactly of the inverses of all the elements contained in $\mathscr{T}_i$.*
   (We shall write $\mathscr{T}_i^*$ for $\mathscr{T}_j$, and τ_i^* for τ_j.)

We call the sets $\mathscr{T}_i$ the *T-classes* of G.

Condition (2) implies the semisimplicity of T ([7], p. 386, footnote). The field $\mathbf{C}$ can easily be replaced by other fields; it is the semisimplicity of T that matters for the theory we are going to develop.

Examples.   1. The group algebra Γ is itself an S-ring on G. The Γ-classes are simply the elements of G.

2. The center Z of Γ is an S-ring on G. The Z-classes are the classes of conjugate elements of G.

3. Let H be any subgroup of G. Then the double cosets HgH, $g \in G$, give rise to an S-ring T_H which we shall call the *double coset S-ring* defined by H on G.

Other examples of S-rings are easily found.

The characters of G are class functions, and the irreducible characters of G form a $\mathbf{C}$-basis of the algebra of all class functions. We are going to generalize this concept.

Given an S-ring T on G, we call a complex valued function on G a *T-class function* if it is constant on every T-class of G. The set of all T-class functions on G becomes a $\mathbf{C}$-algebra $T\#$ by the following definitions:

$$(f_1+f_2)(g) = f_1(g)+f_2(g), \qquad (f_1f_2)(g) = f_1(g)f_2(g), \qquad (cf)(g) = cf(g)$$

for all $f, f_1, f_2 \in T\#$, all $c \in \mathbf{C}$, and all $g \in G$.

Any $\mathbf{C}$-basis

$$\beta_i = \frac{1}{|G|} \sum_{g \in G} f_i(g^{-1})g \quad (i = 1, \dots, t)$$

of T yields a $\mathbf{C}$-basis

$$f_i \colon g \to f_i(g) \qquad (i = 1, \dots, t)$$

of $T\#$, and vice versa. Our aim is to find special $\mathbf{C}$-bases of $T\#$ which are related to the irreducible representations of T, and we shall see that the irreducible representations of G itself provide a connecting link. To outline our procedure we look at the S-ring Γ.

Let $\Delta \colon D_1, \dots, D_n$ be a complete set of pairwise inequivalent irreducible representations of G over $\mathbf{C}$. Write

$$D_\nu(g) = (d_{\kappa\lambda}^{(\nu)}(g))_{\kappa,\lambda = 1, \dots, x_\nu}, \qquad x_\nu = \deg D_\nu, \quad g \in G.$$

Then, as it is well known, the elements

$$\epsilon_{\kappa\lambda}^{(\nu)} = \frac{1}{|G|} \sum_{g \in G} x_\nu d_{\lambda\kappa}^{(\nu)}(g^{-1})g \qquad (\kappa, \lambda = 1, \dots, x_\nu; \nu = 1, \dots, n)$$

are "matrix units" of Γ, which means that

$$\epsilon_{\kappa\lambda}^{(\mu)}\epsilon_{\rho\sigma}^{(\nu)} = \delta_{\mu\nu}\delta_{\lambda\rho}\epsilon_{\kappa\sigma}^{(\nu)},$$

where $\delta_{\mu\nu}$, $\delta_{\lambda\rho}$ are Kronecker-symbols. Can one do something similar with the coefficients of the irreducible representations of T?

Let $\Phi\colon F_1, \ldots, F_r$ be a complete set of pairwise inequivalent irreducible representations of T over $\mathbf{C}$. Write

$$F_\rho(\tau) = (f_{\alpha\beta}^{(\rho)}(\tau))_{\alpha, \beta = 1, \ldots, y_\rho}, \qquad y_\rho = \deg F_\rho, \quad \tau \in T.$$

Denote by

$$t_i = |\mathscr{T}_i| \qquad (i = 1, \ldots, t)$$

the numbers of elements in the T-classes of G.

Theorem 1 ([5], Theorem 1.1 and Proposition 2.1). *If*

$$\beta_{\alpha\gamma}^{(\rho)} = \frac{1}{|G|} \sum_{i=1}^{t} \frac{1}{t_i} f_{\gamma\alpha}^{(\rho)}(\tau_i^*)\tau_i \qquad (\alpha, \gamma = 1, \ldots, y_\rho; \rho = 1, \ldots, r),$$

then there exist rational integers z_ρ $(\rho = 1, \ldots, r)$ such that

$$\beta_{\alpha\gamma}^{(\rho)}\beta_{\mu\nu}^{(\sigma)} = \delta_{\rho\sigma}\delta_{\gamma\mu}\frac{1}{z_\rho}\beta_{\alpha\nu}^{(\rho)},$$

where $\delta_{\rho\sigma}$, $\delta_{\gamma\mu}$ are Kronecker-symbols.

As consequences of Theorem 1 we state:

(i) *The $\beta_{\alpha\gamma}^{(\rho)}$ $(\alpha, \gamma = 1, \ldots, y_\rho; \rho = 1, \ldots, r)$ form a $\mathbf{C}$-basis of T.*

Hence by the remark made above:

(ii) *The functions*

$$\varphi_{\alpha\beta}^{(\rho)}\colon g \to \varphi_{\alpha\beta}^{(\rho)}(g) = \frac{z_\rho}{t_i} f_{\alpha\beta}^{(\rho)}(\tau_i) \quad \text{for } g \in \mathscr{T}_i,$$

$\alpha, \beta = 1, \ldots, y_\rho$ *and* $\rho = 1, \ldots, r$ *are a $\mathbf{C}$-basis of $T\#$.*

(iii) *The elements*

$$\eta_{\alpha\beta}^{(\rho)} = \frac{1}{|G|} \sum_{g \in G} \varphi_{\beta\alpha}^{(\rho)}(g^{-1})g \qquad (\alpha, \beta = 1, \ldots, y_\rho; \rho = 1, \ldots, r)$$

are "matrix units" of T:

$$\eta_{\alpha\beta}^{(\rho)}\eta_{\gamma\epsilon}^{(\sigma)} = \delta_{\rho\sigma}\delta_{\beta\gamma}\eta_{\alpha\epsilon}^{(\rho)},$$

where $\delta_{\rho\sigma}$, $\delta_{\beta\gamma}$ are Kronecker-symbols.

(iv) *The subspaces*

$$R_\alpha^{(\rho)} = \sum_{\beta=1}^{y_\rho} \mathbf{C}\eta_{\alpha\beta}^{(\rho)} \qquad (\alpha = 1, \ldots, y_\rho; \rho = 1, \ldots, r)$$

are minimal right ideals of T. The right multiplication of $R_\alpha^{(\rho)}$ by the elements $\tau \in T$ yields, if related to the basis $\eta_{\alpha\beta}^{(\rho)}$ ($\beta = 1, \ldots, y_\rho$), just the irreducible representation F_ρ of T.

We call the $\varphi_{\alpha\beta}^{(\rho)}$ ($\alpha, \beta = 1, \ldots, y_\rho$) the T-class functions of G related to F_ρ. They are the main subject of this theory. It is important that they can be expressed in terms of the irreducible representations of G. To state this explicitly we introduce the following notations:

$$f_{\alpha\beta}^{(\rho)}: \tau \to f_{\alpha\beta}^{(\rho)}(\tau), \qquad \tau \in T,$$

shall be the complex valued function on T given by the (α, β)-coefficient of F_ρ. We denote by $d_{\kappa\lambda}^{(\nu)}$ the linear extension of the function $g \to d_{\kappa\lambda}^{(\nu)}(g)$ onto Γ, and by

$$\hat{d}_{\kappa\lambda}^{(\nu)}: \tau \to d_{\kappa\lambda}^{(\nu)}(\tau), \qquad \tau \in T,$$

its restriction onto T.

Theorem 2 ([5], Theorem 1.1). *If the irreducible representations of G in $\Delta: D_1, \ldots, D_n$ are such that the representations*

$$\tau \to D_\nu(\tau), \qquad \tau \in T \quad (\nu = 1, \ldots, n)$$

of T are completely reduced with the F_ρ ($\rho = 1, \ldots, r$) as their irreducible constituents, then

$$\varphi_{\alpha\beta}^{(\rho)}(g) = \sum_{\hat{d}_{\kappa\lambda}^{(\nu)} = f_{\alpha\beta}^{(\rho)}} x_\nu d_{\kappa\lambda}^{(\nu)}(g)$$

for all $\alpha, \beta = 1, \ldots, y_\rho$, all $\rho = 1, \ldots, r$, and all $g \in G$, and

$$z_\rho = \sum_{\hat{d}_{\kappa\lambda}^{(\nu)} = f_{\alpha\beta}^{(\rho)}} x_\nu \qquad (\rho = 1, \ldots, r)$$

(where z_ρ is independent of $\alpha, \beta = 1, \ldots, y_\rho$).

By Theorem 2 there exist numerical relations between the $\varphi_{\alpha\beta}^{(\rho)}$'s and the $f_{\alpha\beta}^{(\rho)}$'s which in the case of $T = Z$ reduce to the orthogonality relations of the irreducible characters of G ([5], Section II). Two consequences of those numerical relations can be found in Theorem 2.8 of [5] and in [1].

Let us next look at the center $Z(T)$ of T. According to (iii) the primitive idempotents of $Z(T)$ are given by

$$\eta_\rho = \sum_{\alpha=1}^{y_\rho} \eta_{\alpha\alpha}^{(\rho)} = \frac{1}{|G|} \sum_{g \in G} \Big(\sum_{\alpha=1}^{y_\rho} \varphi_{\alpha\alpha}^{(\rho)}(g^{-1}) \Big) g \qquad (\rho = 1, \ldots, r).$$

Their coefficient functions

$$\psi_\rho = \sum_{\alpha=1}^{y_\rho} \varphi_{\alpha\alpha}^{(\rho)} \qquad (\rho = 1, \ldots, r)$$

will be called the *irreducible T-characters* of G, and z_ρ will be called the *degree of* ψ_ρ. From the definition of the $\varphi_{\alpha\beta}^{(\rho)}$'s in (ii) we get

$$\psi_\rho(g) = \frac{z_\rho}{t_i} \text{ trace } F_\rho(\tau_i) \quad \text{for } g \in \mathcal{T}_i.$$

The irreducible Γ-characters, for example, are the $x_\nu \chi_\nu$, where χ_ν $(\nu = 1, \ldots, n)$ are the irreducible characters of G.

In order to get a theory which really deserved the name of a character theory one would like the ψ_ρ $(\rho = 1, \ldots, r)$ to be a basis of a subalgebra of $T^\#$ which one then would call the *T-character algebra* of G. It is an open question whether this always holds or not, though there are some cases where this is obviously true (for instance if T itself is commutative). We now set out to investigate this problem more closely.

Two elements g and h of G are called T-conjugate if and only if $\psi_\rho(g) = \psi_\rho(h)$ for all $\rho = 1, \ldots, r$.

Obviously this is an equivalence relation on G. Denote by $\mathcal{S}_1, \ldots, \mathcal{S}_s$ the T-conjugacy classes of G, and set

$$\sigma_j = \sum_{g \in \mathcal{S}_j} g \qquad (j = 1, \ldots, s).$$

Since the ψ_ρ's are T-class functions each $\mathcal{S}_j$ is the union of certain T-classes $\mathcal{T}_i$ and consequently each σ_j is the sum of certain τ_i's. Hence $\sigma_j \in T$ for all $j = 1, \ldots, s$.

The mapping

$$\gamma = \sum_{g \in G} c_g g \to \gamma^* = \sum_{g \in G} \bar{c}_g g^{-1},$$

where $c_g \in \mathbf{C}$ and $\bar{c}_g$ is the conjugate complex of c_g, is an antiautomorphism of Γ. Because of condition (2) it induces an antiautomorphism

24—к.

of T, and hence an automorphism of $Z(T)$. Therefore every η_ρ^* is again a primitive idempotent of $Z(T)$. This means that:

Every function $\psi_\rho^*\colon g \to \overline{\psi_\rho(g^{-1})}$ is again an irreducible T-character of G.

Applying this remark to T-conjugacy we obtain:

The set $\mathscr{S}_j^*$ of all g^{-1} for $g \in \mathscr{S}_j$ is again a T-conjugacy class.

This is property (2) of an S-ring. The primitive idempotents can now be written as linear combinations of the σ_j's.

$$\eta_\rho = \frac{1}{|G|} \sum_{j=1}^{s} \psi_\rho(g_j^{-1})\sigma_j, \qquad g_j \in \mathscr{S}_j \quad (\rho = 1, \ldots, r).$$

It follows that $r \le s$, i.e., *the number of irreducible T-characters is not greater than the number of T-conjugacy classes.*

The case where $r = s$ is equivalent to the fact that every σ_j is a linear combination of the η_ρ's and therefore lies in the center of T. The σ_j's then form a **C**-basis of $Z(T)$, that is $Z(T)$ is an S-ring itself, and the $\mathscr{S}_j$'s are then the $Z(T)$-classes of G.

If on the other hand $Z(T)$ is an S-ring, we can apply the whole theory to $Z(T)$ instead of T, and we easily obtain the ψ_ρ's as a **C**-basis of $Z(T)\#$ which is then the T-character algebra of G.

Finally let us assume that the T-characters ψ_ρ $(\rho = 1, \ldots, r)$ are a **C**-basis of a subalgebra X of T which is then the T-character algebra of G. If we extend the mapping

$$\psi_\rho \to \psi_\rho(g_j), \qquad g_j \in \mathscr{S}_j \quad (j = 1, \ldots, s)$$

linearly onto X we get s different irreducible representations of X. Since X consists of all complex valued functions on G which are constant on every T-conjugacy class $\mathscr{S}_j$, it is a semisimple commutative **C**-algebra of dimension r and therefore has r irreducible representations over **C**. Hence we have $s \le r$. Since $r \le s$ always holds we get $r = s$, and $Z(T)$ is an S-ring itself. Thus we have proved

Theorem 3. *The following statements are equivalent:*
I. $\psi_\rho\psi_\sigma = \sum_{\tau=1}^{r} a_{\rho\sigma\tau}\psi_\tau,\ a_{\rho\sigma\tau} \in$ **C**, *for all $\rho, \sigma = 1, \ldots, r$.*
II. *The center $Z(T)$ of T is an S-ring itself.*

III. $r = s$, *i.e., the number of irreducible T-characters of G (which is the same as the number of inequivalent irreducible representations of T) is equal to the number of T-conjugacy classes of G.*

II. Double coset S-rings

Let G be a finite group, and H a subgroup of G. Denote by

$$\mathscr{T}_i = Hg_iH, \qquad g_i \in G \quad (i = 1, \ldots, t)$$

the double cosets of H in G. They define the double coset S-ring T_H on G. In this special case the theory of Section I can be easily obtained by direct proof.

Let $D_1, \ldots, D_r$ be those irreducible representations of G which occur as irreducible constituents in the representation 1_H^G of G which is induced by the 1-representation 1_H of H. Take the D_ρ's such that in the restriction

$$D_\rho|H = \left.\left(\begin{array}{ccc|c} 1_H & & & \\ & \ddots & & 0 \\ & & 1_H & \\ \hline & 0 & & * \end{array}\right)\begin{array}{l}\left.\vphantom{\begin{array}{c}1\\1\\1\end{array}}\right\}y_\rho\\ \ \end{array}\right\}x_\rho$$

of D_ρ onto H all the 1-representations of H occur in the upper left corner. Their number y_ρ is equal to the multiplicity of D_ρ in 1_H^G. Under this assumption on the D_ρ's we get

$$D_\rho(h) = \begin{pmatrix} E_{y_\rho} & 0 \\ 0 & * \end{pmatrix} \quad \text{for all } h \in H,$$

and from this fact it is quite obvious that $d_{\alpha\beta}^{(\rho)}(g) = d_{\alpha\beta}^{(\rho)}(g_i)$ for $g \in Hg_iH$, all $\alpha, \beta = 1, \ldots, y_\rho$ and all $\rho = 1, \ldots, r$, and

$$D_\rho(\tau_i) = \sum_{g \in Hg_iH} D_\rho(g) = \begin{pmatrix} F_\rho(\tau_i) & 0 \\ 0 & 0 \end{pmatrix}.$$

Because $t = \sum_{\rho=1}^r y_\rho^2$ ([8], 29.2) it follows:
(i) *The complex valued functions*

$$d_{\alpha\beta}^{(\rho)} \qquad (\alpha, \beta = 1, \ldots, y_\rho; \rho = 1, \ldots, r)$$

on G are a $\mathbf{C}$-basis of the algebra $T_H^\#$ of all T_H-class functions.

(ii) *The mappings*

$$F_\rho: \tau \to F_\rho(\tau) = (d_{\alpha\beta}^{(\rho)}(\tau))_{\alpha,\beta = 1,\dots,y_\rho} \quad (\rho = 1,\dots,r)$$

are a complete set of pairwise inequivalent irreducible representations of T_H.

We set

$$\varphi_\rho = \sum_{\alpha=1}^{y_\rho} d_{\alpha\alpha}^{(\rho)} \quad (\rho = 1,\dots,r).$$

If we compare these results with Section I we get:

$\varphi_{\alpha\beta}^{(\rho)} = x_\rho d_{\alpha\beta}^{(\rho)}$ ($\alpha, \beta = 1,\dots,y_\rho$) *are the* T_H-*class functions related to* F_ρ.
$\psi_\rho = x_\rho \varphi_\rho$ ($\rho = 1,\dots,r$) *are the irreducible* T_H-*characters of* G *and* x_ρ *is the degree of* ψ_ρ.

Even for double coset S-rings it is an open question whether the statements of Theorem 3 hold. But in this special case we can give the T_H-conjugacy classes explicitly. We denote by $K_1,\dots,K_n$ the classes of conjugate elements of G.

Theorem 4.   *For* $g, h \in G$ *the following statements are equivalent:*
I. g *and* h *are* T_H-*conjugate, i.e.,* $\varphi_\rho(g) = \varphi_\rho(h)$ *for all* $\rho = 1,\dots,r$.
II. $\dfrac{|K_\alpha \cap HgH|}{|HgH|} = \dfrac{|K_\alpha \cap HhH|}{|HhH|}$ *for all* $\alpha = 1,\dots,n$.
III. $|K_\alpha \cap Hg| = |K_\alpha \cap Hh|$ *for all* $\alpha = 1,\dots,n$.

Since the φ_ρ's are linked with the T_H-classes HgH and with the conjugacy classes K_α one would expect the irreducible characters χ_ρ to be linked with them too. This is shown by the following

Theorem 5.

$$\frac{|K_\alpha| |HgH|}{|G|} \sum_{\rho=1}^{r} \chi_\rho(g_\alpha)\varphi_\rho(g^{-1}) = |K_\alpha \cap HgH|$$

for all $g \in G$, $g_\alpha \in K_\alpha$ *and all* $\alpha = 1,\dots,n$.

In the special case $H = \langle 1 \rangle$, φ_ρ is equal to χ_ρ and Theorem 5 becomes the orthogonality relations for the irreducible characters of G.

Furthermore the χ_ρ's and the φ_ρ's can be expressed by each other in the following way:

$$\varphi_\rho(g_i) = \sum_{\alpha=1}^{n} \chi_\rho(g_\alpha)\frac{|K_\alpha \cap Hg_iH|}{|Hg_iH|},$$

$$y_\rho\chi_\rho(g_\alpha) = \sum_{i=1}^{t} x_\rho\varphi_\rho(g_i)\frac{|K_\alpha \cap Hg_iH|}{|K_\alpha|}.$$

Finally we note another connection of the conjugacy classes with the double cosets.

Theorem 6. *Set* $\epsilon = 1/|H| \sum_{h \in H} h$ *and* $\kappa_\alpha = \sum_{g \in K_\alpha} g$ $(\alpha = 1, \ldots, n)$. *Then each element of the center* $Z(T_H)$ *of* T_H *is a linear combination of the elements* $\epsilon \kappa_\alpha$ $(\alpha = 1, \ldots, n)$. *The mapping*

$$\zeta \to \epsilon\zeta, \qquad \zeta \in Z,$$

is an epimorphism of the center Z *of* Γ *onto* $Z(T_H)$.

Theorem 6 is easily proved by showing that the primitive idempotents η_ρ of $Z(T_H)$ can be expressed by the primitive idempotents

$$\epsilon_\rho = \frac{1}{|G|} \sum_{g \in G} x_\rho \chi_\rho(g^{-1}) g$$

of Z as

$$\eta_\rho = \epsilon\epsilon_\rho \qquad (\rho = 1, \ldots, r).$$

In addition we get

$$\eta_\rho = \frac{1}{|G|} \sum_{\alpha=1}^{n} (x_\rho \chi_\rho(g_\alpha^{-1})) \epsilon \kappa_\alpha,$$

$$\epsilon \kappa_\alpha = \sum_{\rho=1}^{r} \frac{|K_\alpha| \chi_\rho(g_\alpha)}{x_\rho} \eta_\rho.$$

References

[1] J. SUTHERLAND FRAME and OLAF TAMASCHKE, Über die Ordnungen der Zentralisatoren der Elemente in endlichen Gruppen, *Math. Z.* **83** (1964), 41–45.

[2] OLAF TAMASCHKE, Ringtheoretische Behandlung einfach transitiver Permutationsgruppen, *Math. Z.* **73** (1960), 393–408.

[3] OLAF TAMASCHKE, Zur Theorie der Permutationsgruppen mit regulärer Untergruppe I, II, *Math. Z.* **80** (1963), 328–354, 443–465.

[4] OLAF TAMASCHKE, S-Ringe und verallgemeinerte Charaktere auf endlichen Gruppen, *Math. Z.* **84** (1964), 101–119.

[5] OLAF TAMASCHKE, S-rings and the irreducible representations of finite groups, *J. Algebra* **1** (1964), 215–232.

[6] HELMUT WIELANDT, Zur Theorie der einfach transitiven Permutationsgruppen *Math. Z.* **40** (1935) 582–587.

[7] HELMUT WIELANDT, Zur Theorie der einfach transitiven Permutationsgruppen, II, *Math. Z.* **52** (1949), 384–393.

[8] HELMUT WIELANDT, *Finite permutation groups*, Academic Press, New York–London, 1964.

Proc. Internat. Conf. Theory of Groups, Austral. Nat. Univ. Canberra,
August 1965, pp. 357–362. © Gordon and Breach Science Publishers, Inc. 1967

On Hughes' H_p problem

G. E. WALL

1. If G is a group and p a prime, let $H_p(G)$ denote the subgroup of G generated by the elements of order $\neq p$. Hughes [1] conjectured that if $H_p(G)$ is a proper, nontrivial subgroup then $H_p(G)$ has index p in G. This is an elementary result for $p=2$ and was proved for $p=3$ by Straus and Szekeres [4]. It is also true, for arbitrary p, when G is finite and not a p-group (Hughes and Thompson [2]) and when G is a finite p-group of class $\leq p$ (Zappa [5]). The purpose of this note is to outline a proof that the conjecture is false for $p=5$. We construct a finite group G of exponent 25 such that $|G:H_5(G)|=25$.

2. The general principle of the construction is given in

Lemma 1. *Let P be a finite group of prime exponent p and Q ($\neq P$) a normal subgroup of P. Let Γ denote the group algebra of P over $\mathbf{Z}_p$. Suppose there exists a linear functional φ on Γ such that*

$$(2.1) \qquad \varphi\left(\sum_{i=0}^{p-1} ab^i\right) = \delta_b \quad \text{for all } a, b \in P,$$

where $\delta_b = 0$ or 1 according as $b \in Q$ or $b \notin Q$. Then there exists a finite group G of exponent p^2 such that $G/H_p(G) \cong P/Q$.

Proof. Let Γ^* denote the group algebra of P over $\mathbf{Z}_{p^2}$, Δ^* the ideal of Γ^* generated by the elements

$$p(a-e), \quad (a-e)^{p-1}+p\delta_a e \qquad (a \in P, \ e \text{ the identity of } P).$$

We take $G = A/D$, where A is the group of all "affine" transformations

$$T_{a,b}x = ax+b \qquad (a \in P; \ x, b \in \Gamma^*)$$

on Γ^* and D is the normal subgroup of A formed by the transformations $T_{e,b}$ ($b \in \Delta^*$).

Now

$$T^p_{a,b} = T_{a^p,(e+a+\cdots+a^{p-1})b}$$
$$\equiv T_{e,(p+\binom{p}{2}(a-e)+\cdots+(a-e)^{p-1})b}$$
$$\equiv T_{e,(1-\delta_a)pb} \quad (\mathrm{mod}\ D).$$

Also, by (2.1), $pe \notin \varDelta^*$. It follows easily that G has exponent p^2 and $H_p(G) = \{T_{a,b}D \mid a \in Q,\ b \in \varGamma^*\}$, which gives the lemma.

3. *From now on we assume that P is a two-generator group and $Q = P'$.* Our present purpose is to recast Lemma 1 in exponential form (Lemma 4).

It is well known that the elements $u - e$ $(u \in P)$ span the radical, $\varOmega$, of $\varGamma$. Each $a \in P$ has a unique exponential representation

$$(3.1) \qquad a = e^\alpha = \sum_{k=0}^{\infty} \frac{\alpha^k}{k!} \qquad (\alpha \in \varOmega,\ \alpha^p = 0).$$

If x, y are generators of P, we can choose commutators in x, y:

$$x_1(= x),\ x_2(=y\ ),\ x_3,\ldots, x_m,$$

arranged in order of nondecreasing weight, such that each $a \in P$ has a unique representation

$$(3.2) \qquad a = x_1^{\lambda_1} x_2^{\lambda_2} \cdots x_m^{\lambda_m} \qquad (0 \le \lambda_i < p).$$

Suppose $x = e^\xi$, $y = e^\eta$, and define ξ_i $(1 \le i \le m)$ as the ring commutator in ξ, η corresponding to the group commutator x_i in x, y. The elements

$$\xi_1^{\lambda_1} \xi_2^{\lambda_2} \cdots \xi_m^{\lambda_m} \qquad (0 \le \lambda_i < p)$$

will be called the *standard monomials*. The *weight* of a standard monomial is defined to be its degree as homogeneous polynomial in ξ, η.

Lemma 2 (Campbell–Hausdorff formula). *In the notation of (3.1) and (3.2),*

$$\alpha = \sum_1^m \mu_i \xi_i + f_p(\lambda_1 \xi_1,\ \lambda_2 \xi_2) + \cdots,$$

where $\mu_i = \lambda_i + a$ polynomial in the λ_j such that weight $\xi_j <$ weight ξ_i $(1 \le i \le m)$, f_p is a homogeneous polynomial of degree p and the omitted terms are all of degree $> p$ in ξ_1, ξ_2.

Lemma 3 (cf. Jennings [3]). *The standard monomials of weight $\ge t$ form a basis of $\varOmega^t$ $(t = 1, 2, \ldots)$.*

By Lemma 3, each $u \in \Omega$ can be expressed in a unique way as a linear combination of standard monomials; this is called the *standard form* of u. By Lemma 2,

$$(a-e)^{p-1} = \alpha^{p-1} \equiv \left(\sum_1^m \mu_i \xi_i\right)^{p-1} + F(\lambda_1 \xi_1, \lambda_2 \xi_2) \pmod{\Omega^{2p-1}},$$

where F is a homogeneous polynomial of degree $2p-2$. Since $\lambda_i^p = \lambda_i$, we may write

$$F(\lambda_1 \xi_1, \lambda_2 \xi_2) = \sum_{i=0}^{p-1} \Theta_i \lambda_1^i \lambda_2^{p-1-i} + \Theta \lambda_1^{p-1} \lambda_2^{p-1},$$

where Θ_i, Θ are homogeneous polynomials in ξ_1, ξ_2 of degree $2p-2$.

Lemma 4. *Let Λ denote the ideal of Γ generated by Ω^{2p-1} and the powers ω^{p-1}, where ω runs over the linear combinations of $\xi_1, \xi_2, \ldots, \xi_m$. If $\Theta \notin \Lambda$, there exists a finite group G of exponent p^2 such that $|G: H_p(G)| = p^2$.*

Notation. We write the multinomial expansion of $(\sum_1^m \mu_i \xi_i)^r$ $(r < p)$ as

$$\left(\sum_1^m \mu_i \xi_i\right)^r = \sum_{\alpha_1 + \cdots + \alpha_m = r} \binom{r}{\alpha_1, \ldots, \alpha_m} |\xi_1^{\alpha_1} \cdots \xi_m^{\alpha_m}| \mu_1^{\alpha_1} \cdots \mu_m^{\alpha_m}.$$

We use the notation $[u, v] = u^{-1} v^{-1} u v$, $[u, v, w] = [[u, v], w]$, etc. for group commutators and the notation $(\alpha, \beta) = \alpha\beta - \beta\alpha$, $(\alpha, \beta, \gamma) = ((\alpha, \beta), \gamma)$, etc. for ring commutators.

4. The actual construction of the counterexample to Hughes' conjecture is given by

Lemma 5. *The condition $\Theta \notin \Lambda$ of Lemma 4 is satisfied when $p = 5$ and P is the two-generator free group of the variety defined by the laws $z^5 = [z_1, z_2, \ldots, z_8] = 1$.*

Outline of proof. Write

$$\delta = (\xi_1, \xi_2), \qquad \delta_i = (\delta, \xi_i), \qquad \delta_{ij} = (\delta_i, \xi_j), \quad \text{etc.,}$$

where $\xi_1 = \xi$, $\xi_2 = \eta$. Then we may take $\xi_3, \ldots, \xi_m$ to be δ, δ_1, δ_2, δ_{11}, δ_{12}, δ_{22}, (δ_1, δ), (δ_2, δ), (δ_{11}, δ), (δ_{12}, δ), (δ_{22}, δ), (δ_2, δ_1), $(\delta_1, \delta, \delta)$, $(\delta_2, \delta, \delta)$, (δ_{12}, δ_1), (δ_{12}, δ_2).

The dots in (4.1)–(4.3) below indicate sums of standard monomials of weight 8 which have the form $\delta^\lambda \delta_1^\mu \cdots$ but not the form $\delta_{ij} \delta_{kl}$; in effect, we are ignoring such terms (and the elements of Ω^9) in our calculations.

The first step in the proof is to show that

(4.1) $$\Theta \equiv \delta_{12}^2 - \delta_{11}\delta_{22} + \cdots \quad (\mathrm{mod}\ \Lambda).$$

This requires a rather lengthy calculation.

If α is an automorphism of P, we define the action of α on Γ by

$$\Big(\sum_{a \in P} \lambda_a a \Big)^\alpha = \sum_{a \in P} \lambda_a a^\alpha.$$

An element ω of Ω^8 is called an *invariant* if, for all such α,

(4.2) $$\omega^\alpha \equiv \omega + \cdots \quad (\mathrm{mod}\ \Omega^9).$$

$\delta_{12}^2 - \delta_{11}\delta_{22}$ is one such invariant. The following are invariants *which belong to Λ*:

$$S_1 = |\xi^3\delta|\eta^3 - 3|\xi^2\eta\delta|\,|\xi\eta^2| + 3|\xi\eta^2\delta|\,|\xi^2\eta| - |\eta^3\delta|\xi^3,$$

$$\begin{aligned}
S_2 = {}& |\xi^3\delta_2|\eta^2 - \{2|\xi^2\eta\delta_2|\,|\xi\eta| + |\xi^2\eta\delta_1|\eta^2\} \\
& + \{|\xi\eta^2\delta_2|\xi^2 + 2|\xi\eta^2\delta_1|\,|\xi\eta|\} - |\eta^3\delta_1|\xi^2,
\end{aligned}$$

$$\begin{aligned}
S_3 = {}& |\xi^3\delta_{22}|\eta - \{2|\xi^2\eta\delta_{12}|\eta + |\xi^2\eta\delta_{22}|\xi\} \\
& + \{|\xi\eta^2\delta_{11}|\eta + 2|\xi\eta^2\delta_{12}|\xi\} - |\eta^3\delta_{11}|\xi,
\end{aligned}$$

$$S_4 = |\xi^2\delta^2|\eta^2 - 2|\xi\eta\delta^2|\,|\xi\eta| + |\eta^2\delta^2|\xi^2,$$

$$S_5 = |\xi^2\delta\delta_2|\eta - \{|\xi\eta\delta\delta_1|\eta + |\xi\eta\delta\delta_2|\xi\} + |\eta^2\delta\delta_1|\xi,$$

$$S_6 = |\xi^2\delta\delta_{22}| - 2|\xi\eta\delta\delta_{12}| + |\eta^2\delta\delta_{11}|,$$

$$S_7 = |\xi^2\delta_2^2| - 2|\xi\eta\delta_1\delta_2| + |\eta^2\delta_1^2|,$$

$$S_8 = |\xi\delta^3|\eta - |\eta\delta^3|\xi,$$

$$S_9 = |\xi\delta^2\delta_2| - |\eta\delta^2\delta_1|.$$

The second step in the proof is to show that this list is complete in the sense that every invariant S in Λ has the form

(4.3) $$S \equiv \sum_1^9 \lambda_i S_i + \cdots \quad (\mathrm{mod}\ \Omega^9).$$

This is proved by examining the structure of $(\Lambda \cap \Omega^8)/\Omega^9$ as module for the group of automorphisms of P.

It remains only to prove that $\delta_{12}^2 - \delta_{11}\delta_{22}$ does *not* have the form (4.3). To do this, we express the S_i in standard form and compile a partial table of coefficients.

		$\eta^2 \times$			$\eta \times$				
	δ_1^2	$\delta\delta_{11}$	(δ_{11}, δ)	$(\delta_2\delta_{11} - \delta_1\delta_{12})$	$\delta^2\delta_1$	(δ_{12}, δ_1)	$\delta(\delta_1, \delta)$	$(\delta_1, \delta, \delta)$	$(\delta_{12}^2 - \delta_{11}\delta_{22})$
S_1	2	-2	2	2	2	-2	-2	2	-1
S_2	-2	2	1	2	2	-1	-1	0	-2
S_3	0	2	-2	-2	0	-2	2	0	1
S_4	-2	-1	1	2	1	-2	-1	1	-2
S_5	-2	-2	1	-2	1	0	-1	1	-1
S_6	0	1	-2	1	0	-1	1	-1	1
S_7	1	0	0	-1	0	2	0	0	-2
S_8	0	0	0	0	1	0	-1	1	0
S_9	0	0	0	0	-1	0	-1	-2	0

The 9×9 coefficient matrix has rank 8 and the final row of the table gives the coefficients of the one linear relation between its columns. This makes the assertion evident.

References

[1] D. R. HUGHES, Research problem 3, *Bull. Amer. Math. Soc.* **63** (1957), 209.

[2] D. R. HUGHES and J. G. THOMPSON, The H_p-problem and the structure of H_p-groups, *Pacific J. Math.* **9** (1959), 1097–1101.

[3] S. A. JENNINGS, The structure of the group ring of a p-group over a modular field, *Trans. Amer. Math. Soc.* **50** (1941), 175–185.

[4] E. G. STRAUS and G. SZEKERES, On a problem of D. R. HUGHES, *Proc. Amer. Soc.* **9** (1958), 157–158.

[5] G. ZAPPA, Contributo allo studio del problema di Hughes sui gruppi, *Ann. Mat. Pura Appl.* (4) **57** (1962), 211–219.

Proc. Internat. Conf. Theory of Groups, Austral. Nat. Univ. Canberra,
August 1965, pp. 363–365. © Gordon and Breach Science Publishers, Inc. 1967

Basic commutators for polynilpotent groups

M. A. WARD

The conventional theory of basic commutators may be considered to be an investigation of the properties of the lower central series $\gamma_c(F)$: $c = 1, 2, \ldots$ of an absolutely free group F, or alternatively of the properties of the corresponding factor groups, which are the free nilpotent groups of various classes: $F/\gamma_{c+1}(F) = F(\mathfrak{N}_c)$. The work described in this paper is a generalization of this theory which allows free polynilpotent groups to be investigated in much the same way. This paper is a summary of a thesis to be presented to the Australian National University, and it is hoped that a detailed account will be published shortly.

It is convenient to replace the notion of "formal expressions" for elements of a group by a free algebra as follows.

Definition. **A** is the free algebra on a set $\{g_i\}_{i<\tau}$ of free generators, where τ is some ordinal, with a nullary operator (the unit element) denoted **1**, a unary operator (inversion) whose action on **x** is denoted $\mathbf{x}^{-1}$, and two binary operators (multiplication and commutation) whose actions on **x**, **y** are denoted **xy** and [**x**, **y**] respectively. The only law of **A** is the associative law of multiplication, so that for instance $\mathbf{x1} \neq \mathbf{x}$ and $\mathbf{xx}^{-1} \neq \mathbf{1}$.

If G is any group with generating set $\{g_i\}_{i<\tau}$ there is a unique epimorphism $\rho \colon \mathbf{A} \to G$ such that $g_i\rho = g_i$ for all $i < \tau$; this epimorphism allows the elements of **A** to describe the elements of G in much the same way as "formal expressions" do.

The following notation is used for polynilpotent groups. Let $K = (k_i)_{i=1}^{\infty}$ be a sequence of integers, each ≥ 2. Then for each r and any group G the subgroup $P_r(G)$ is defined recursively: $P_0(G) = G$ and

$P_r(G) = \gamma_{k_r}(P_{r-1}(G))$, $(r > 0)$. This is the polynilpotent series of G of type K. The corresponding varieties are denoted $\mathfrak{P}_r$.

Definition.

(A) A "semiweight range" is a set W upon which an order $\leq$ and addition $+$ is defined satisfying:

  (i) $\leq$ well-orders W, which has a least element 1 and a greatest element ∞.

  (ii) W is closed under addition and $W - \{\infty\}$ is generated by 1 under addition.

  (iii) Addition is commutative but not necessarily associative.

  (iv) $\alpha, \beta < \infty \Rightarrow \alpha + \beta < \infty$,
  $\alpha < \infty \Rightarrow \alpha < \alpha + \beta$,
  $\alpha + \infty = \infty$.

  (v) $\alpha_1 < \alpha_2$ and $\beta < \infty \Rightarrow \alpha_1 + \beta < \alpha_2 + \beta$,

  (vi) $\alpha \leq \beta \leq \gamma \Rightarrow (\gamma + \beta) + \alpha = (\gamma + \alpha) + \beta \leq (\beta + \alpha) + \gamma$.

(B) A corresponding mapping $\sigma \colon \mathbf{A} \to W$ (the "semiweight" associated with W) is defined recursively: $\sigma(1) = \infty$, $\sigma(\mathbf{g}_i) = 1$, $\sigma(\mathbf{x}^{-1}) = \sigma(\mathbf{x})$, $\sigma(\mathbf{xy}) = \min\{\sigma(\mathbf{x}), \sigma(\mathbf{y})\}$, and $\sigma([\mathbf{x}, \mathbf{y}]) = \sigma(\mathbf{x}) + \sigma(\mathbf{y})$.

(C) For each group G and each element $\alpha \in W$ the subgroup $W_\alpha(G)$ is the set $\{\mathbf{x}\rho \colon \sigma(\mathbf{x}) \geq \alpha\}$ for some epimorphism $\rho \colon \mathbf{A} \to G$. This is a verbal subgroup of G and is independent of the particular epimorphism ρ chosen. $\mathfrak{W}_\alpha$ is the variety of all groups G for which $W_\alpha(G) = \{1\}$.

(D) λ is a "limiting element" of W if $\xi < \lambda \Rightarrow \xi + 1 < \lambda$.

(E) The set of "W-basic commutators" is defined and ordered recursively over their semiweight. The W-basic commutators of semiweight 1 are just the generators $\{\mathbf{g}_i\}_{i < \tau}$ of $\mathbf{A}$ and $\mathbf{g}_i < \mathbf{g}_j \Leftrightarrow i < j$. The W-basic commutators of semiweight $\alpha > 1$ are the elements $[\mathbf{x}, \mathbf{y}]$ where $\mathbf{x}, \mathbf{y}$ are W-basic commutators, $\sigma(\mathbf{x}) + \sigma(\mathbf{y}) = \alpha$, $\mathbf{x} > \mathbf{y}$ and if $\mathbf{x} = [\mathbf{x}_1, \mathbf{x}_2]$ then $\mathbf{x}_2 \leq \mathbf{y}$. The set of W-basic commutators of weight exactly α are ordered as follows: $[\mathbf{b}_1, \mathbf{a}_1] \leq [\mathbf{b}_2, \mathbf{a}_2]$ if either $\mathbf{b}_1 < \mathbf{b}_2$ or $\mathbf{b}_1 = \mathbf{b}_2$ and $\mathbf{a}_1 \leq \mathbf{a}_2$. If $\mathbf{x}_1$ and $\mathbf{x}_2$ are W-basic commutators and $\sigma(\mathbf{x}_1) < \sigma(\mathbf{x}_2)$ then $\mathbf{x}_1 < \mathbf{x}_2$. The set of "$W$-basic expressions mod α" are expressions of the form $\mathbf{1}$ or $\mathbf{b}_1^{\beta_1}\mathbf{b}_2^{\beta_2} \cdots \mathbf{b}_k^{\beta_k}$ where $\mathbf{b}_1, \mathbf{b}_2, \ldots, \mathbf{b}_k$ are W-basic commutators, $\mathbf{b}_1 < \mathbf{b}_2 < \cdots < \mathbf{b}_k$, each $\sigma(\mathbf{b}_i) < \alpha$ and each β_i is a nonzero integer.

The set of positive integers with the usual order and addition extended to encompass an extra element ∞ in the fashion indicated above provides an example of a semiweight range. The associated semiweight is then the usual "weight" and the subgroups $W_\alpha(G)$ are then the members of the lower central series of G. A semiweight range W may also be chosen so that the subgroups $W_\alpha(G)$ contain amongst them the members of any particular polynilpotent series $P_r(G)$. The following results have been established.

(1) Let $G = F(\mathfrak{W}_\beta)$, $x \in W_\alpha(G)$ and suppose there is no limiting element λ such that $\alpha < \lambda \leq \beta$. Then there exists a unique W-basic expression $\mathbf{x}$ mod β which describes x, that is, $\mathbf{x}\rho = x$. This is established by a collecting process.

(2) If F is an absolutely free group, $W_\alpha(F)/W_{\alpha+1}(F)$ is a free abelian group, freely generated by the W-basic commutators of semiweight $< \alpha + 1$ and $\geq \alpha$.

(3) Let $G = F(\mathfrak{W}_\alpha)$ be of rank at least 2. Then $\zeta(G) = W_{\alpha-1}(G)$ where $\alpha - 1 = \min\{\xi \colon \xi + 1 \geq \alpha\}$. Now let $G = F(\mathfrak{W}_\alpha \cap \mathfrak{N}_c)$ be of rank at least 2. Then $\zeta(G) = W_{\alpha-1}(G) \cdot \gamma_c(G)$. In particular, if P_r is a polynilpotent variety of (polynilpotent) length other than 1, then $\zeta(F(\mathfrak{P}_r))$ is trivial and the upper and lower central series of $F(\mathfrak{P}_r \cap \mathfrak{N}_c)$ coincide.

(4) The group $G = F(\mathfrak{W}_\alpha)$ is residually nilpotent. Alternatively, let $x \in G$, $x \neq 1$. Then there exists $\xi < \alpha$ such that $x \in W_\xi(G) - W_{\xi+1}(G)$. This makes it possible to obtain information about elements of an absolutely free group F which it is known do not belong to one of the subgroups $W_\alpha(F)$.

(5) It is possible in some cases to obtain several different expressions for the subgroups $W_\alpha(G)$ in ordinary group-theoretic terms. This makes possible the calculation of certain nontrivial subgroup identities. For example, suppose r and m are positive integers, $r \leq m$, and F is an absolutely free group. Then

$$[\gamma_m(F), \gamma_r(F)][\gamma_{m-1}(F), \gamma_{r+1}(F)] \cdots [\gamma_r(F), \gamma_m(F)]$$
$$= \gamma_{m+r}(F) \cap [\gamma_r(F), \gamma_r(F)].$$

Proc. Internat. Conf. Theory of Groups, Austral. Nat. Univ. Canberra,
August 1965, pp. 367–371. © Gordon and Breach Science Publishers, Inc. 1967

Critical and basic p-groups

PAUL M. WEICHSEL*

Introduction. The study of finite p-groups can, in a sense, be reduced
to the study of finite critical p-groups (a group is said to be critical if it
is not contained in the variety generated by its proper subgroups and
proper factor groups). That is to say, every finite p-group is either
critical or can be constructed from its critical factors (factor groups of
subgroups) by the operations of forming finite direct products, taking
subgroups and taking factor groups. In fact the set of all critical
groups is an unnecessarily large class of groups to consider. For there
exist critical p-groups which generate varieties that can be obtained as
the join of proper subvarieties. For example, if G is the two-generator
p-group of order p^3, exponent p^2, with cyclic center $(p > 2)$, then the
variety generated by G is the join of the varieties generated by A, the
cyclic group of order p^2 and B, the two-generator p-group of order p^3,
exponent p and cyclic center. Hence we may restrict ourselves to those
critical p-groups which generate join-irreducible varieties. Such groups
are called basic ([2], p. 96).

In this paper we will study properties of finite basic p-groups and
obtain a characterization of these p-groups for class ≤ 3 and $p > 3$.

I. Marginal subgroups and regular basic p-groups

A useful tool for analyzing basic groups is the concept of a marginal
subgroup introduced by P. Hall [1].

Definition 1.1. Let $V = \{f_1, \ldots, f_n\}$ be a collection of functions
(words) on a group G. (A function f is a word in a finite number of

---

* This work was supported in part by the National Science Foundation,
NSF GP–4616.

25—к.

variables for which we substitute elements of G. $f(G)$ is the subgroup of G thus generated and is called the verbal subgroup of G.) An element g of G is said to be marginal with respect to a set of words V, if

$$f(x_1, \ldots, x_n) = f(gx_1, x_2, \ldots, x_n) = \cdots = f(x_1, \ldots, x_{n-1}, gx_n)$$

for all $f \in V$ and all $x_i \in G$. The set of all marginal elements of G is a subgroup, called the marginal subgroup, $M_V(G)$.

The role of marginal subgroups in the study of basic groups is given by Theorem A below whose proof appeared in [2] (pp. 97–98).

Theorem A. *If G is a basic p-group, then $M_V G \subseteq \Phi(G)$, the Frattini subgroup of G, for all sets of functions V such that $V(G) \neq 1$.*

Thus if a p-group G is basic then it can have no marginal generators.

Before we proceed to our main results concerning basic p-groups, we need to restate the definition of a regular p-group.

Definition 1.2. A finite p-group G is called regular if for any $a, b \in G$ and any positive integer α,

$$(ab)^{p^\alpha} = a^{p^\alpha} b^{p^\alpha} c_1^{p^\alpha} \cdots c_s^{p^\alpha}$$

where the c_i are elements of the derived group of the group generated by $\{a, b\}$.

We list below without proof a number of properties of regular p-groups which we need later.

(1) Let G be a p-group of class c. If $p > c$, then G is regular.

Thus for fixed class c, and with the possible exception of a finite number of primes, all p-groups of class c or less are regular.

(2) Let $(x_1, \ldots, x_n)$ be a simple commutator defined inductively by $(x_1, x_2) = x_1^{-1} x_2^{-1} x_1 x_2$, $(x_1, \ldots, x_r) = ((x_1, \ldots, x_{r-1}), x_r)$. Then if $x_i \in G$ $(i = 1, \ldots, n)$ and G is regular, the order of $(x_1, \ldots, x_n)$ cannot exceed the order of x_i, all $i = 1, \ldots, n$. It is easy to show that the c_i in Definition 1.2 can be considered to be simple commutators.

(3) It is an easy consequence of (2) that if $a^{p^\alpha} = b^{p^\alpha}$ for a, b elements of the regular p-group G, then $(ab^{-1})^{p^\alpha} = 1$, and conversely.

In the following three theorems, G is a basic regular p-group of class c and $\Phi(G)$ is the Frattini subgroup of G.

Theorem 1.1. *If a and b are elements of G and $a, b \notin \Phi(G)$, then $|a| = |b|$.*

Proof. Assume that $|a| > |b| = p^\beta$, and consider $f(x) = x^{p^\beta}$. Since $|a| > p^\beta$, $f(a) \neq 1$ and thus $f(G) \neq 1$.

Now $f(bx) = (bx)^{p^\beta} = b^{p^\beta} x^{p^\beta} c_1^{p^\beta} \cdots c_r^{p^\beta}$. But $b^{p^\beta} = 1$ and each c_i is a simple commutator containing at least one occurrence of b. It follows from (2) that $f(bx) = f(x)$ and b is marginal, a contradiction. Thus $|a| = |b|$.

Theorem 1.2. *Let a, b be elements of G not in $\Phi(G)$ and let $Z(G)$ be the center of G. Then $a^{p^\alpha} \in Z(G)$ if and only if $b^{p^\alpha} \in Z(G)$. That is, $|aZ(G)| = |bZ(G)|$.*

Proof. Suppose that $|aZ(G)| > |bZ(G)| = p^\beta$ and $g(x, y) = (x, y)^{p^\beta}$.

(i) If $g(a, y) = 1$ for all $y \in G$, then, by (2), $a^{p^\beta} \in Z(G)$, a contradiction. Thus for some $y \in G$, $g(a, y) \neq 1$ and thus $g(G) \neq 1$.

(ii) $g(bx, y) = (bx, y)^{p^\beta} = [(b, y)(b, y, x)(x, y)]^{p^\beta} = (b, y)^{p^\beta} (b, y, x)^{p^\beta} (x, y)^{p^\beta} R$ where R is a product of p^β-th powers of simple commutators, each containing a b-entry. Hence it follows from (2) that $(b, y)^{p^\beta} = (b, y, x)^{p^\beta} = R = 1$ and thus $g(bx, y) = g(x, y)$. Similarly $g(x, by) = g(x, y)$. Therefore b is a marginal element, a contradiction; and $|aZ(G)| = |bZ(G)|$.

It is not difficult to show that the above result can be generalized by replacing $Z(G)$ by any marginal subgroup $M_V(G)$ with V a set of words satisfying $V(G) \neq 1$.

Theorem 1.3. *The exponent of G is equal to the exponent of G_2, the derived group of G.*

Proof. It follows from (2) that if $(a, b)^{p^\beta} = 1$ for fixed a and all $b \in G$, then $a^{p^\beta} \in Z(G)$. Hence if the exponent of $G_2 = p^\beta < p^\alpha = $ exponent of G, then there exists $a \in G$, $a \notin \Phi(G)$ such that $a^{p^\beta} = 1$ and $(a, x)^{p^\beta} = 1$ for all $x \in G$. Hence $a^{p^\beta} \in Z(G)$ and, from Theorem 1.2, $b^{p^\beta} \in Z(G)$ for all generators $b \in G$. But since G is basic, G is critical and $Z(G)$ is cyclic. Now let b be a generator independent of a ($b \notin a\Phi(G)$). Then since $|a| = |b|$, $1 \neq a^{p^\beta} = (b^n)^{p^\beta}$ with $(n, p) = 1$. It follows from (3) that $|a^{-1}b^n| = p^\beta < p^\alpha$ and since $a^{-1}b^n \notin \Phi(G)$ this contradicts Theorem 1.1. Hence G and G_2 have the same exponent.

II. Basic groups of class ≤ 3, $p > 3$

Theorem 2.1. *For each prime p (> 3), positive integer α, and $c = 1, 2, 3$, there is exactly one basic p-group of exponent p and class c. We denote such a group by $B_p(c, \alpha)$. They are defined as follows:*

$$B_p(1, \alpha) = \{a: a^{p^\alpha} = 1\},$$

$$B_p(2, \alpha) = \{a, b: a^{p^\alpha} = b^{p^\alpha} = (a, b)^{p^\alpha} = (a, b, b) = (a, b, a) = 1\},$$

$$B_p(3, \alpha) = \{a, b: a^{p^\alpha} = b^{p^\alpha} = (a, b)^{p^\alpha} = (a, b, a)^{p^\alpha}$$

$$= (a, b, b) = (a, b, a, a) = (a, b, a, b) = 1\}.$$

Proof. If $c = 1$, then clearly the basic p-group of exponent p^α is the cyclic group of order p^α.

If $c = 2$ and the group G is basic, then $e(G) = e(G_2) = p^\alpha$ by Theorem 1.3 [$e(G)$ is the exponent of G], and there is precisely one such critical p-group which must be a two-generator group whose relations are given by $B_p(2, \alpha)$. $B_p(2, \alpha)$ is a basic group because the variety generated by $B_p(2, \alpha)$ has as a basis for its identical relation $x^{p^\alpha} = 1$, $(u, v)^{p^\alpha} = 1$, and $(r, s, t) = 1$. But every proper subvariety must satisfy the relation $(u, v)^{p^{\alpha-1}} = 1$ and hence the variety generated by $B_p(2, \alpha)$ is join-irreducible and $B_p(2, \alpha)$ is basic.

If $c = 3$ and the group G is basic, then again $e(G) = e(G_2) = p^\alpha$. If $G_3 = (G_2, G)$ and $e(G_3) = p^\gamma$ with $\gamma < \alpha$, then the variety generated by G is the join of the proper subvarieties generated by groups $B_p(2, \alpha)$ and the unique critical group H satisfying $e(H) = e(H_2) = e(H_3) = p^\gamma$. The existence of the group H is a consequence of Lemma 5.3 of [2]. The fact that the variety generated by G is the join of the proper subvarieties mentioned follows by noting that the set of identical relations which characterize the variety of G is in fact the intersection of those which characterize $B_p(2, \alpha)$ and H. (See Lemma 5.3 and Theorem 5.4 of [2].) Thus if G is basic, it is the unique critical group given by the relations of $B_p(3, \alpha)$, again by Lemma 5.3 of [2]. $B_p(3, \alpha)$ is basic because every subvariety of the variety generated by $B_p(3, \alpha)$ satisfies the relation $(x, y, z)^{p^{\alpha-1}} = 1$.

Corollary 2.2. *Let G be a finite p-group ($p > 3$) of class ≤ 3, such that $e(G) = p^\alpha$, $e(G_2) = p^\beta$ and $e(G_3) = p^\gamma$. Then G is contained in the variety generated by $B_p(1, \alpha)$, $B_p(2, \beta)$, $B_p(3, \gamma)$.*

Proof. This is a direct consequence of Theorem 2.1 and Lemma 5.1 of [2].

The corollary above can be thought of as a "Basis Theorem" for finite p-groups, $p > 3$, of class ≤ 3. For we have completely specified a relatively small collection of groups, the $B_p(c, \alpha)$, and proved that every p-group $(p > 3)$ of class ≤ 3 can be constructed from at most three of these using the operations of forming finite direct products, taking subgroups and taking factor groups. Further we have shown that the $B_p(c, \alpha)$ are in this sense indecomposable.

References

[1] P. HALL, Verbal and marginal subgroups, *J. Reine Angew. Math.* **182** (1940), 130–141.

[2] P. M. WEICHSEL, On critical p-groups, *Proc. London Math. Soc.* (3) **14** (1964), 83–100.

Proc. Internat. Conf. Theory of Groups, Austral. Nat. Univ. Canberra,
August 1965, pp. 373–378. © Gordon and Breach Science Publishers, Inc. 1967

On a useful theorem for commutator calculation and the theory of associative rings

KENNETH W. WESTON

Introduction

There have been very useful connections made between associative rings and group theory. We present here a theorem which has not only proved quite useful in exploring commutator identities, but has also been used to provide results in the theory of associative rings, particularly for nil rings. The theorem is the following.

Theorem 1. *For every group G, there is a right module M over a ring R, a homomorphism $x \to x^*$ from G to R^+, and a mapping $x \to x'$ from G to M such that, for $x_{1,i} \in G_1 - G_2$,*

$$\prod_i (x_{1,i}, \ldots, x_{m,i})^{e_i} \equiv 1 \mod G_{m+1}$$

$$\Leftrightarrow \sum_i e_i x'_{1,i} \prod_{j=2}^m x^*_{j,i} = \sum_i x'_{1,i} e_i \prod_{j=2}^m x^*_{j,i} = 0.$$

It is still unknown whether M is unique, but it seems unlikely. Theorem 1 then gives the translation of group commutator identities into module notation. This in practice not only seems to simplify the commutator calculation, but occasionally clarifies the group identities as well. We shall give an example of a commutator identity which upon translation reduces to a binomial expansion in R.

Theorem 1 has another interesting feature. Commutator calculations can be imitated in M, so that occasionally only R is needed. Thus a group commutator theorem may result in a theorem in associative rings. The application of Theorem 1, sighted in this discussion, allowed for the copying of [1] in the notation of M, and the group commutator calculations[1] in [1] yielded the following theorems.

Theorem 2. *If R is an associative ring with a set of generators S such that*

(a) *S forms a group under addition;*
(b) *$s_1[s_2, s_3] = 0$ for $s_1, s_2, s_3 \in S$;*
(c) *$s^m = 0$ for $s \in S$;*

then

$$\prod_{i=1}^{m} i! R^{m+1} = 0.$$

Theorem 3. *If R is a commutative ring with a set of generators S such that*

(a) *S forms a group under addition,*
(b) *$s^m = 0$ for $s \in S$,*

then

$$\prod_{i=1}^{m} i! R^{m} = 0.$$

Proof of Theorem 1

We need the following notation. Suppose that G is a group. Let (x, y) for $x, y \in G$ designate the commutator $x^{-1}y^{-1}xy$, and let $(x, y, z) = ((x, y), z)$. Designate the lower central series of G by $G = G_1 \supseteq G_2 \supseteq \cdots$. If H and K are subgroups of G and $H \triangleleft K$, then let $[x]_H^K$ designate the coset in K/H which is represented by $x \in K$.

Let $M = \sum_{\alpha \in \theta} G_\alpha / G_{\alpha+1}$ where θ represents the collection of ordinals. For $y \in G$, define the mapping y^* of M into M by

$$\sum_{\alpha \in \theta} [x_\alpha]_{G_{\alpha+1}}^{G_\alpha} \xrightarrow{y^*} [1]_{G_2}^{G_1} + \sum_{\alpha \in \theta} [(x_\alpha, y)]_{G_{\alpha+1}}^{G_\alpha}.$$

That y^* is a well defined endomorphism of M follows directly from the identity

(1) $$(xz, y) = (x, y)(x, y, z)(z, y).$$

---

[1] The editors understand that [1] will appear in a substantially revised and extended form, so that these references may no longer be pertinent.

Let R be the endomorphism ring generated by elements y^* for $y \in G$. From the identity

$$(2) \qquad (x, zy) = (x, y)(x, z)(x, z, y)$$

we see that $y \to y^*$ is a homomorphism of G into the additive group R^+ of R.

We define the mapping $x \to x'$ of G into M by

$$x' = \begin{cases} \sum_{\beta \neq \alpha} [1]_{G_\beta+1}^{G_\beta} + [x]_{G_\alpha+1}^{G_\alpha} & \text{if } x \in G_\alpha - G_{\alpha+1}, \\ 0 & \text{otherwise.} \end{cases}$$

The proof of Theorem 1 follows directly from the equations

$$e_i x'_{1,i} \prod_{j=2}^{m} x^*_{j,i} = x'_{1,i} e_i \prod_{j=2}^{m} x^*_{j,i} = \sum_{\beta \neq m} [1]_{G_\beta+1}^{G_\beta} + [(x_{1,i}, \ldots, x_{m,i})^{e_i}]_{G_m+1}^{G_m}.$$

Properties of M

The following are direct consequences of Theorem 1.

P_1 $G_\alpha/G_{\alpha+1}$ is periodic for $i \leq \alpha < \omega$ with periods dividing q if and only if $qx'R^{i-1} = 0$ for all $x \in G_1 - G_2$.

P_2 $(x,z,w,y) \equiv (x,z,y,w) \bmod G_5 \Leftrightarrow x'z^*[w^*,y^*] = 0$ for $x \in G_1 - G_2$.

P_3 $(x, \underbrace{y, \ldots, y}_{m}) \equiv 1 \bmod G_{m+2} \Leftrightarrow x'(y^*)^m = 0$ for $x \in G_1 - G_2$.

P_4 $x'y^* = -y'x^*$ for $x, y \in G_1 - G_2$.

P_5 $x'y^*z^* + y'z^*x^* + z'x^*y^* = 0$ for $x, y, z \in G_1 - G_2$.
        (This is a restatement of P. Hall's relation $(x, y, z)(y, z, x)(z, y, x) \equiv 1 \bmod G_4$.)

P_6 $x'x^* = 0$.

Application of Theorem 1

We wish to prove the following statement:

Theorem 4. *If G is a metabelian group satisfying the mth Engel congruence, then $G_\alpha/G_{\alpha+1}$ is periodic for $m+2 \leq \alpha < \omega$ where the periods divide $k = (m-1)! \prod_{i=1}^{m-1} i!$ and G_{m+1}/G_{m+2} has exponent dividing $(m+1)k$.*

This was first proved by N. D. Gupta and M. F. Newman in [1], with "congruence" replaced by "identity" and "$G_\alpha/G_{\alpha+1}$," "G_{m+1}/G_{m+2}"

replaced by "G_α" and "G_{m+1}", respectively. We will use the calculations in [1], but carry them out completely inside M. If a comparison is made between the calculations in [1] and here, their simplification using the notation of M will become apparent. Not only are the calculations simplified, but all calculations in M, without the use of P_4–P_6, yield a result in the theory of associative rings.

The proof of Theorem 4 results from the following list of identities. If I_k is the kth identity here and (j) is the jth identity[2] in [1], we shall write $I_k(j)$ to indicate the correspondence between the two.

$I_1(8)$ *If R is an associative ring and $a[a, b] = b[a, b] = 0$ for $a, b \in R$ then, for $j \geq 1$,*

$$(a+b)^j = a^j + ba^{j-1} + (j-1)a^{j-1}b + \sum_{\alpha,\beta} \gamma_{\alpha,\beta} a^\alpha b^\beta + \sum_{\alpha',\beta'} \gamma'_{\alpha',\beta'} b^{\beta'} a^{\alpha'}$$

where $\alpha + \beta = \alpha' + \beta' = j$ and $\beta, \beta' \geq 2$. (This is a binomial expansion in R.)

$I_2(9)$ *If R is an associative ring with generators S where*

   (a) *S forms a group under addition,*
   (b) *$s_1[s_2, s_3] = 0$ for $s_1, s_2, s_3 \in S$,*
   (c) *$s^{n+1} = 0$ for $s \in S$,*

then, for any $s_1, s_2 \in S$,

$$f(i+1)s_1^i s_2^{n-i+1} + if(i)s_2^{n-i+1}s_1^i = d_i$$

where $d_i = \sum_{\alpha,\beta} \gamma_{\alpha,\beta} s_1^\beta s_2^\alpha + \sum_{\alpha',\beta'} s_2^{\alpha'} s_1^{\beta'}$ with $\alpha + \beta = \alpha' + \beta' = n+1$, $\beta, \beta' \geq i+1$, and $f(i) = n!/(n-i+1)!$, $0 \leq i \leq n+1$.

The proof of $I_2(9)$ follows from an induction on i, using $I_1(8)$ and, for $i = 0$, $d_0 = s_1^{n+1} + (s_1 + s_2)^{n+1} = 0$.

$I_3(10)$ *If R is an associative ring with generators S and S satisfies (a), (b), and (c), then, for any $s_1, s_2 \in S$,*

$$n!(s_1^n s_2 + n s_2 s_1^n) = 0,$$

(This follows from $I_2(9)$ by setting $i = n$.)

Proof of Theorem 3. $I_3(10)$ yields directly that $m! s_1 s_2^{m-1} = 0$ for $s_1, s_2 \in S$ and Theorem 3 follows by induction on m.

Proof of Theorem 2.   By $I_3(10)$ we have $m! s_1 s_2 s_3^{m-1} = 0$ for $s_1, s_2, s_3 \in S$. But since S satisfies $s_1[s_2, s_3] = 0$ for $s_1, s_2, s_3 \in S$, then $r_1[r_2, r_3] = 0$ for $r_1, r_2, r_3 \in R$ and thus R can be regarded as a commutative ring over $m! R^2$. The proof then follows by Theorem 3.

---

[2] See footnote 1.

I_4 *If G is a metabelian group satisfying the mth Engel congruence,*
then

$$x' \prod_{i=1}^{m} i! R^{m+1} = 0 \quad \textit{for all } x \in G_1 - G_2.$$

(The proof of I_4 follows from Theorem 2 by P_2 and P_3.)
I_5 *If G is a metabelian group satisfying the mth Engel congruence,*
then

$$y'(m-1)!w^*z^*(x^*)^{m-1} = 0 \quad \text{for all } x, y, z, w \in G_1 - G_2.$$

By P_2, P_3, and $I_3(10)$ we have

$$(m-1)!z'y^*(x^*)^{m-1} + (m-1)!(m-1)z'(x^*)^{m-1}y^* = 0.$$

Hence by P_4

(3) $$-(m-1)!y'z^*(x^*)^{m-1} + (m-1)!(m-1)z'(x^*)^{m-1}y^* = 0.$$

But, by Theorem 1, (3) is equivalent to

(3') $$(y, z, _{m-1}x)^{-(m-1)!}(z, _{m-1}x, y)^{(m-1)!(m-1)} \equiv 1 \bmod G_{m+2}.$$

If we replace y by (y, w) in (3'), we have

$$(y, w, z, _{m-1}x)^{-(m-1)!} \equiv 1 \bmod G_{m+3},$$

and thus, by Theorem 1,

$$-(m-1)!y'w^*z^*(x^*)^{m-1} = 0.$$

(Notice that only in this argument do we need to refer to the group. This is because there is no equivalent notion in M for metabelian, i.e., M equivocates $G'' = 1$ to (x, y, z, w) (x, y, w, z) mod G_5.)

From I_5 we have $x'(m-1)!R^2y^{m-1} = 0$ for all $x \in G_1 - G_2$. But R has commutative action on $x'(m-1)!R^2$ and thus by Theorem 3

$$x'(m-1)! \prod_{i=1}^{m-1} i! R^{m+1} = 0 \quad \text{for all } x \in G_1 - G_2.$$

Hence by P_1 we have that $G_\alpha/G_{\alpha+1}$ is periodic for $m+2 \le \alpha < \omega$ and the periods divide $(m-1)! \prod_{i=1}^{m-1} i!$.

In order to complete the proof of Theorem 4, we need the following identities.

$I_6(14)$ *If G is metabelian, then*

$$x'(y^*)^{m-1}z^* - z'(y^*)^{m-1}x^* + z'x^*(y^*)^{m-1} = 0 \quad for \ x, y, z \in G_1 - G_2.$$

The proof of $I_6(14)$ follows from $(z'x^*y^* + y'z^*x^* + x'y^*z^*)(y^*)^{m-2} = 0$ by P_5, P_2, and P_4.

$I_7(15)$ *If G is metabelian and satisfies the mth Engel congruence, then* $(m+1)(m-1)!z'x^*(y^*)^{m-1} = 0$ *for* $x, y, z \in G_1 - G_2$.

By $I_6(14)$ we have

$$(m-1)(m-1)!(x'(y^*)^{m-1}z^* - z'(y^*)^{m-1}x^* + z'x^*(y^*)^{m-1}) = 0.$$

Therefore by $I_3(10)$, P_4, and $I_7(15)$ we have $x'(m+1)(m-1)!Ry^{m-1} = 0$ for all $x \in G_1 - G_2$. Thus by Theorem 3

$$x'(m+1)(m-1)! \prod_{i=1}^{m-1} i!R^m = 0$$

and Theorem 4 is proved.

Corollary. *If G is a metabelian group with prime exponent p, then* $G_{p+1} = G_{p+2}$ (Meier–Wunderli).

Since G has prime exponent, G satisfies the $(p-1)$th Engel congruence. Thus, by Theorem 4, G_{p+1}/G_{p+2} has exponent dividing $k = (p-2)! \prod_{i=1}^{p-2} i!$. But obviously k and p are relatively prime.

As is evident from the numbering, some of the commutator identities used in [1] have been eliminated. Also, some of the commutator identities have been made clearer. For example, identity (8) in [1] is

$$(a, {}_jbc) = (a, {}_jc)[(a, {}_{j-1}c, b)^{j-1}(a, b, {}_{j-1}c)]^c d^{(j)}$$

where $d^{(j)}$ is a product of commutators of the form $(a, a_1, \ldots, a_t)$ where $t \geq j$ and each element of $\{a_1, \ldots, a_t\}$ belongs to $\{b, c\}$, also at least two elements from $\{a_1, \ldots, a_t\}$ belong to $\{b\}$. Identity (8) translates into $I_1(8)$, where $I_1(8)$ is the expansion of the binomial $(a+b)^j$ in an associative ring satisfying $r_1[r_2, r_3] = 0$.

We do not wish to imply that every commutator argument can be copied to prove the same result. This certainly is not possible. But we do state that sometimes commutator arguments lend themselves well to module notation and that a corresponding theorem is provable.

References

[1] N. D. Gupta and M. F. Newman, On metabelian groups, *J. Austral. Math. Soc.* [to appear].

Proc. Internat. Conf. Theory of Groups, Austral. Nat. Univ. Canberra,
August 1965, pp. 379–388. © Gordon and Breach Science Publishers, Inc. 1967

On the structure of composite groups

HELMUT WIELANDT

This lecture is concerned with the subgroup problem: to obtain a survey
of all subgroups of a given group. Unfortunately, there is no hope for a
general solution; the investigation has to be restricted to important
subgroups. To find a reasonable interpretation of this term one may
take a look at that part of group theory which is developed best, i.e.,
the theory of finite groups, and see what kinds of subgroups appear in
the basic theorems. No doubt they are the theorems of Jordan–Hölder
and Sylow. The subgroups of a group G which occur in the Jordan–
Hölder theorem are those which appear in the various subnormal
series of G; and Sylow's theorem refers, broadly speaking, to subgroups
of G which are maximal in G under certain conditions on their structure.
This suggests to investigate two types of important subgroups which
may be termed *subnormal subgroups* and *relatively maximal subgroups*.
The first type has been investigated intensively; a sketch of the development and present state of this theory will form the main part of this
lecture. Little is known about the second type. In fact it is not at all
clear how to define it adequately for general groups, or even for finite
groups. For the latter case, a suggestion will be discussed briefly in the
second part of this lecture.

1. Subnormal subgroups

A subgroup A of G is called subnormal if it occurs in some subnormal
series of G, that is, $A = G_k$ in a finite chain

$$(1.1) \qquad G = G_0 \rhd G_1 \rhd \cdots \rhd G_k \rhd \cdots \rhd G_n = 1$$

where each term is a normal subgroup of the preceding one. Although
Jordan's theorem on composition series is nearly a hundred years old
the theory of subnormal subgroups was initiated only 30 years ago.

Robert Remak mentioned in a seminar that in an attempt to dualize his theory of subdirect decompositions of finite groups he had run into the question whether the group generated by two subnormal subgroups is subnormal itself. I found an affirmative answer and later developed the theory for finite groups. In infinite groups additional difficulties arise. Only recently significant results have been obtained, in particular by D. S. Robinson and J. E. Roseblade. In order to appraise what has been accomplished we begin by recalling the main results in the finite case [16, 17, 18].

(1.2) *Let G be a group of finite order, and let A, B be subnormal subgroups of G. Then*
  (a) *the intersection $A \cap B$ is subnormal in G;*
  (b) *the join $\langle A, B \rangle$ is subnormal in G;*
  (c) *each composition factor group of $\langle A, B \rangle$ is isomorphic to a composition factor group of A or of B;*
  (d) *if the commutator indices $|A : A'|$ and $|B : B'|$ are relatively prime, then A and B commute as wholes: $AB = BA$;*
  (e) *if $A \cap B = 1$, and if A, B have no prime composition factor in common, then A and B commute elementwise;*
  (f) *if $G \neq 1$ then $\bigcap N_G(S) \neq 1$ where S runs over all subnormal subgroups of G and $N_G(S)$ denotes the normalizer of S in G;*
  (g) *for any subgroup X of G, define the subnormal closure $\bar{X} = \bigcap S$ where S runs over those subnormal subgroups of G which contain X. Then $XY = YX$ implies $\bar{X}\bar{Y} = \bar{Y}\bar{X}$.*

Before we turn to generalizations to infinite groups a few comments may be in order. Statements (a) and (b) mean that the set $S(G)$ of all subnormal subgroups of G is a sublattice of the lattice $L(G)$ of all subgroups. $S(G)$ is more easily accessible than $L(G)$. It is remarkable that in most finite groups where interesting results on the lattice of all subgroups have been obtained (see Suzuki's report [14]) the two lattices coincide. (This condition characterizes the nilpotent groups.) The lattice $S(G)$ has been investigated mainly by Tamaschke [15] and Zappa [21]. $S(G)$ is modular if and only if any two subnormal subgroups of G commute. This fact may motivate search for conditions which imply that two given subnormal subgroups A and B of G commute. Apparently this occurs often. For instance (1.2d) shows that the condition $A = A'$ is sufficient, with no condition on B.

In many ways, subnormal subgroups are close to being normal. For instance, (1.2f) states that the "common normalizer" of all subnormal subgroups of G is "large": it is never trivial unless G is trivial.

Finally, (1.2g) shows that subnormal subgroups may be used to investigate arbitrary subgroups.

How far do these statements remain valid for infinite groups? It is easily seen that (a) remains true; that is, $S(G)$ is closed with respect to intersections. The corresponding question for joins is more difficult. In fact, it remained open for twenty years till Zassenhaus [22] gave an example to show that $S(G)$ is not, in general, closed with respect to joins. This gave rise to the *join problem*: When is the join of any two (hence, of finitely many) subnormal subgroups of G subnormal? To formulate recent answers, let $\mathfrak{S}$ resp. $\mathfrak{S}^\infty$ ($\subseteq \mathfrak{S}$) denote the class of groups G in which all joins of finitely many (resp. arbitrarily many) subnormal subgroups are subnormal. It was recognized early that for investigation of $\mathfrak{S}$ and $\mathfrak{S}^\infty$ chain conditions are relevant. Let $\mathfrak{M}_{\hat{S}}$ resp. $\mathfrak{M}_{\hat{L}}$ denote the class of all groups G such that $S(G)$ resp. $L(G)$ satisfies the maximum condition, define similarly $\mathfrak{M}_S^\vee$ by the minimum condition on $S(G)$, and denote by $\mathfrak{M}_S$ the class of those groups G which possess a subnormal series (1.1) in which each quotient group G_{k-1}/G_k belongs to either $\mathfrak{M}_{\hat{S}}$ or $\mathfrak{M}_S^\vee$.

The following list of results indicates major steps in the application of chain conditions:

(1.3) $\mathfrak{M}_{\hat{S}} \subseteq \mathfrak{S}^\infty$ (Wielandt [16]).

(1.4) $\mathfrak{M}_S^\vee \subseteq \mathfrak{S}$ (Roseblade [10]).

(1.5) $\mathfrak{M}_S \subseteq \mathfrak{S}$; *more precisely,* $\mathfrak{M}_S\mathfrak{S} = \mathfrak{S}$ *and* $\mathfrak{M}_S\mathfrak{S}^\infty = \mathfrak{S}^\infty$.
Also $\mathfrak{S}\mathfrak{M}_{\hat{L}} = \mathfrak{S}$ *and* $\mathfrak{S}^\infty\mathfrak{M}_S = \mathfrak{S}^\infty$ (Robinson [8, 9]). (After P. Hall, we use $\mathfrak{X}\mathfrak{Y}$ to denote the class of all groups which possess a normal $\mathfrak{X}$-subgroup with quotient group in $\mathfrak{Y}$.)

Of course, chain conditions cannot be the only relevant restrictions in the join problem. For instance we have obviously $\mathfrak{T} \subseteq \mathfrak{S}^\infty$ where $\mathfrak{T}$ denotes the class of those groups G in which normality is transitive (that is, in which each subnormal subgroup is normal). The class $\mathfrak{T}$ has been used to build up a large subclass of $\mathfrak{S}$:

(1.6) $$\mathfrak{S}^\infty\mathfrak{T} \subseteq \mathfrak{S} \quad \text{(Robinson [9])}.$$

Another version of the join problem is to ask for conditions on two *given* subnormal subgroups A, B of G which imply that $\langle A, B \rangle$ is subnormal in G. It has been known for a long time that the condition $AB = BA$ is sufficient; a proof can be found in Robinson [8]. I would like to point out that a weaker permutability condition is sufficient:

(1.7)   *If* $A \in S(G)$, $B \in S(G)$, *and* $\langle A, B \rangle = ABA$ *then* $\langle A, B \rangle \in S(G)$.

One thing about this theorem is annoying: the assumption is unsymmetric although the assertion is symmetric. It might be worth investigating whether already the existence of a natural number n such that $\langle A, B \rangle = (AB)^n$ is sufficient for $\langle A, B \rangle \in S(G)$. This would be a satisfactory symmetric, weak permutability condition.

Instead of assumptions on the behavior of A and B in their join, conditions on the interior structure of A and B can be used. (This is somewhat surprising since the statement $\langle A, B \rangle \in S(G)$ concerns the way the join is embedded in G.) Results of this type have turned up in attempts to generalize theorem (1.2c) which deals with the structure of $\langle A, B \rangle$. Starting from an idea of Zassenhaus [22], Roseblade [10] introduced the following concepts: A class $\mathfrak{X}$ of abstract groups is called a *subnormal coalition class* if, for any group G, the join of any two subnormal $\mathfrak{X}$-subgroups of G is again a subnormal $\mathfrak{X}$-subgroup of G. Normal coalition classes are defined correspondingly. For example, the class of all nilpotent groups is known to be a normal coalition class.

These concepts lead to wide generalizations of (1.2c) which at the same time contribute to the join problem:

(1.8)   *If* $\mathfrak{X}$ *is a normal coalition class contained in either* $\mathfrak{M}_S^{\wedge}$ *or* $\mathfrak{M}_S^{\vee}$ *then* $\mathfrak{X}$ *is a subnormal coalition class.* (Roseblade [10, 12]).

Let us illustrate this result by two examples.

In any group G two subnormal subgroups which are nilpotent and finitely generated have a subnormal join with the same properties. This corollary of (1.8) is equivalent to a well known result of Baer on the nil radical ([1], p. 419).

In any group G two subnormal subgroups which possess composition series generate a subnormal subgroup which has a composition series and does not have any new composition factor groups. In this sense, (1.2c) remains true for infinite groups G.

As for (1.2d), it is not obvious what a generalization to arbitrary groups might look like since the condition on commutator indices makes sense only if they are finite. A wide generalization has been given by Roseblade [11]:

(1.9)   Let A and B be subnormal in G. Consider the commutator quotient groups A/A' and B/B' as modules over the ring of rational integers. Assume their tensor product is trivial: $A/A' \otimes B/B' = 0$. Then $AB = BA$ (and hence AB is subnormal in G).

For instance, if a subnormal subgroup A of G is perfect $(A = A')$ then it commutes with every subnormal subgroup of G. Refinements of (1.9) might be sought along the lines of Hainzl's investigation [2] for finite groups.

With respect to (1.2e), I would like to point out that finiteness conditions can be discarded completely:

(1.10)   Let A and B be subnormal in G; let $A \cap B = 1$. Then either A and B commute elementwise, or there are a prime number p and subnormal subgroups $A_1 \lhd A_2$ of A and $B_1 \lhd B_2$ of B such that $|A_2 : A_1| = |B_2 : B_1| = p$.

A generalization of (1.2f) has been given by Roseblade [10]:

(1.11)   If the set S of all subnormal subgroups of G satisfies the minimal condition then $N_G(S)$, the common normalizer of all subnormal subgroups of G, has finite index in G.

As for (1.2g), no generalization to infinite groups seems to be known.

These samples of results may suffice to give an idea of what is going on with subnormal subgroups. I would like to add a few words about more general concepts. Various generalizations of subnormal subgroups have been investigated recently, in particular descendant and ascendant subgroups of G (e.g., Hall [5], Kegel [7], Robinson [8], Specht [13]). They are defined as the members of well-ordered subnormal series which are either descending from G, or ascending to G. In general neither of these two families of subgroups is closed with respect to joins. A drastical counterexample has been given by Robinson [8]: There exists a group G (solvable with derived length 3 and exponent 8) which contains two subnormal subgroups whose join is neither descendant nor ascendant in G. Sufficient conditions in order that the ascendant

26—K.

subgroups of a group G form a lattice have been given by Kegel [7]. A generalization of descendant subgroups has been investigated by Heineken [6].

I would like to conclude the first part of this lecture by pointing out three problems:

(1.12) *Let $\mathfrak{A}$ be a given set of subnormal subgroups of G, and let $\mathfrak{L}$ be the smallest sublattice of the lattice of all subgroups of G which contains $\mathfrak{A}$. What restrictions on the structure of the subgroups in $\mathfrak{L}$ can be derived from knowledge on the structure of the groups in $\mathfrak{A}$?*

(1.13) *Does every group $G \neq 1$ possess a finite subnormal series in which some quotient group is simple?*

(1.14) *Investigate arbitrary subgroups by means of their subnormal closures and their subnormal cores.*

2. Relatively maximal subgroups

The problem we are concerned with now is how to generalize the concept of a Sylow p-subgroup of G. For finite solvable groups G, this has been achieved by P. Hall [3]. His basic result may be stated as follows:

(2.1) *Let ω be a set of prime numbers. Let G be a finite solvable group, and let $L_\omega(G)$ denote the set of all ω-subgroups of G (whose orders contain at most prime factors in ω), ordered by inclusion. Then any two maximal elements of $L_\omega(G)$ are conjugate subgroups of G.*

By this theorem, knowledge of a maximal element M of $L_\omega(G)$ suffices to give a complete description of all ω-subgroups of G: they are just all the subgroups of all the conjugates of M.

Combining the theorems of Hall and Sylow, one may characterize the maximal ω-subgroups of a solvable finite group G of order g as those subgroups of G whose order is the maximal ω-divisor of g. The *Hall ω-subgroups* of G defined by this latter property have been studied also for nonsolvable groups G. However, they cannot help much in the general subgroup problem even in finite groups, because Hall [4] has shown that for any finite nonsolvable group G there is a set ω of primes such that G does not contain a Hall ω-subgroup. So one has to look for a different generalization of Sylow p-groups. The obvious suggestion is to consider the maximal elements of $L_\omega(G)$ which at least have the advantage of existing for any choice of G. Surprisingly, they have not appeared at all in the vast literature on finite groups, the reason probably being that they behave badly with respect to homomorphisms, and hence are difficult to handle. Nevertheless they surely occupy a key position in structure theory; and there is a way to attack them. Although my results are restricted essentially to finite groups I would like to sketch the method for arbitrary groups.

The basic idea is to study a given subgroup A of a group G with given subnormal series (1.1) by looking at certain subgroups of the quotient groups $G^{(k)} = G_{k-1}/G_k$. For each k, we define the *projection of A into $G^{(k)}$* to be

$$(2.2) \qquad A \dashv G^{(k)} = (A \cap G_{k-1})G_k/G_k \qquad (k = 1, 2, \ldots, n).$$

The elements of $A \dashv G^{(k)}$ are those cosets of G_k in G_{k-1} which contain elements of A. Given the subnormal series G_k, each subgroup of G determines a set of projections, but not vice versa. There is an existence problem I would not like to talk about today, and there is a uniqueness problem: just how far is a group A determined by its n projections? A partial answer can be derived from the obvious isomorphism

$$(2.3) \qquad A \dashv G^{(k)} \cong (A \cap G_{k-1})/(A \cap G_k).$$

This shows that the projections of A certainly determine the order $|A|$ of A:

$$(2.4) \qquad |A| = \prod_k |A \dashv G^{(k)}|.$$

So two groups with coinciding projections have the same order, and one might conjecture that they even have to be isomorphic. But this is not true; the abelian group G of order 8 and exponent 4 is a counterexample.

Equation (2.4) shows that a finite subgroup A of G is an ω-group if and only if each projection $A \dashv G^{(k)}$ is an ω-group. So projections are an appropriate tool for investigating $L_\omega(G)$. In fact, the method carries farther. Let $\mathfrak{X}$ be a class of groups which is closed with respect to normal subgroups, quotient groups, and extensions, so that a group H with a normal subgroup N is in $\mathfrak{X}$ if and only if $N \in \mathfrak{X}$ and $H/N \in \mathfrak{X}$. For each group G, let $\mathfrak{X}(G)$ denote the set of all $\mathfrak{X}$-subgroups of G. Because of the isomorphism (2.3) we have

$$(2.5) \quad A \in \mathfrak{X}(G) \quad \textit{if and only if} \quad A \rightharpoondown G^{(k)} \in \mathfrak{X}(G^{(k)}) \quad (k = 1, \ldots, n).$$

Now let $\mathscr{M}\mathfrak{X}(G)$ denote the set of maximal elements of $\mathfrak{X}(G)$ with respect to inclusion. Then

$$(2.6) \quad A \rightharpoondown G^{(k)} \in \mathscr{M}\mathfrak{X}(G^{(k)}) \quad (k = 1, \ldots, n) \quad \textit{implies } A \in \mathscr{M}\mathfrak{X}(G).$$

Although the converse is not true, (2.6) justifies hope that projections are useful for investigating maximal $\mathfrak{X}$-subgroups. We would like to decide whether two maximal $\mathfrak{X}$-subgroups A, B of G with given projections are conjugate. Now it is obviously impossible to do that unless any two maximal $\mathfrak{X}$-subgroups with identical projections are conjugate. So one has to start by investigating this restricted question, and that is what I have done so far. For arbitrary $\mathfrak{X}$, the answer is no. But one can give conditions on $\mathfrak{X}$ which are, in the finite case, essentially necessary and sufficient:

(2.7)   *Assume* (i) $\mathfrak{X}$ *is a class of finite groups which is closed with respect to subgroups, quotient groups, and extensions.*

(ii) *G is a finite group and* (1.1) *is a subnormal series of G.*

(iii) *A and B are maximal $\mathfrak{X}$-subgroups of G.*

(iv) *$A \rightharpoondown G^{(k)} = B \rightharpoondown G^{(k)}$ $(k = 1, \ldots, n)$.*

(v) *Either A is solvable, or the outer automorphism groups of all simple epimorphic images of subgroups of G are solvable.*

Then A and B are conjugate in their join.

Note that (i) requires $\mathfrak{X}$ to be closed with respect to arbitrary subgroups, not just normal ones. This condition might seem inadequately strong. However it is crucial for the validity of the theorem: If a class $\mathfrak{X}$ violates this condition then there is a triple G, A, B such that all other conditions are satisfied but A and B are not conjugate.

(v) probably is no restriction at all (Schreier's conjecture).

The proof of (2.7) is based on the fact that for any normal Hall ω-subgroup of a finite group there is a complement, and that any two complements are conjugate (hence the proof relies, at present, on the Feit–Thompson theorem). Theorem (2.7) may be expected to carry over to classes of infinite groups where a corresponding splitting and conjugacy theorem is available.

To conclude, I would like to mention a refinement of (2.7):

(2.7*) *In theorem (2.7) the conclusion is valid if assumption (iv) is replaced by*

(iv*) $A \dashv G^{(k)} = B \dashv G^{(k)}$ *whenever* $G^{(k)}$ *is not solvable.*

This theorem includes Hall's basic theorem (2.1) on solvable groups. Indeed, if $\mathfrak{X}$ is the class of all ω-groups, A, B are maximal ω-subgroups of a solvable group G and if $\{G_k\}$ denotes any subnormal series of G then the conditions of (2.7*) are trivially satisfied, hence A and B are conjugate.

It may be hoped that Theorem (2.7) opens a way for extending the rich theory of relations between subnormal subgroups and Hall ω-subgroups [19, 20] to $\mathfrak{X}$-subgroups.

References

[1] REINHOLD BAER, Nilgruppen, *Math. Z.* **62** (1955), 402–437.

[2] JOSEF HAINZL, Über Seminormalität in Gruppen und Verbänden, *Math. Z.* **80** (1963), 358–362.

[3] P. HALL, A note on soluble groups, *J. London Math. Soc.* **3** (1928), 98–105.

[4] P. HALL, A characteristic property of soluble groups, *J. London Math. Soc.* **12** (1937), 198–200.

[5] P. HALL, On non-strictly simple groups, *Proc. Cambridge Philos. Soc.* **59** (1963), 531–553.

[6] HERMANN HEINEKEN, Eine Verallgemeinerung des Subnormalteilerbegriffs, *Arch. Math.* **11** (1960), 244–252.

[7] O. H. KEGEL, Über den Normalisator von subnormalen und erreichbaren Untergruppen, *Math. Ann.* [to appear].

[8] DEREK J. S. ROBINSON, Joins of subnormal subgroups, *Illinois J. Math.* **9** (1965), 144–168.

[9] DEREK J. S. ROBINSON, On the theory of subnormal subgroups, *Math. Z.* **89** (1965), 30–51.

[10] J. E. ROSEBLADE, On certain subnormal coalition classes, *J. Algebra* **1** (1964), 132–138.

[11] JAMES E. ROSEBLADE, The permutability of orthogonal subnormal subgroups, *Math. Z.* **90** (1965), 365–372.

[12] JAMES E. ROSEBLADE, A note on subnormal coalition classes, *Math. Z.* **90** (1965), 373–375.

[13] WILHELM SPECHT, *Gruppentheorie*, Springer, Berlin–Göttingen–Heidelberg, 1956.

[14] MICHIO SUZUKI, *Structure of a group and the structure of its lattice of subgroups*, Ergebnisse der Math. (2) **10**, Springer, Berlin–Göttingen–Heidelberg, 1956.

[15] OLAF TAMASCHKE, Gruppen mit reduziblem Subnormalteilerverband, *Math. Z.* **75** (1961), 211–214.

[16] HELMUT WIELANDT, Eine Verallgemeinerung der invarianten Untergruppen, *Math. Z.* **45** (1939), 209–244.

[17] HELMUT WIELANDT, Vertauschbare nachinvariante Untergruppen, *Abh. Math. Sem. Univ. Hamburg* **21** (1957), 55–62.

[18] HELMUT WIELANDT, Über den Normalisator der subnormalen Untergruppen, *Math. Z.* **69** (1958), 463–465.

[19] H. WIELANDT, Entwicklungslinien in der Strukturtheorie der endlichen Gruppen, *Proc. Internat. Congress Math.* 1958, pp. 268–278; Cambridge Univ. Press, New York, 1960.

[20] H. WIELANDT, Arithmetische Struktur und Normalstruktur endlicher Gruppen, *Conv. Internaz. di Teoria dei Gruppi Finiti (Firenze*, 1960), pp. 56–65; Edizioni Cremonese, Rome, 1960.

[21] GUIDO ZAPPA, Sui gruppi finiti per cui il reticolo dei sottogruppi di composizione è modulare, *Boll. Un. Mat. Ital.* (3) **11** (1956), 315–318.

[22] HANS ZASSENHAUS, *The theory of groups*, 2nd ed., Chelsea, New York, 1958.

Proc. Internat. Conf. Theory of Groups, Austral. Nat. Univ. Canberra,
August 1965, pp. 389–393. © Gordon and Breach Science Publishers, Inc. 1967

On automorphisms of doubly transitive permutation groups

HELMUT WIELANDT

Let G be a permutation group on a set Ω; that is, let G be a subgroup
of the symmetric group S of all permutations of Ω. Then every element
of the normalizer of G in S will induce an automorphism of G. We
denote the group of all automorphisms of G which arise in this manner
by $P(G)$. Usually these "permutation automorphisms" of G are easier
to handle than other automorphisms of G, hence it is a reasonable
question to ask for conditions that a given automorphism α of the
permutation group G be contained in $P(G)$.

If G is transitive on Ω, then there is a simple and probably well
known answer to this question. To formulate it, we pick a point o in
Ω and form its stabilizer G_o in G. Then we have, for finite Ω:

Theorem 1. $\alpha \in P(G)$ if and only if G_o^α has a fixed point.

It is the purpose of this note to point out that weaker conditions
may suffice if G is doubly transitive. We shall prove:

Theorem 2. Let G be a doubly transitive permutation group of finite
degree n. Let α be an automorphism of G such that α has odd order r and
that G_o^α has an orbit whose length k is relatively prime to n. Then $\alpha \in P(G)$,
and $k = 1$ or $k = n - 1$.

Remark. The condition that α be of odd order cannot be omitted.
The simple group G of degree 7 and order 168 has an automorphism
$\alpha \notin P(G)$ of order 2 such that the orbit lengths of G_o^α are 3 and 4, hence
relatively prime to 7.

Proof of Theorem 2. We may assume $n > 2$. We denote by $M(g)$
the $n \times n$ permutation matrix associated with the permutation $g \in G$.

and by $M'(g)$ the $n \times n$ permutation matrix belonging to that permutation representation of G which is induced by G_o^α. A simple argument due to Ito (*Acta Sci. Math. Szeged* vol. 21, 1960) which is based on Frobenius' reciprocity theorem shows that G_o^α has precisely two transitive constituents, and that the representations M and M' of G are equivalent. The latter fact implies that the representation $g \rightarrow M(g^\alpha)$ is equivalent to g, hence there is a nonsingular $n \times n$ matrix K such that

(1) $$M(g)K = KM(g^\alpha) \quad \text{for all } g \in G.$$

This equation means that certain of the entries $k_{\alpha\beta}$ of K have to have the same value, hence there is a matrix $K \neq 0$ satisfying (1) with the additional property $k_{\alpha\beta} = 1$ or 0, and since the "automorphism group" of this incidence matrix K acts transitively on the rows as well as on the columns, every row and every column of K will contain the same number, say l, of 1's. Specializing equation (1) to the elements $g \in G_o$ we find that G_o^α maps a set of l points of Ω onto itself, hence $l = k$ or $l = n - k$. We may assume, after a change of notation, that $l = k$ and $k \le n/2$, and, since $(k, n) = 1$ and $n > 2$, even $k < n/2$; we have still $(n, k) = 1$. So we have

(2) $$1 \le k < \frac{n}{2}, \qquad (k, n) = 1$$

and

(3) $$Ke = K^*e = ke \quad \text{for } e = \begin{pmatrix} 1 \\ 1 \\ \vdots \\ 1 \end{pmatrix}$$

where K^* denotes the complex conjugate transposed matrix of K.

From (1) we may deduce two $n \times n$ matrixes which commute with every $M(g)$, $g \in G$. Firstly, application of the operation $*$ to (1) yields

$$K^*M(g^{-1}) = M(g^{-\alpha})K^* \quad \text{for all } g \in G,$$

hence

(4) $$M(g)KK^* = KK^*M(g) \quad \text{for all } g \in G.$$

Secondly, replacing g by $g^\alpha, g^{\alpha^2}, \ldots, g^{\alpha^{r-1}}$ where r denotes the odd order of α we find from (1)

$$M(g^\alpha)K = KMg(\alpha^2), \ldots, M(g^{\alpha^{r-1}})K = KM(g)$$

which leads to

(5) $$M(g)K^r = K^r M(g) \quad \text{for all } g \in G.$$

Since every matrix which commutes with the doubly transitive group $M(G)$ elementwise is a linear combination of the $n \times n$ identity matrix I and the $n \times n$ matrix J consisting entirely of 1's, there exist $a, b, c, d \in \mathbf{Z}$ such that

(6) $$KK^* = aI + bJ,$$

(7) $$K^r = cI + dJ.$$

From this we want to deduce conditions on a, b, c, d by investigating the eigenvalues of K. Multiplication by the eigenvector e of K yields by (3)

(8) $$k^2 = a + bn,$$

(9) $$k^r = c + dn.$$

Furthermore (6) implies that K is a normal matrix, that is, $KK^* = K^*K$. Hence K possesses an eigenvector f which is orthogonal to e. If we denote the corresponding eigenvalue by κ then $Kf = \kappa f$ and (because of the normality of K) $K^*f = \bar{\kappa}f$. Multiplication of (6) and (7) by f yields

(10) $$\kappa\bar{\kappa} = a,$$

(11) $$\kappa^r = c.$$

Now (10) and (11) imply $|\kappa| \in \mathbf{Q}$ since r is odd; but as κ is an eigenvalue of the matrix K which has coefficients in $\mathbf{Z}$, κ is an algebraic integer. Hence $|\kappa| \in \mathbf{Z}$. The rational integer $|\kappa|$ satisfies, by (8) and (9), the congruences

(12) $$k^2 \equiv |\kappa|^2 \mod n,$$

(13) $$k^r \equiv \pm |\kappa|^r \mod n.$$

Since k is relatively prime to the modulus n we find

(14) $$k \equiv \pm |\kappa| \mod n.$$

Furthermore, since (6) implies $b \geq 0$, we find from (8) and (10)

$$(15) \qquad |\kappa|^2 \ = \ a \ = \ k^2 - bn \ \leq \ k^2,$$

hence $|\kappa| \leq k$; and from (2) we find

$$0 \ \leq \ |\kappa| \ \leq \ k \ < \ \frac{n}{2}.$$

This together with (14) obviously implies $k = |\kappa|$, hence by (15) $b = 0$. Now (6) reduces to $KK^* = aI$, which implies that any two distinct rows of K are orthogonal. Since the elements of K are nonnegative and each column of K contains precisely k 1's, we have $k = 1$. This again means that K is a permutation matrix, and from (1) it is obvious that the corresponding permutation on Ω is in the normalizer of G and induces the automorphism α. This finishes the proof of Theorem 2.

To give an application of Theorem 2, let us denote by $A^*(G)$ the group generated by those automorphisms of G which have odd orders. Obviously $A^*(G)$ is a normal subgroup of 2-power index in the automorphism group $A(G)$. We prove

Theorem 3. *Let G be a doubly transitive permutation group of finite degree n such that G_o is a Hall subgroup of G (that is, n relatively prime to $|G|n^{-1}$). Then*

$$A^*(G) \ \leq \ P(G).$$

Proof. By Theorem 2 it suffices to show that for every automorphism α of G the length of that orbit of G_o^α which contains o is relatively prime to n. To show this, note that l divides $|G_o^\alpha|$, which is equal to $|G|n^{-1}$ and hence, by assumption, relatively prime to n.

As a consequence of Theorem 3, we settle a special case of a well known conjecture of Schreier:

Theorem 4. *If G is a simple finite group which possesses a subgroup of prime index p then $A(G)/I(G)$ is solvable; $I(G)$ denotes the group of inner automorphisms of G.*

Proof. If $|G| = p$ then the statement is trivial. If $|G| \neq p$ then G is not soluble, hence by a famous theorem of Burnside that faithful permutation representation of degree p which is induced by the given subgroup (and which we shall denote by G again) is doubly transitive.

Theorem 3 is applicable and yields $A^*(G) \le P(G)$. By Sylow's conjugacy theorem, applied to a Sylow p-subgroup of G, we find easily that $P(G)/I(G)$ is cyclic of an order which divides $p-1$. Hence the same is true for $A^*(G)/I(G)$ since $A^*(G) \le P(G)$. So $A^*(G)/I(G)$ is soluble, and since $A(G)/A^*(G)$ is a 2-group, $A(G)/I(G)$ is soluble.

Remark. Under the assumptions of Theorem 4, one can show by a more elaborate argument that $A(G)/I(G)$ is cyclic of an order which divides $p-1$.

Proc. Internat. Conf. Theory of Groups, Austral. Nat. Univ. Canberra, August 1965, pp. 395–397. © Gordon and Breach Science Publishers, Inc. 1967

Sur les S-partitions de Hall dans les groupes finis

G. ZAPPA

Soit G un groupe et soit S un sous-groupe de G. Un ensemble Π de sous-groupes de G est appelé une *S-partition* de G si, pour tout élément $x \in G$, $x \notin S$, il y a un sous-groupe et un seul $H \in \Pi$ tel que $x \in SH$. La S-partition Π est appelée *triviale* si il y a au plus un seule $H \in \Pi$ tel que $H \nsubseteq S$. Si, pour tout $H \in \Pi$, on a $S \subseteq H$, Π est appelée S-partition *étroite*.

Soit G fini. Alors Π est appelée S-partition *de Hall* si S et tous les sous-groupes $H \in \Pi$ sont des sous-groupes de Hall de G. Si, pour tout $H \in \Pi$, l'ordre de $H \cap S$ est le plus grand commun diviseur des ordres de H et S, Π est appelée *demi-étroite*. Évidemment, toute S-partition étroite d'un groupe fini est demi-étroite.

Si $S = 1$, les S-partitions se réduisent aux partitions de G, qui ont été étudiées par plusieurs auteurs (Baer, Kegel, Suzuki, etc.).

Les S-partitions des groupes ont intérêt dans la théorie des plans projectifs [2].

Je vais donner les résultats que j'ai obtenu jusqu'à ce moment dans une recherche sur les S-partitions de Hall des groupes finis.

1. Soit G un groupe et S un sous-groupe normal de G. Alors un ensemble Π de sous-groupes de G est une S-partition de G si et seulement si l'ensemble des images homomorphes des sous-groupes $H \in \Pi$ dans l'homomorphisme naturel de G sur G/S est une partition de G/S. Donc, si S est normal dans G, l'étude des S-partitions de G peut être reconduit à l'étude des partitions de G/S; par conséquent c'est intéressant de donner des conditions sur les S-partitions de G qui entraînent la normalité de S.

On a le théorème suivant:

*Soit G un groupe fini résoluble et soit S un sous-groupe de Hall de G.
S'il existe une S-partition étroite de Hall non triviale de G, S est normal
dans G.*

Le corollaire suivant descend directement de ce théorème.

*Soit G un groupe fini résoluble, soit S un sous-groupe de Hall de G, et
soit Π une S-partition de Hall non triviale de G. Condition nécessaire et
suffisante à fin que S soit normal dans G est que Π soit demi-étroite et que
tout H ∈ Π soit permutable avec S.*

En supposant seulement que Π soit une partition de Hall non triviale
demi-étroite sans imposer que tout $H \in \Pi$ soit permutable avec S, on
ne peut pas déduire que S soit normal dans G. En effet, si $G = SL$
$(|S|, |L|) = 1$, et si L a une partition Π de Hall non triviale, Π est aussi
une S-partition de Hall non-triviale de G; et il n'est pas dit que S soit
normal. Par exemple, si $G = \{a, b\}$, $a^6 = b^7 = 1$, $a^{-1}ba = b^3$, $S = \{a^2\}$,
$H_i = \{a^3b^{i-1}\}$ $(i = 1, \ldots, 7)$, $H_8 = \{b\}$, l'ensemble Π des sous-groupes H_i
$(i = 1, \ldots, 8)$ est une S-partition demi-étroite non triviale de Hall de G,
et S n'est pas normal dans G.

Tous les résultats précédents sont démontrés dans [3].

Maintenant le problème de voir si toutes les S-partitions non triviales
de Hall des groupes résolubles sont de la façon précédente se pose
naturellement.

Jusqu'à ce moment j'ai atteint seulement le théorème suivant, assez
particulier.

*Soit G un groupe supersoluble d'ordre $p^\alpha q^\beta r^\gamma$ (p, q, r nombres premiers
différents) et soit S un sous-groupe de G d'ordre p^α. Supposons qu'il
existe une S-partition demi-étroite non triviale de Hall de G. Alors
$G = SL$, $S \cap L = 1$, où L est un sous-groupe possédant une partition non
triviale de Hall.*

2. L'étude des S-partitions de Hall non demi-étroites est bien plus
difficile, et je peux donner maintenant seulement des cas très particuliers.

Soit G un groupe diédral d'ordre $2m$ (m impair > 1) c'est à dire
$G = \{a, b\}$, $a^2 = b^m = 1$, $a^{-1}ba = b^{-1}$. Alors, si $S = \{a\}$, l'ensemble des sous-
groupes de G conjugués à S et $\neq S$ est une S-partition de Hall non
triviale (et non demi-étroite) de G.

Soit G un groupe non abélien d'ordre pq (p, q nombres premiers,
$q \equiv 1 (\mod p)$), et soit S un sous-groupe de G d'ordre p. Une S-
partition non triviale de G doit se composer de sous-groupes conjugués

à S, et n'est pas demi-étroite. Mais, il n'y a pas toujours une telle S-partition. Des simples calculs arithmétiques prouvent que s'il y a une S-partition non triviale de G, il doit être $q \equiv 1 \pmod{p(p-1)}$. Dans ce cas, le problème de trouver une S-partition non triviale de G est reconduit à la question de trouver, dans le groupe multiplicatif Γ des classes de restes $(\bmod\ q)$ un complément à l'ensemble A des éléments $1, 1+x, 1+x+x^2, \ldots, 1+x+\cdots+x^{p-2}$ où x est tel que $x^p \equiv 1 \pmod{q}$, $x \not\equiv 1 \pmod{q}$; c'est à dire un ensemble B tel que pour tout élément $c \in \Gamma$ il y a un seul $a \in A$ et un seul $b \in B$ pour lesquels $c = ab$. Si $p-1$ est une puissance de 2 (c'est à dire, p est un nombre premier de Gauss), grâce à un résultat de Sands [1], B doit être périodique, c'est à dire, il y a un élément $g \in \Gamma$, $g \neq 1$ tel que $Bg = B$.

Si $p = 3$, on prouve qu'il y a toujours une S-partition non triviale de G. À la même conclusion on arrive si $p = 5$ (et naturellement, $q \equiv 1 \pmod{20}$), au moins dans le cas $1 + x \equiv 0 \pmod 4$ (par exemple, pour $q = 41$, $q = 61$).

Bibliographie

[1] A. D. Sands, On the factorisation of finite abelian groups, *Acta Math. Acad. Sci. Hungar.* **8** (1957), 65–86.

[2] G. Zappa, Sugli spazi generali quasi di traslazione, *Matematiche (Catania)* **19** (1964), 127–143.

[3] G. Zappa, Sulle S-partizioni di Hall di un gruppo finito, *Atti Accad. Naz. Lincei Rend. Cl. Sci. Fis. Mat. Nat.* (8) **38** (1965), 755–759.